Essential Concepts in
Sediment Transport Processes

Essential Concepts in Sediment Transport Processes

Edited by **Alice Grenouille**

R Callisto Reference

New York

Published by Callisto Reference,
106 Park Avenue, Suite 200,
New York, NY 10016, USA
www.callistoreference.com

Essential Concepts in Sediment Transport Processes
Edited by Alice Grenouille

International Standard Book Number: 978-1-63239-315-9 (Hardback)

Printed in the United States of America.

Contents

Preface

The purpose of the book is to provide a glimpse into the dynamics and to present opinions and studies of some of the scientists engaged in the development of new ideas in the field from very different standpoints. This book will prove useful to students and researchers owing to its high content quality.

The essential concepts in sediment transport processes are elucidated in this all-inclusive book. It presents a variety of topics related to sediment transport. It extensively discusses sediment dynamics, thereby helping in understanding effective management of aquatic environment. Economic and environmental consequences, especially in cases associated with anthropogenic involvement, have also been discussed. The fate of sediment movement and transport is often judged by tools like numerical models due to their efficiency in estimating temporal and spatial fluxes. Nature of the local sediments - whether fully cohesive, non-cohesive or mixture of both - is a crucial factor in determining physical sedimentary processes. This is the reason why engineers, mathematicians, scientists and hydrologists have been conducting researches for a long time about the various factors affecting sediment transport. These vary from processes such as erosion, scour and deposition to applications of sediment process observations in transport modeling frameworks. Researchers specializing in transport of sediments and supporting issues have been continually conducting researches in applied sediment transport in various aquatic environments. The inferences and outcomes of these crucial researches have been included in this book.

At the end, I would like to appreciate all the efforts made by the authors in completing their chapters professionally. I express my deepest gratitude to all of them for contributing to this book by sharing their valuable works. A special thanks to my family and friends for their constant support in this journey.

Editor

Sediment Transport Processes

Longshore Sediment Transport Measurements on Sandy Macrotidal Beaches Compared with Sediment Transport Formulae

Adrien Cartier, Philippe Larroudé and
Arnaud Héquette

Additional information is available at the end of the chapter

1. Introduction

In a context of global climate change, local sea level rise could affect the different coastal processes as erosion, transport and deposition which are responsible in maintaining the coastline. The study of sediment transport processes is one of the key for a better understanding of the coastal evolution which is needed for effective design of coastal engineering or to protect anthropogenic activities and population from marine submersion. One of the main processes that control coastal evolution is sediment transport. A number of studies have been focused on this topic, but they were mostly restricted to micro- to mesotidal beaches [1-3] and field investigations on sandy macrotidal beaches appear to be more limited, notably because these environments are less common along the worldwide coastline [4].

Only a few studies have been conducted for quantifying sediment flux on macrotidal beaches [5, 6] where sediment transport results from the complex interactions of tidal currents with longshore currents generated by obliquely incident breaking waves, this complexity being further increased by the large variations in water level that induce significant horizontal translations of the surf zone. Although a number of studies were recently conducted on the morphodynamics of the barred macrotidal beaches of Northern France [7-12], relatively little effort has been dedicated to measuring longshore sediment transport on these beaches, even though it is largely recognized that they are affected by significant longshore transport that plays a major role in the morphodynamics of the intertidal zone [10, 13]. Apart from some attempts to make estimates of longshore sediment transport from fluorescent tracers [10, 14-16] and to infer transport directions using grain-size trend analysis techniques [17,

18] no studies were conducted up to now for trying to quantify accurately longshore sand transport on these sandy macrotidal beaches.

Very recent field experiments conducted on macrotidal beaches of Northern France showed that, at a very short time scale (minutes), cross shore sediment flux is generally higher than longshore flux, suggesting that shore-perpendicular sediment transport associated with wave oscillatory currents probably represents a major factor controlling the cross-shore migration of intertidal bars [19]. Further analysis highlighted the strong dependence of longshore sediment transport (LST) on instantaneous hydrodynamic conditions that are extremely variable from one hydrodynamic zone to the other, notably between the non-breaking zone of wave shoaling and the surf zone [20]. Although such field experiments can provide very useful results that contribute to a better understanding of beach morphodynamic and sediment transport dynamics, *in situ* experiments are hard to undertake due to a series of technical and environmental factors. During the last decades, numerical modeling of coastal sediment transport and morphodynamics has grown substantially and is now largely used by the coastal scientific community. As a first step, models have to be calibrated in order to correspond as close as possible to natural phenomenae. Thus, a major focus of nearshore research is to relate (measured and predicted) sediment transport rates to morphological change, with the aim of improving our understanding and modeling capabilities of beach morphodynamics.

This study is based on previous field investigations conducted on sandy barred macrotidal beaches of northern France by Cartier and Héquette [19-22] during which longshore sediment fluxes were estimated using streamer traps, following the method proposed by Kraus [23]. Longshore sediment transport rates were compared with several sediment transport formulae integrated in a numerical model. This numerical model is characterized by a coupling of three codes consisting in enchained Artemis for swells, Telemac2d for the currents and Sisyphe for the morphodynamic evolution [24]. The aim of this contribution is to present the results of the field measurements carried out on these multi-barred sandy beaches and to discuss the abilities of numerical models to predict longshore sediment transport on these macrotidal environments.

2. Study area

This study has been conducted on three sandy barred beaches of Northern France from November 2008 to March 2010. The first field experiment site (Zuydcoote, ZY) is located near the Belgian border, facing the North Sea; the second site (Wissant Bay, WI) is on the shore of the Dover Strait, while the third study site (Hardelot, HA) is located on the eastern English Channel coast (Figure 1).

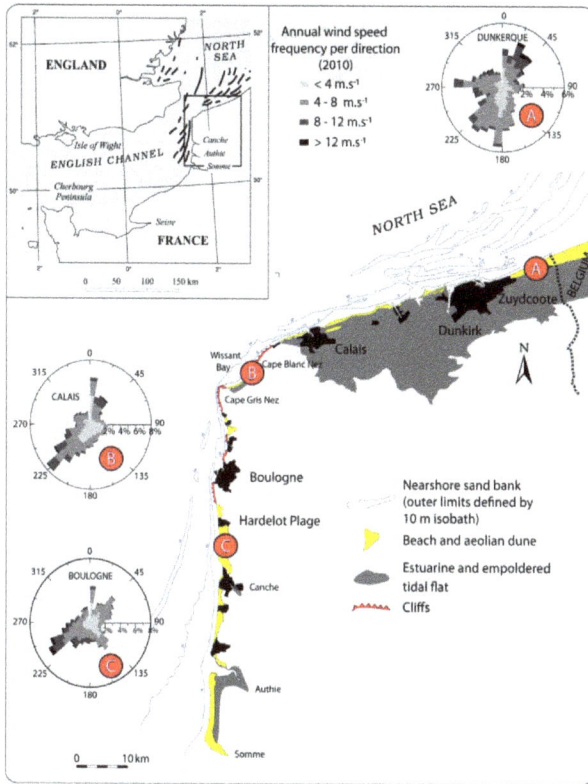

Figure 1. Location of the three study sites along the coast of Northern France: A) Zuydcoote; B) Wissant Bay; C) Harde-lot Beach.

The study sites consist of 300 to 800 m wide dissipative beaches characterized by extensive intertidal bar-trough systems (Figure 2). The coasts of Northern France are exposed to relatively low-energy waves that are refracted by numerous offshore sand banks. Dominant wave directions are from southwest to west, originating from the English Channel followed by waves from the northeast to north, generated in the North Sea. Offshore modal significant wave heights are similar for all the study sites and are less than 1.5 m, but may exceed 4 m during storms [25, 26]. The presence of several sand banks on the shoreface and the inner shelf and the gentle beach slopes that characterize the coasts of Northern France are responsible for strong wave energy dissipation, resulting in modal significant wave heights lower than 0.6 m in the intertidal zone [11, 25]. Wave heights can nevertheless reach 2 m on the foreshore during extreme events [27]. During such high wave energy conditions, substantial volumes of sediment can be transported on these beaches as revealed by the formation and migration of large megaripples across the intertidal zone [10].

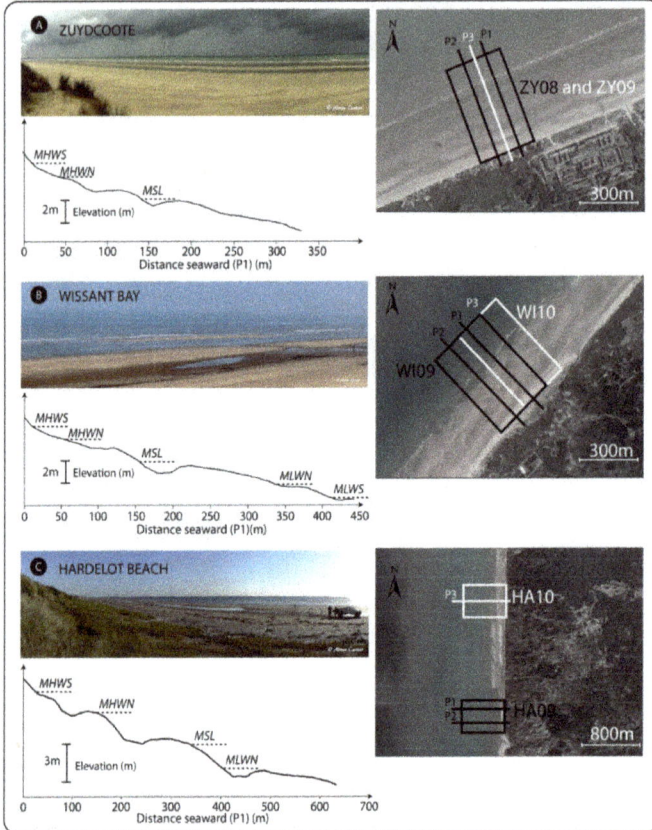

Figure 2. Shore perpendicular beach profiles, panoramic and aerial vertical photographs (©Orthophoto, 2005) of each study site (See Figure 1 for location).

The study sites are affected by semi-diurnal tides with mean spring tide ranging from about 6 m at Zuydcoote to almost 10 m at Hardelot (Figure 3). This high tidal range is responsible for relatively strong tidal currents that flow almost parallel to the shoreline in the coastal zone, at Wissant and Zuydcoote, the ebb is directed westward and the flood is flowing eastward, while at Hardelot Beach, the ebb and flood are directed southward and northward respectively. The reversing of tidal currents does not occur at high or low tide, but typically after a delay of two to three hours. Current measurements conducted in previous studies revealed that the speeds of flood currents exceed those of the ebb, resulting in a flood-dominated asymmetry responsible for a net regional sediment transport to the east-northeast [9, 28].

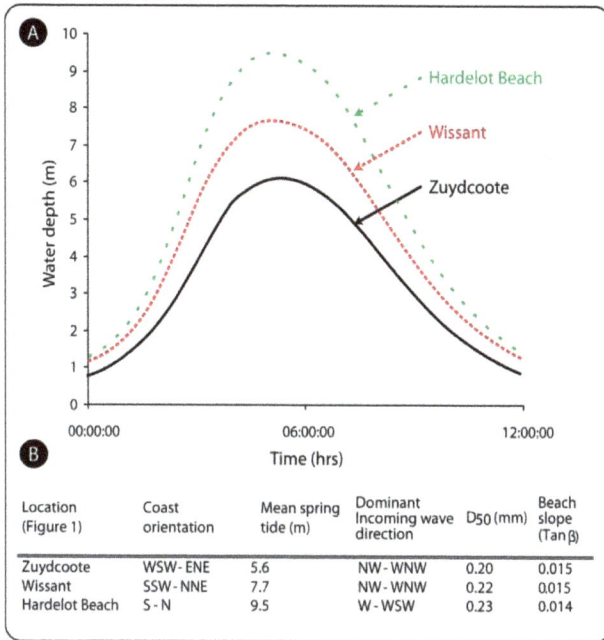

Figure 3. General characteristics of each field site. A) Water level variations for mean spring tide. B) Summary of hydrodynamic and morphodynamic characteristics at each field sites (see Figure 1 for location)

3. Field measurements

3.1. Field methodology

In order to determine the ability of numerical models to predict sediment transport and morphodynamics over sandy beds, several field measurements of sediment transport have been carried out on three different sandy macrotidal beaches. Although sediment fluxes can be estimated using acoustic or optical backscatter instruments, this study was based on direct sediment transport measurements using sediment traps rather than these techniques. Previous studies highlighted that acoustic or optical backscatter sensors can often be problematic in the coastal zone due to bubbling in the breaker and surf zones [29] and/ or to the presence of organic matter in the water column and to grain size variability [30]. Thus, streamer traps appeared to be the most adapted and the most accurate method to measure longshore sediment transport in the shoaling, breaking and surf zone on these macrotidal beaches.

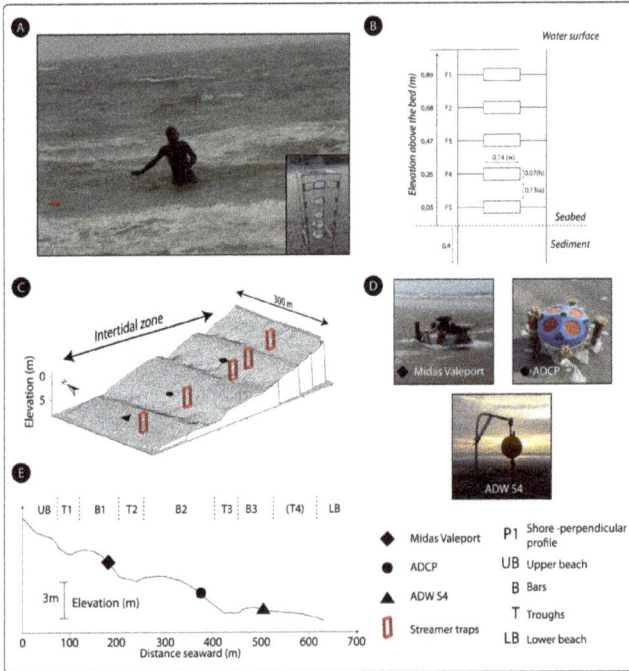

Figure 4. Field methodology. A) *In situ* measurements of longshore sediment transport during low to moderate wave energy conditions. B) Schematic representation of streamer trap used during the experiments. C) Locations of sediment trap deployment along a shore-perpendicular transect. D) Hydrographic instruments used during the field experiments. E) Shore-perpendicular profile showing the location of the hydrographic instruments. Codes refer to beach morphology where UB and LB are upper and lower beach respectively; B and T correspond to bars and troughs. Elevations are relative to the French topographic datum (IGN69).

Longshore sediment fluxes were estimated using streamer traps, following the method proposed by Kraus [23]. The sediment traps consisted of a vertical array of five individual streamer traps with 63 μm mesh size sieve cloth that collects sand-size particles at different elevations above the bed. First streamer trap (F5) is located 0.05 m above the bed and the last one (F1) is at approximately 0.90 m (Figure 4, B). Measurements of LST with the sediment traps were undertaken during 10 minutes.

Calculations of the sediment flux from sand traps were carried out according to the procedure of Rosati and Kraus [31]. The sediment flux Q(f), in kg.s^{-1}.m^{-2}, at a streamer trap (f) is equivalent to:

$$Q(f) = \frac{S(f)}{w * h * t}$$

Where S(f) is the dry weight of sediment collected in the streamer (f), h is the height of the streamer opening (0.07 m), w is the streamer width (0.14 m), and t is the sampling period (≈

10 minutes). The sediment flux between neighboring streamers QE(f) corresponds to the linear interpolation between two adjacent traps:

$$QE(f)=0,5*(F(f)+F(f+1))$$

The depth integrated flux (Q) in kg.s^{-1}.m^{-1} is:

$$Q=h*\sum_{i=1}^{N}Q(f)+\sum_{i=1}^{N}a(f)*QE(f)$$

Where h is the height of the streamer opening in meters, Q(f) is the sediment flux at a streamer f, a(f) is the distance between neighboring streamers, QE(f) is the sediment flux between neighboring streamers and N is the total number of streamers (N = 5).

Measurements of LST were carried out at several locations across the intertidal zone during rising and falling tides in order to obtain estimates of longshore sediment flux from the lower to the upper beach during flood and ebb (Figure 4, C).Although the sediment traps were usually deployed in similar water depths, ranging from 0.8 to 1.4 m, sediment transport measurements took place in various hydrodynamic zones including shoaling, breaker and surf zones, depending on wave activity during sampling. For safety reasons, sand transport measurements were conducted only under low to moderate wave energy conditions (H_s max $\approx$ 0.7 m).

Two field experiments were conducted on each study site: Zuydcoote in November 2008 (ZY08) and December 2009 (ZY09), Wissant in March 2009 (WI09) and March-April 2010, (WI10), and Hardelot in June 2009 (HA09) and January-February 2010 (HA10), resulting in the collection of 172 depth-integrated sediment flux measurements.

Coastal hydrodynamics were measured at different locations across the bar-trough morphology along cross shore transects using various hydrographic instrument (Figure 4, D). Waves and currents were measured using three different hydrographic instruments: an Acoustic Doppler Current Profiler (ADCP), and two electromagnetic wave and current meters (Midas Valeport©, and InterOcean ADW S4). All instruments operated during 9 minutes intervals every 15 minutes at a frequency of 2 Hz, providing almost continuous records of significant wave height (H_s), wave period and direction, longshore current velocity (V_l), and mean current velocity (V_m) and direction. Current velocity was measured at different elevations above the bed depending on each instrument. The ADW S4 and Valeport current meters recorded current velocity at 0.4 m and 0.2 m above seabed respectively, while the ADCP measured current velocity at intervals of 0.2 m through the water column from 0.4 m above the bed to the water surface. Current velocity at 0.2 m above the bed was estimated using the ADCP data by applying a logarithmic regression curve to the measured velocities obtained at different elevations in the water column.

Beach morphology plays a major role in the variation of sediment transport rates, especially on bar-trough topography [9, 10, 13, 32, 33]. Thus, during each field experiment, beach morphology was surveyed using a very high resolution Differential Global Positioning System (DGPS) with horizontal and vertical error margins of ± 2 cm and ± 4 cm respectively. A 300 m wide zone of the beach was systematically surveyed on each study site whereas the cross-shore extent of the surveyed area was variable depending on tidal range.

3.2. Sediment transport: role of the main physical forcings

3.2.1. Hydrodynamic conditions during the field experiments

The field experiments have been conducted under different conditions of wave energy and tidal range at each study site. Because sand transport measurements were restricted to moderate to low wave energy conditions, the range of wave heights recorded during sediment trapping is relatively constrained. A classification of wave energy conditions during the field experiments was adopted in which $H_s < 0,2$ m represents low wave energy conditions, $0,2 \leq H_s < 0,4$ m refers to moderate wave energy conditions and $H_s \geq 0,4$ m represents higher energetic conditions (Figure 5).

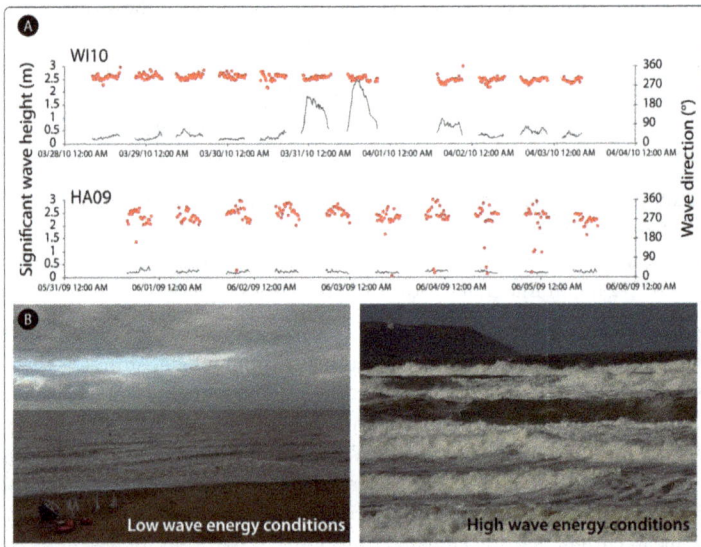

Figure 5. Examples of hydrodynamic conditions during some field experiments. A) Time series of significant wave height (m) and wave direction (°) during WI10 and HA09 experiments. B) Photographs of low and high wave energy conditions during the experiments.

The lowest wave energy conditions took place during the HA09 experiment. 99% of the significant wave heights were under 0.4 m and 46% were under 0.2 m for a mean longshore current velocity of 0.2 m.s^{-1}. High wave energy conditions occurred during several field experiments with a maximum wave height of 2.4 m reached at Wissant in 2010 (WI10) while longshore current velocity reached 2 m.s^{-1}. Such high energy conditions lasted over only two tidal cycles however during this field experiment (Figure 5). In comparison, the maximum significant wave height during the HA10 experiment was 2.1 m, but the duration of high energy conditions was considerably longer as 80% of the recorded wave heights was higher than 0.4

m for an average H_s of 0.70 m, which is remarkably high for the relatively low-energy coasts of Northern France where wave heights are generally lower in the intertidal zone [11, 13].

3.2.2. Longshore sediment transport rates

During the six field experiments, more than 700 sediment samples were collected, which were used to compute 172 depth-integrated sediment fluxes. Among these, 79 depth-integrated sediment fluxes were obtained close to a hydrographic instrument, allowing a comparison between LST and hydrodynamic parameters. Throughout the 6 field experiments, most of the longshore sediment transport rates (> 50%) ranged from 1 x 10^{-4} kg.s^{-1}.m^{-1} to 1 x 10^{-3} kg.s^{-1}.m^{-1} (Figure 6A). Sediment transport rates show a high variability depending on the study site and during each field experiment due to variations in hydrodynamic conditions (Figure 6B). Longshore sediment flux reached values up to 2.1 x 10^{-1} kg.s^{-1}.m^{-1} during the ZY09 experiment, which was the most energetic event during which sediment sampling took place. Lower rates of sediment transport were measured in the vicinity of current meters where sediment flux nevertheless reached approximately 1.6 x 10^{-1} kg.s^{-1}.m^{-1} for a mean flow velocity of 0.5 m.s^{-1}. Significantly higher transport rates were observed during the most energetic conditions, however, notably during the HA10, ZY09 and WI09 field experiments (Figure 6B).

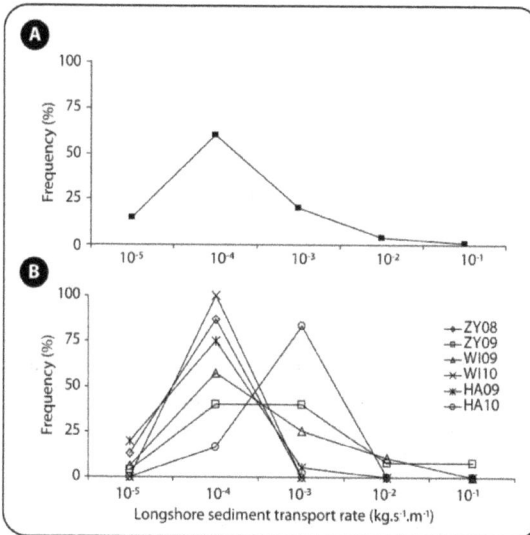

Figure 6. Range of LST for (A) all field experiments and (B) each field experiment.

Longshore sediment transport rates measured during this study are in the same order of magnitude as other studies conducted on microtidal beaches [23, 31, 34-37] as well as on macrotidal beaches [5] (Figure 7). The fact that ranges of values are similar whatever the tidal conditions are, suggests that tidally-induced currents are not the main forcing and do not act significantly on longshore sediment transport magnitude.

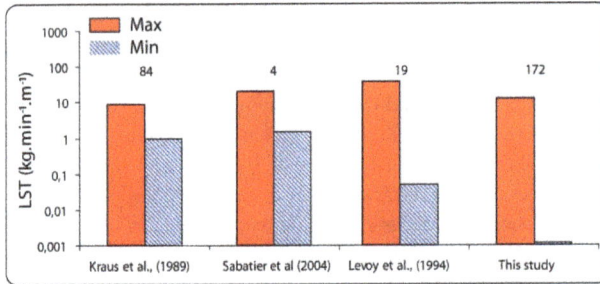

Figure 7. Maximum and minimum values of LST measured using streamer traps on microtidal beaches[35, 38] and macrotidal beaches [5]. Numbers of samples are located just above the bar charts.

3.2.3. Relationship between longshore sediment transport and hydrodynamics

Comparisons of transport rates with hydrodynamic data showed that longshore sediment transport increases with both significant wave height and mean current velocity (Figure 8). Low sediment transport ($< 1.0 \times 10^{-4}$ kg.s^{-1}.m^{-1}) is mainly associated with small wave heights (< 0.3 m), but only a small increase in wave height appears to induce significantly larger sediment transport. However, high sediment transport values can also be associated with low wave conditions, such as during the WI09 experiment when a sediment transport rate of 2.4×10^{-2} kg.s^{-1}.m^{-1} was measured with a significant wave height of about 0.2 m for example. The variability in sediment flux values obtained during conditions of equivalent wave heights, and the fact that similar transport rates may be associated with different wave heights, even on the same beach and during the same field experiment (e.g., WI09), suggest that waves do not represent the only factor controlling sediment transport.

Least-square regression analyses show that longshore sediment flux is better correlated with current velocity than with wave height as revealed by higher determination coefficients (Figure 8). Similarly to what was observed with significant wave heights, high sediment transport rates can also be observed with lower current velocities. Further analysis detailed in Cartier and Héquette [20] highlighted that variations in LST are better explained by these two forcing parameters in the surf zone, where currents generated by obliquely incident breaking waves are acting, than in the shoaling zone.

Longshore sediment transport in the nearshore zone is commonly related to the longshore wave energy flux (P_l) evaluated at the breaker zone, sand transport being expressed as an immersed-weight transport rate (I_l) and related to P_l [39]. Conversely to what was observed in other studies [40], however, previous analyses of our data showed no relationship between LST and wave breaking angle, which is directly involved in the computation of P_l [19,

20, 22]. These results may be explained by the influence of tidal currents that interact with wave-induced longshore currents on these macrotidal beaches

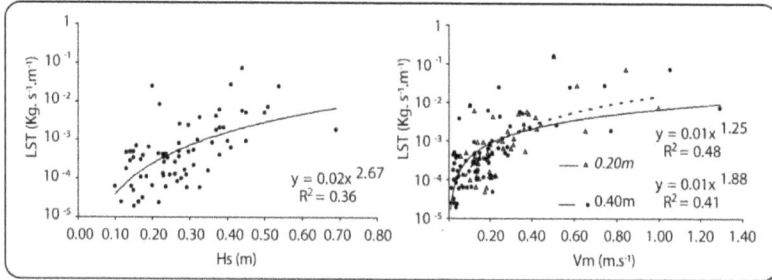

Figure 8. Relationship between LST and significant wave height (H_s), and with mean current velocity (V_m) at 0.2 m (triangle) and 0.4 m (circles) above the bed using all the samples collected near a current meter during all field experiments.

4. Sediment transport modelling

4.1. Methodology

The methodology used during field experiments did not allow us to compare sand transport rates with well known formulae such as the CERC formula [41] or the Kamphuis's formula [42] for example. These formulations, which are essentially based on a wave energy flux approach, provide estimates of total sediment transport rates across the entire surf zone. A large amount of studies have compared that kind of numerical model with in situ measurements [29, 43-47], or with laboratory measurements [48-50], and even between several numerical models in order to understand their behaviour with virtual data [51-53]. However, our measurements can not be compared with these numerical models because sand trapping took place at only one location in the surf or shoaling zone while these formulae estimate the total longshore sediment transport across the surf zone.

There is, nevertheless, a number of sediment transport formulae that can be compared with localized measurements of sand flux. In the present study, calculations of sediment transport rates have been realized using a coupling of three codes [54]. The sedimentary evolution is modeled under the action of the oblique incident waves and is coupled with different numerical tools dedicated to the other process involved in the nearshore zone. We can mention the following modules:

The wave module takes into account the surge energy dissipation (hyperbolic equation of extended Berkhoff). The Artemis code (Agitation and Refraction with Telemac2d on a Mild Slope) solves the Berkhoff equation taken from Navier-Stokes equations with some other hypotheses (small wave steepness of the surface wave, small slope...). The main results are, for every node of the mesh, the height, the phase and the incidence of the waves. Artemis can take into account the reflection and the refraction of waves on an obstacle, the bottom friction and the breakers. One of the difficulties with Artemis is that a fine mesh must be used to have good results whereas Telemac2d does not need such a fine mesh.

The hydrodynamic module calculates currents induced by means of the surge of the waves, from the concept of radiation constraints obtained according to the module of waves. Telemac2d is designed to simulate the free surface flow of water in coastal areas or in rivers. This code solves Barré Saint-Venant equations taken from Navier-Stokes equations vertically averaged. Then, the main results are, for every node of the mesh, the water depth and the velocity averaged over the water column. Telemac2d is able to represent the following physical phenomena: propagation of long periodic waves, including non-linear effects, wetting and drying of intertidal zone, bed friction, turbulence…

The sedimentary module integrates the combined actions of the waves and the wave currents (2D or 3D) on the transport of sediment [24].

The Sisyphe code solves the bottom evolution equation which expresses the mass conservation by directly using a current field result file given by Telemac2d (Figure 9). Several of the most currently used empirical or semi-empirical formulas are already integrated in Sisyphe.

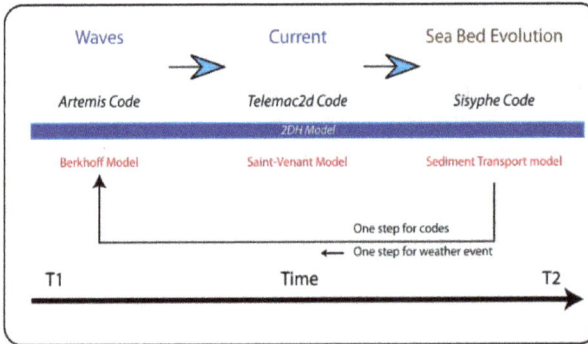

Figure 9. Diagram of the model ATS (Artemis-Telemac-Sisyphe) used in our simulations, showing the principle of external coupling to make a loop over one hydro-meteorological event time step (between t1 and t2).

A hydrodynamic simplified model (called Multi1DH) uses the following assumptions: a random wave approach and a 1DH (cross-shore) direction. An offshore wave model (shoaling + bottom friction + wave asymmetry) is used with the break point estimation. The waves in the surf zone are modeled with the classic model of Svendsen (1984) with an undertow model (roller effect) [55, 56]. The longshore current model is the Longuet-Higgins's model [57]. The model is included in the Sysiphe code to calculate the sea bed evolution with several sediment transport formulas.

It is generally accepted that the estimate is acceptable when the flux is between 0.5 and 2 times the in situ measurement [58]. Thus, in the following figures, three lines symbolize the extent of data that are significant at $0.5\ Q_{sm}<Q_{sc}<2\ Q_{sm}$. The standard error ($S_{rms}$) was also calculated using the following equation to characterize the dispersion of data, where the higher the values, the higher the data are scattered.

$$S_{rms} = \sqrt{\frac{\sum_{i=1}^{N}\left[Log(Q_{Sc}) - Log(Q_{Sm})\right]^2}{N-2}}$$

where Q_{sc} is the flux calculated, Q_{sm} the measured flux and N the number value.

4.2. Results

4.2.1. Potential effects of water depth

Sediment flux measurements were performed in a water depth between approximately 1 m and 1.5 m, but because of the excursion of the tide, these measurements were made at several locations on the foreshore (Figure 4). Calculation of sediment load at the same location on the field requires a lot of computer manipulations. So initially, the sediment loads are modelled in the middle of the digital domain. The water level in the middle of the simulated field rarely matches the exact water level measured during sampling. A test was therefore carried out to assess the potential effects of water depth on computed sediment flux using the data measured during the two field experiments at Hardelot beach (2009 and 2010). For the numerical simulations, bathymetry has been considered as stable during and between the two field campaigns.

The following graphs show the control of the water level on the accuracy of modeled flux (Figure 10). The example shows calculations based on the expression of Bijker [59], similar observations having been observed for the other formulations.

The results show that sediment fluxes tend to be better estimated with decreasing water depth, which is consistent with the depths at which measurements were made. The correlations are even better for water depths just below the water level of the in situ measurement (<1 m). In a water depth between 2.70 m and 1.74 m, the computed values seem to line up around $1 \times 10^{-3} kg.s^{-1}.m^{-1}$. There is very little change in calculated flux in comparison with those measured in situ. As soon as the water level is similar to that of measurement at the time of trapping (1.51 m), the distribution of points tends to align the right $Q_{sc} = Q_{sm}$. However, when the water column is much lower than reality, many errors appear in the calculation and the estimation of sediment flux becomes completely erroneous. In fact, the RMS errors are quite high when the water level is far from the actual water depth (Figure 10). It should be emphasized, however with all the initial approximations, the mere fact of positioning the water level at the same level as during the sampling resulted in relatively better results. In particular, when h = 0.91 m, which resulted in a percentage of acceptable values of 32%. Overall calculated sediment fluxes are nevertheless generally overestimated.

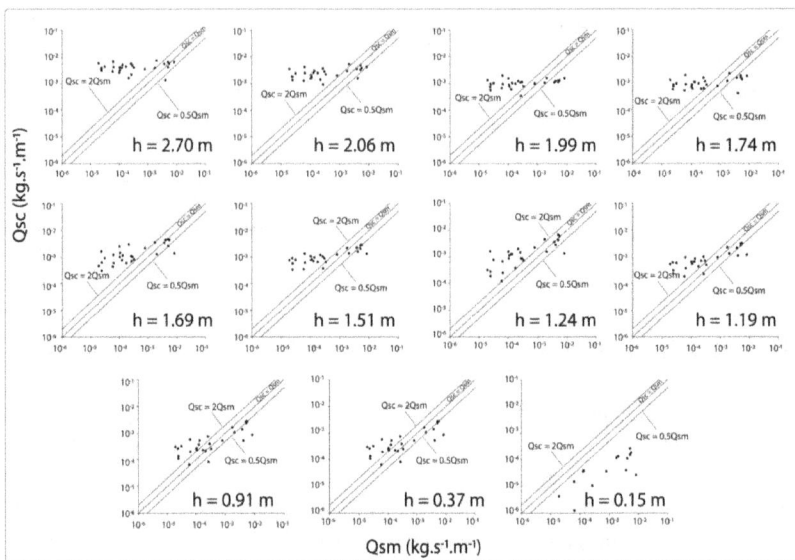

Figure 10. Comparison between measured (Q_{sm}) and computed (Q_{sc}) transport rates using Bijker's formula [59] for different water depths (h).

The water level acts directly on the current profile and associated sediment transport mechanisms. Thus, when the water column is higher than that at the time of measurement, the transport in suspension is favoured in the modelling, leading to an overestimation of the integrated flux in the water column. In contrast, a shallower water column could lead to higher sediment fluxes due to increased bed shear. However, it appears that the formulas are struggling to express sediment transport in very shallow water. Despite very simplified initial conditions, the results obtained in this study show that the coupling of these three codes is not so far from reality when we take into account the water level measured in situ. More precise calibrations will likely improve the accuracy of the results in the future.

4.2.2. Calculated sediment flux against in situ data

Using the water levels measured during the field experiments, the calculations of sediment transport were calibrated for each case to be as close as possible to the in situ measurement. Sediment transport modeling was carried out on the beaches of Wissant and Hardelot, and sediment loads calculated by several formulas have been extracted and compared to field data. The calculations concerning the Zuydcoote field site are not presented because of problems related to the computational domain. Because the bathymetry input was not broad enough, the waves spread by the code Artemis underwent many artifacts and measurement errors that did not allow reliable sediment transport calculations.

Sediment fluxes were compared with the expressions of Bijker [59], noted BI68, and a coupling of Van Rijn formula [60] with an expression modelling the transport in suspension [61], this coupling being called the Soulsby-Van Rijn formula noted SVR97 (Figure 11). The reader is referred to [58, 62]for details on the different formulae.

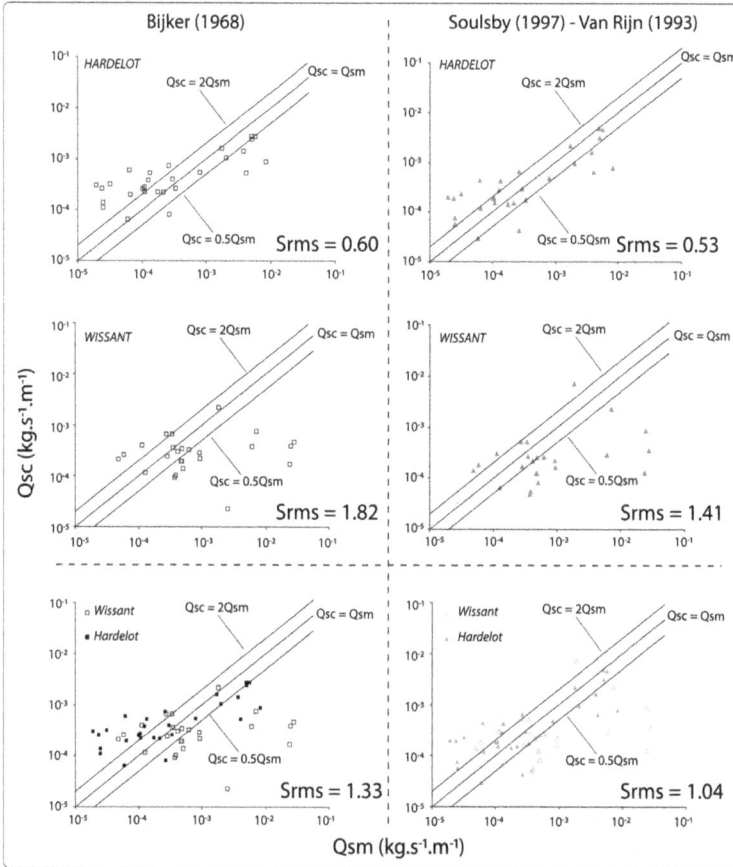

Figure 11. Comparisons between in situ (Q_{sm}) sand transport rates measured during HA09, HA10, WI09 and WI10, with calculated sediment fluxes (Q_{sc}) following Bijker and Soulsby – Van Rijn formulae.

The results were analyzed according to the study site and the mathematical expression used. The results are better on the site of Hardelot for both formulas, the error S_{rms} being less than 1.0 with a minimum of 0.53 with SVR97. Moreover, the percentage of significant values reaches 32% and 46% for SVR97 and BI68, respectively. On the site of Wissant, it does not exceed 30% whatever the formula. RMS errors are associated with a greater dispersion of data readily observable in the graphs (Figure 11). When considering all the data, the formula for SVR97 is the expression that is most satisfactory with a S_{rms} of only 1.04 and 36% of acceptable values. The proportion of sediment flux of low intensities ($<1 \times 10^{-3}$ kg.s^{-1}.m^{-1}) is higher during the campaigns carried out in Hardelot than those held in Wissant. Conversely, the flux measurements acquired during the field investigations in the Bay of Wissant proved to be higher. Morphological changes were also more significant in Wissant than during the campaigns conducted in Hardelot.

5. Discussion

In the present study, measurements of sediment flux were carried out at successive positions between the lower and the upper beach during rising or falling tide in order to measure sediment transport in comparable water depths. Our results showed that sand transport was mainly dependent on the mean flow, especially above a velocity threshold of approximately 0.4 m.s^{-1} (Figure 8). Further results have shown that sediment transport was also controlled by wave action, but correlation analyses between LST and significant wave height showed a better relationship in the surf zone than in the non-breaking zone [20]. These observations highlight the mere fact that sediment transport processes are strongly different from a hydrodynamic zone to another and it underlines the need to use appropriate formulae in order to model sediment transport properly. Although a large amount of studies compared sediment transport calculation against in situ data, few of them have been undertaken using sediment transport measurements from macrotidal beaches.

Results obtained in a previous study by Camenen and Larroudé [62] showed that the Bijker formula generally tends to underestimate the sediment transport when there is interaction of waves and current, which is generally the case in the coastal zone. In this formula, the swell is considered to be the only mechanism responsible for suspending sediment. Therefore, when wave height is low, the simulated sediment transport remains insignificant even if the average current velocity is high. Although the strength of currents in the intertidal zone is usually related to the conditions of agitation, it may sometimes be forced by wind or induced by the combination of tidal currents and those generated by the incident swell [10]. The suspended particles are then provided by currents and waves. Our results show, however, that the higher sediment fluxes ($> 1 \times 10^{-3}$ kg.s^{-1}.m^{-1}) are underestimated, which may be due to low wave height and mean current that are powerful enough to induce substantial sediment transport, which cannot be modeled by this expression. Conversely, the lowest sediment transport rates occur when the swell and the mean current are of low intensity. The direction of tidal currents is also directly involved in the magnitude of the sediment flux since it can easily reduce or conversely increase wave-induced currents, depending on the phase of the tidal cycle and the direction of the longshore current. Such types of case are not considered in the model, however, which may explain a part of the variability observed between measured and computed sediment fluxes.

The formula of Soulsby-Van Rijn comes from the coupling of the Van Rijn formula [60] with an expression modeling the transport in suspension. It takes into account many physical parameters for estimating the bed load and suspended load. Even though the calculations are more complicated and time consuming, the estimate is generally better, but significant errors may occur when the wave direction is opposite to that of the current [62].

In this study, the results are particularly satisfactory for the data obtained during the field experiments at Hardelot (Figure 11) where beach morphology changes were more limited than at Wissant. When taking into account all values, it clearly appears that the most significant fluxes can nevertheless be largely underestimated since some values may be up to four times lower than the measured transport rates. During high wave energy conditions, beach morphology and bed roughness change rapidly due to an increase in bed load sediment transport. The impact of these bottom changes on the distribution of sediment in the water column is crucial and plays an important role in the mechanisms of suspended sediment transport. Because bed roughness is variable, largely depending on the local morphology of the beach, it is necessary to incorporate changes in beach morphology in the process of calculation which has not been done yet in this first attempt to model longshore transport on macrotidal beaches.

6. Conclusion

Despite several simplifications in the modeling procedure, comparisons of longshore sediment transport fluxes measured on sandy macrotidal beaches with computed sand fluxes gave encouraging results. It was shown that water depth is one of the major parameter affecting modeled sediment transport rates, as calculated sand fluxes were more comparable with in situ measurements when simulated water depth was similar to the actual water depth measured in the field. The height of the water column therefore represents a key term to consider in modeling sediment transport on these beaches. The best modeling results were obtained with the data collected during low energy conditions at Hardelot beach where the beach morphology was the most stable. A limitation of the modeling approach used in this study is related to the fact that beach morphology changes are not taken into account in the calculations, which should be considered in future modeling studies. Because sediment traps mainly collect sediments transported in suspension, future investigations using this data set will be aimed at de-coupling suspended and bed load transport calculation in the different sediment transport formulae in order to evaluate only the suspended sediment flux, which should result in more accurate comparisons between modeled and measured sediment transport rates.

Acknowledgements

This work was supported by French Research National Agency (ANR) through the Vulnerability Milieu and Climate program (project VULSACO, n° ANR-06-VMC-009) and by the

French Centre National de la Recherche Scientifique (CNRS) through the PLAMAR and MI-CROLIT projects of the Programme "Relief de la Terre". Additional funding was provided by the "Syndicat Mixte de la Côte d'Opale" through a doctoral scholarship to Adrien Carti-er. The authors would like to thank Aurélie Maspataud, Vincent Sipka and Antoine Tresca as well as all the students and permanent staff for their help during the field experiments.

Author details

Adrien Cartier[1*], Philippe Larroudé[2] and Arnaud Héquette[3]

*Address all correspondence to: adrien.cartier@univ-littoral.fr

1 LOG - UMR CNRS 8187, University of Littoral Côte d'Opale,, France

2 LEGI-UMR 5519 UJF, University of Grenoble,, France

3 LOG - UMR CNRS 8187, University of Littoral Côte d'Opale,, France

References

[1] Wang, P., Kraus, N. C., & Davis, R. A. (1998). Total longshore sediment transport rate in the surf zone-field measurement and empirical predictions. *J Coast. Res.*, 14, 269-282.

[2] Bayram, A., Larson, M., Miller, H. C., & Kraus, N. C. (2001). Cross-shore distribution of longshore sediment transport: comparison between predictive formulas and field measurements. *Coast. Eng.*, 44(2), 79-99.

[3] Kumar, V. S., Anand, N. M., Chandramohan, P., & Naik, G. N. (2003). Longshore sediment transport rate-measurement and estimation, central west coast of India. *Coast. Eng.*, 48, 95-109.

[4] Davies, A. G. (1980). Geographical variations in coastal development. London, Long-man, 212.

[5] Levoy, F., Montfort, O., & Rousset, H. (1994). Quantification of longshore transport in the surf zone on macrotidal beaches. Fields experiments along the western coast of Cotentin (Normandy, France). Paper presented at 24 th International Conference on Coastal Engineering., Kobé, Japon. 2282-2296.

[6] Corbau, C., Ciavola, P., Gonzalez, R., & Ferreira, O. (2002). Measurements of Cross-Shore Sand fluxes on a Macrotidal Pocket Beach (Saint-Georges Beach, Atlantic Coast, SW France). *J. Coast. Res.* [36].

[7] Reichmüth, B., & Anthony, E. J. (2008). Dynamics of intertidal drainage channels on a multi barred macrotidal beach. *Earth Surf. Process. Landf.*, 33(1), 142 -151 .

[8]   Reichmüth, B., & Anthony, E. J. (2002). The variability of ridge and runnel beach morphology: examples from Northern France. *J. Coast. Res.* [36].

[9]   Anthony, E. J., Levoy, F., & Montfort, O. (2004). Morphodynamics of intertidal bars on a megatidal beach, Merlimont, Northern France. *Mar. Geol.*, 208, 73-100.

[10] Sedrati, M., & Anthony, E. J. (2007). Storm-generated morphological change and longshore sand transport in the intertidal zone of a multi-barred macrotidal beach. *Mar. Geol.*, 244, 209-229.

[11] Héquette, A., Ruz, M. H., Maspataud, A., & Sipka, V. (2009). Effects of nearshore sand bank and associated channel on beach hydrodynamics: implications for beach and shoreline evolution. *J. Coast. Res.* [56].

[12] Maspataud, A., Ruz, M. H., & Héquette, A. (2009). Spatial variability in post-storm beach recovery along a macrotidal barred beach, southern North Sea. *J. Coast. Res.* [56].

[13] Reichmüt, B., & Anthony, E. J. (2007). Tidal influence on the intertidal bar morphology of two contrasting macrotidal beaches. *Geomorphology*, 90(1- 2), 101-114.

[14] Levoy, F., Montfort, O., & Larsonneur, C. (1997). Quantification des débits solides sur les plages macrotidales à l'aide de traceurs fluorescents, application à la côte ouest du Cotentin. *Oceanol. Acta*, 20(6), 811-822.

[15] Voulgaris, G., Simmonds, D., Michel, D., Howa, H., Collins, M. B., & Huntley, D. (1998). Measuring and modelling sediment transport on a macrotidal ridge and runnel beach: an intercomparison. *J. Coast. Res.*, 14(1), 315 .

[16] Stépanian, A., Vlaswinkel, B., Levoy, F., & Larsonneur, C. (2001). Fluorescent tracer experiment on a macrotidal ridge and runnel beach : a case study at Omaha beach, North France. *Coastal dynamics'01 Lund, Suède*, 1017-1027.

[17] Reichmüth, B. (2003). Contribution à la connaissance de la morphodynamique des plages à barres intertidales: Approche expérimentale, Côte d'Opale, Nord de la France. *Thèse de doctorat Université du Littoral Côte d'Opale*, 247.

[18] Héquette, A., Hemdane, Y., & Anthony, E. J. (2008). Determination of sediment transport paths in macrotidal shoreface environments: a comparison of grain-size trend analysis with near-bed current measurements. *J. Coast. Res.*, 24, 695-707.

[19] Cartier, A., & Héquette, A. (2011). Estimation of longshore and cross shore sediment transport on sandy macrotidal beaches of Northern France. Paper presented at Proceedings Coastal sediments'11, Miami, Florida, USA. 2130-2143.

[20] Cartier, A., & Héquette, A. (2011). *Longshore and cross shore variation in sediment transport on barred macrotidal beaches, Northern France.*

[21] Cartier, A. (2011). Evaluation des flux sédimentaires sur le littoral du Nord Pas-de-Calais: Vers une meilleure compréhension de la morphodynamique des plages macrotidales. *Thèse de Doctorat, Université du Littoral Côte d'Opale.*

[22] Cartier, A., & Héquette, A. (2011). Variation in longshore sediment transport under low to moderate conditions on barred macrotidal beaches. *J. Coast. Res.* [64], 45 -49 .

[23] Kraus, N. C. (1987). Application of portable traps for obtaining point measurements of sediment transport rates in the surf zone. *J. Coast. Res.*, 3, 139-152.

[24] Hervouet, J. M. (2007). Hydrodynamics of Free Surface Flows: Modelling With the Finite Element Method. John Wiley & Sons, 360, 10.1002/9780470319628.

[25] Clique, P. M., & Lepetit, J. P. (1986). Catalogue sédimentologique des côtes françaises, côtes de la Mer du Nord et de la Manche. 133.

[26] Ruz, M. H., Héquette, A., & Maspataud, A. (2009). Identifying forcing conditions responsible for foredune erosion on the northern coast of France. *J. Coast. Res.* [56].

[27] Maspataud, A., Idier, D., Larroudé, Ph., Sabatier, F., Ruz, M. H., Charles, E., Lecacheux, S., & Héquette, A. (2010). L'apport de modèles numériques pour l'étude morphodynamique d'un système dune-plage macrotidal sous l'effet des tempêtes : plage de la dune Dewulf, Est de Dunkerque, France. XIèmes. *Journées Nationales Génie Civil-Génie Côtier Les Sables d'Olonne*, 353-360.

[28] Héquette, A., Hemdane, Y., & Anthony, E. J. (2008). Sediment transport under wave and current combined flows on a tide-dominated shoreface, northern coast of France. *Mar. Geol*, 249, 226-242.

[29] Esteves, L. S., Lisniowski, M. A., & Wiliams, J. J. (2009). Measuring and modelling longshore sediment transport. *Est. Coast. Shelf Sci.*, 83, 47-59.

[30] Battisto, G. M., Friedrichs, C. T., Miller, H. C., & Resio, D. T. (1999). Response of OBS to mixed grain-size suspensions during Sandyduck'97. *Coastal sediment 99 Long Island, New York*, 1, 297-312.

[31] Rosati, J. D., & Kraus, N. C. (1989). Development of a portable sand trap for use in the nearshore Department of the army, U.S. Corps of Engineers. *Technical report CERC* [89-91], 181.

[32] Sipka, V., & Anthony, E. J. (1999). Morphology and Hydrodynamics of a macrotidal ridge and runnel beach under modal low conditions. *J. Res. Oceanogr.*, 24, 25-31.

[33] Cartier, A., Larroudé, Ph, & Héquette, A. (2012). Comparison of sediment transport models with in-situ sand flux measurements and beach morphodynamic evolution. ICCE, Santander 2012. Submitted

[34] Kraus, N. C., & Dean, J. L. (1987). Longshore sediment transport rate distributions measured by trap. *Coastal sediment'87 ACE*, 881-898.

[35] Kraus, N. C., Gingerish, K. J., & Rosati, J. D. (1989). Duck 85 surf zone sand transport experiment. *US Army Corps of Engineers, Waterways Experiment Station, Coastal Engineering Research Center, Vicksburg, Mississippi.*, 48.

[36] Sabatier, F. (2001). Fonctionnement et dynamiques morpho-sédimentaires du littoral du delta du Rhone. *Thèse de Doctorat Université Aix-Marseille III.*

[37] Sabatier, F., Stive, M. J. F., & Pons, F. (2004). Longshore variation of depth of closure on a micro-tidal wave-dominated coast. Paper presented at International Conference of Coastal Engineering American Society of Civil Engineering, Lisboa. 2329-2339.

[38] Sabatier, F., Samat, O., Chaibi, M., Lambert, A., & Pons, F. (2004, 7-9 septembre 2004). Transport sédimentaire de la dune à la zone du déferlement sur une plage sableuse

soumise à des vents de terre. Paper presented at VIIIèmes Journées Nationales Génie Civil- Génie Côtier Compiègne,. 223-229.

[39] Komar, P. D. (1998). Beach processes and sedimentation. *Pearson Education*, 544.

[40] Komar, P. D., & Inman, D. L. (1970). Longshore sand transport on beaches. *Journal of Geophysical Research*, 75(30), 5514-5527.

[41] Manual, S. P. (1984). Coastal engineering Research Center, in: U.S Army Corps of Engineers, t.E.W., DC 20314 (Ed.).

[42] Kamphuis, J. W., Davies, M. H., Nairn, R. B., & Sayao, O. J. (1986). Calculation of littoral sand transport rate. *Coast. Eng.*, 10, 1-21.

[43] Davies, A. G., Ribberink, J., Temperville, A., & Zyserma, J. A. (1997). Comparisons between sediment transport models and observations made in wave and current flows above plane beds. *Coast. Eng.*, 31, 163-198.

[44] Miller, H. C. (1999). Field measurements of longshore sediment transport during storms. *Coast. Eng.*, 36, 310-321.

[45] Eversole, D., & Fletcher, C. H. (2002). Longshore sediment transport rates on a reef-fronted beach: field data and empirical models Kaanapali Beach, Hawaii. J. Coast. Res. In press, 19.

[46] Zengh, J., & Hu, J. (2003, November 9-11). Calculation of longshore sediment transport in Shijiu Bay. Paper presented at International Conference on Estuaries and Coasts,, Hangzhou, China.

[47] Rogers, A. L., & Ravens, T. M. (2008). Measurement of Longshore Sediment Transport Rates in the Surf Zone on Galveston Island, Texas. *J. Coast. Res.* [2], 62-73.

[48] Smith, E. R., & Wang, P. (2001). Longshore sediment transport as a function of energy dissipation. Paper presented at ASCE Conf. Proc. Ocean Wave Measurement and Analysis.

[49] Wang, P., Ebersole, B. A., Smith, E. R., & Johnson, B. D. (2002). Temporal and spatial variations of surf-zone currents and suspended sediment concentration. *Coast. Eng.*, 46, 175-211.

[50] Singh, A. K., Deo, M. C., & Kumar, V. S. (2007). Prediction of littoral drift with artificial neural networks. *Hydro. Earth Syst. Sciences Disc.*, 4, 2497-2519.

[51] Schoonees, J. S., & Theron, A. K. (1995). Evaluation of10 cross-shore sediment transport morphological models. *Coast. Eng.*, 25, 1-41.

[52] Davies, A. G., van Rijn, L. C., Damgaard, J. S., van de Graaff, J., & Ribberink, J. S. (2002). Intercomparison of research and practical sand transport models. *Coast. Eng*, 46(1), 1-23.

[53] Van Maanen, B., Ruiter, d. P. J., & Ruessink, B. G. (2009). An evaluation of two alongshore transport equations with field measurements. *Coast. Eng.*, 56, 313-319.

[54] Larroudé, Ph. (2008). Methodology of seasonal morphological modelisation for nourishment strategies on a Mediterranean beach. *Mar. Pollut. Bull.*, 67, 45-52.

[55] Dally, W. R., Dean, J. L., & Dalrymple, R. A. (1984). A model for breaker decay on beaches. Paper presented at In 19th Coastal Engineering Conference Proceedings. 82-88.

[56] Svendsen, I. A. (1984). Mass flux and undertow in the surf zone. *Coast. Eng.*, 8, 347-365.

[57] Longuet-Higgins, M. S. (1970). Longshore currents generated by obliquely Incident waves 1. *Journal of Geophysical Research*, 75, 6778-6789.

[58] Camenen, B. (2003). Comparison of sediment transport formulae for the coastal environment. *Coast. Eng.*, 48, 111-132.

[59] Bijker, E. (1968). Littoral drift as function of waves and current. Paper presented at Coastal Engineering Conference Proceedings, London, UK. 415-435.

[60] Van Rijn, L. C. (1993). *Principles of sediment transport in rivers, estuaries and coastal seas.*, Publ. Aqua Publications, Zwolle, the Netherlands.

[61] Soulsby, R. (1997). *Dynamics of marine sands, a manual for practical applications.*, Thomas Telford,, Wallingford, England.

[62] Camenen, B. (2000). Numerical comparison of sediment transport formulae. Sandwave Dynamics Workshop Lille, France. , 37-42.

Chapter 2

Sediment Transport Patterns Inferred from Grain-Size Trends: Comparison of Two Contrasting Bays in Mexico

Alberto Sanchez and Concepción Ortiz Hernández

Additional information is available at the end of the chapter

1. Introduction

The analysis and description of geological processes in sedimentary environments have been widely defined by the frequency distribution of grain size (Friedman, 1961) whereas the change in the sediment textural characteristics has been used to evaluate the net sediment transport (Mc Laren and Bowles, 1985). The evaluation as a grain-size trend analysis (GSTA) is defined in Gao and Collins (1991, 1992) and LeRoux (1994a,b). The use of textural characteristics (grain size, sorting, and skewness) to infer the sediment transport was originally shown by a decrease in particle size in the direction of flow. In Sunamura and Horikawa (1971) a combination of grain-size and sorting identified four possible examples where it is possible to infer the direction of the sediment transport. In the early 1980s, the use of grain size, sorting, and skewness were proposed to infer sediment transport on the basis of the statistical analysis of sediment-transport paths along the transect (Friedman, 1961). Later bidimensional models of sediment transport proposed by Gao and Collins (1992), LeRoux (1994a) and Poizot et al. (2008) are supported by analytic geometry, vector analysis, and statistics to obtain more robust results of the magnitude and direction of the transport vectors. The GSTA is an excellent approach for establishing sediment transport in a variety of environments such as rivers, beaches, harbors, estuaries, continental shelf, and submarine canyons (Carriquiry and Sánchez, 1999; Carriquiry et al., 2001; Sanchez et al., 2008, 2009, 2010; Sanchez and Carriquiry, 2012).

1.1. Bahia Chetumal

Bahia Chetumal is located in the Mexican Caribbean at the mouth of the Rio Hondo, which defines the border between Mexico and Belize and is one of the few surface runoffs from the Yucatan Peninsula. Organic and inorganic wastes of the extensive cane crops adjacent to the river are discharged directly into the Rio Hondo (Ortiz Hernández and Sáenz Morales, 1999) and subsequently are transported and deposited into the bay. Although studies have been made in Bahia Chetumal concerning the sedimentology (De Jesús Navarrete et al., 2000; Sanchez et al., 2008), the distribution of metals (García Ríos and Gold Bouchot, 2003; Díaz López et al., 2006; Buenfil Rojas and Flores Cuevas, 2007), and aromatic hydrocarbons (Álvarez Legorreta and Sáenz Morales, 2005) only in last study there is a good relationship between fine sediments and the concentration of aromatic hydrocarbons in the deepest parts of the bay reported.

1.2. Bahia Magdalena

The lagoon complex Bahia Magdalena-Almejas is an ecosystem with high biodiversity, fisheries, and tourism on the peninsula of Baja California. Recently, a synthesis and integration of studies on biology, ecology, physical oceanography, and social sciences in this lagoon complex was made for the purpose of providing accessible information for making decisions on the use and sustainable exploitation of the natural resources and to identify possible areas for ecological protection (Funes Rodríguez et al., 2007). However, the sedimentary processes in this bay were not integrated into this synthesis, but they are of importance in the assessment of the ecological risk and the recovery and rehabilitation of marine environments (Carriquiry and Sánchez, 1999; Sanchez et al., 2008). In contrast to Bahia Chetumal, in the lagoon complex there is no evidence of impacts derived from human activities, at least for metal contamination (Shumilin et al., 2005).

In our work, the transport and dispersion of the surface sediments are compared in these two contrasting bays. The interpretation of spatial trends of textural characteristics and their comparison with hydrographic records may form the basis of a framework for implementing environmental monitoring schemes in both bays, in addition to expanding our knowledge of the biogeochemical processes and their relationship with the sedimentary dynamics that determine the functioning of the ecosystem.

2. Setting

2.1. Bahia Chetumal

Bahia Chetumal is semielongated (~ 110-km long and ~ 20-km wide) with a maximum of 49 km in its central area and a minimum of 5 km at the head (Fig. 1). The bathymetry of the bay is relatively shallow (4-m on average) with a center channel from 6- to 8-m depth with a SW direction. There are some narrow, deep depressions known locally as pozas. The Rio Hondo and some smaller streams that flow into the bay cause estuarine conditions with a salinity of 10 to 18, decreasing from the mouth to the head of the bay. In the summer, rainfall is highest and accentuates the estuarine conditions in the bay (Carrillo et al., 2009). In this Caribbean region, the tides are mixed semidiurnal with a microtidal range between 10 and 20 cm (Kjerfve, 1981). Local winds are dominated by easterly and southeasterly trade winds and by large-scale perturbations, such as tropical storms, hurricanes, and cold fronts (Gallegos et al., 1997; Mooers and Maul, 1998). The winds are easterlies-south easterlies with a mean speed of 3.1 m s^{-1} and the maximum air temperature (> 34.6 °C) is during August (Carrillo et al., 2009). The dry season lasts from late March to early June and the winds are mainly from the southeast. The mean annual river discharge into the bay is about 1,500 million m^3. Recently, from observations during 2005–2006, it was estimated that the Rio Hondo discharge varies from about 9 to 24 to over 78 m^3 s^{-1} during the dry and the wet seasons. It can be as high as 220 m^3 s^{-1} during the wet season (SARH-CNA, 1987).

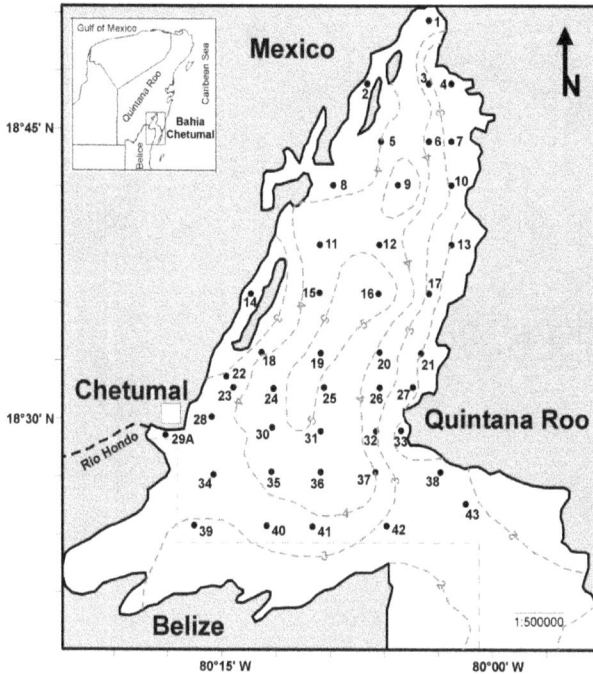

Figure 1. Study area and sampling stations in Bahia Chetumal. The dotted lines denote the bathymetry of the bay.

2.2. Bahia Magdalena

Bahia Magdalena is on the southwestern coast of the peninsula of Baja California (Fig. 2). The bay is characterized by a relatively shallow area with marshes, lagoons, and channels with a depth < 10 m. In the central part of the bay, the depth is greater than 20 m, with a channel that connects the bay with the ocean (Álvarez Borrego et al., 1975). In the ocean there is an area of seasonal upwelling (Zaitsev et al., 2003) that transports nutrients into the bay during spring tides. Inside the bay, the speed of the tidal current is ~23 cm s^{-1} during the flood tide and 20 cm s^{-1} during the ebb tide (Acosta Ruíz and Lara Lara, 1978), with a maximum tidal-current speed of 1.09 m s^{-1} in the mouth (Obeso Nieblas et al., 1999). In Morales Zárate et al. (2006) is described the circulation and passive transport of particles in Bahia Magdalena. Seeded particles tend to concentrate in the shallow and internal bay, along Isla Margarita and specific areas in the northern part of the bay. The maximum particle concentration occurred off the northern part of Isla Margarita, and was associated with transport generated by wind and the residual tidal flow.

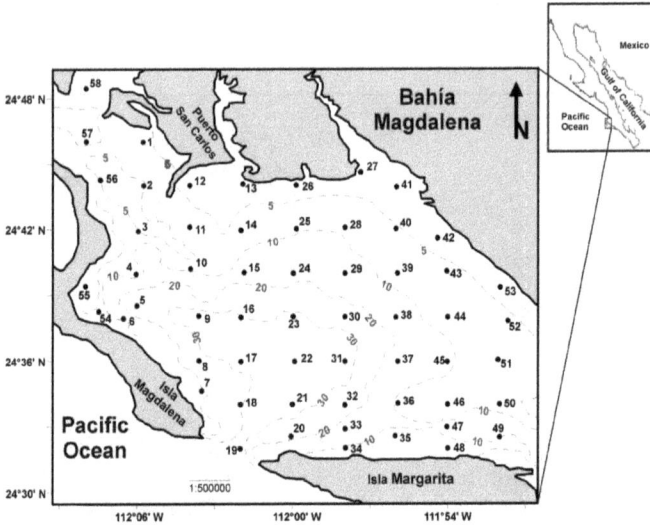

Figure 2. Study area and sampling stations in Bahia Magdalena. The dotted lines denote the bathymetry of the bay.

3. Methods

3.1. Sampling and analysis of surface sediments

In September 1998, the first 2 cm of the surface sediment was collected by scuba diving and Van Veen grab samples at the 43 sampling stations in Bahia Chetumal and the 58 in Bahia Magdalena (Fig. 1, 2). The sieve analysis was made by the sieving method (Ingram, 1971). The textural characteristics (grain size, sorting, and skewness) were calculated using granulometric data (Folk, 1974).

3.2. Multivariate statistical analysis

The principal component analysis (PCA) is a classical statistical method. This linear transform has been widely used in data analysis (Carriquiry et al., 2001). If X is a n X m data-matrix (n samples of m variables, here we choose grain size, sorting coefficient, and skewness, and X is demeaned and the covariance matrix is R, then a set of orthogonal eigenvectors $U - [u_1; u_2;...; u_n]$ exists:

RU−UΛ

Define:

Y - XU

Then Y – $[y_1; y_2; \ldots; y_n]$ also forms an orthogonal set. The $\{y_i\}$ are the principal components (PC), the $\{u_i\}$ are the eigenvectors of the covariance matrix, and the proportion of the total variance that each eigenvector "accounts for" is given by the magnitude of the eigenvalue. Here, we do not provide the principal components, but show the principal component weights (eigenvector) of the first PC that indicates the dominate relation among the three grain-size characteristics (Davis, 1986).

3.3. Sediment-transport model

Sediment-transport models have allowed coastal oceanographers to infer the residual-sediment transport based on spatial trends of sediment (Mc Laren and Bowles, 1985; Gao and Collins, 1992; LeRoux, 1994a; Poizot et al., 2008). In our study, the model proposed by LeRoux, (1994a,b), based on the principles of analytic geometry and vector analysis of textural data, was used. With this method, the magnitude and direction of the vector of transport were obtained by comparison of the textural characteristics of five neighboring sampling stations (one central and four satellites). The general considerations of the model are (1) textural trends resulting from the hydrodynamic conditions of the environment, (2) applicable in the coastal zone and shelf where sediment transport is unidirectional, (3) the gradient between textural parameters is constant in the area where we compared the five sampling stations, (4) textural parameters in the model have the same weight and importance, and (5) the distance between the five stations (interseasonal) is not critical, especially if there is a clear textural gradient between stations.

4. Results

4.1. Grain size trends

4.1.1. Bahia Chetumal

The average grain size was 1.6 ϕ with maximum grain size of -0.43 ϕ and minimum of 2.4 ϕ. In general, the spatial trend of grain size is to decrease toward the central part and head of the bay and in the area around the mouth of the Rio Hondo and the city of Chetumal, whereas the central area of the coast has a finer grain size than the eastern coastal area (Fig. 3A). The sediments with better sorting are associated with stations where the grain size is fine and vice versa (Fig. 3B). The coefficient of determination between these variables (grain size vs. sorting) is $R^2 = 0.52$ ($F_{1, 41, \alpha = 0.05} = 43.7$, $P = 0.0000$). The skewness (Fig. 3C) of the surface sediments is negative (toward coarse sediments) throughout the bay, except in the coastal margin, south of Isla Tamalcab (Fig. 3C), where skewness is moderately positive (toward fine sediments). The surface sediments are dominated 87% on average by sand (minimum 76% and maximum 96%) and a minor proportion of mud 13% on average (minimum 4% and maximum 24%).

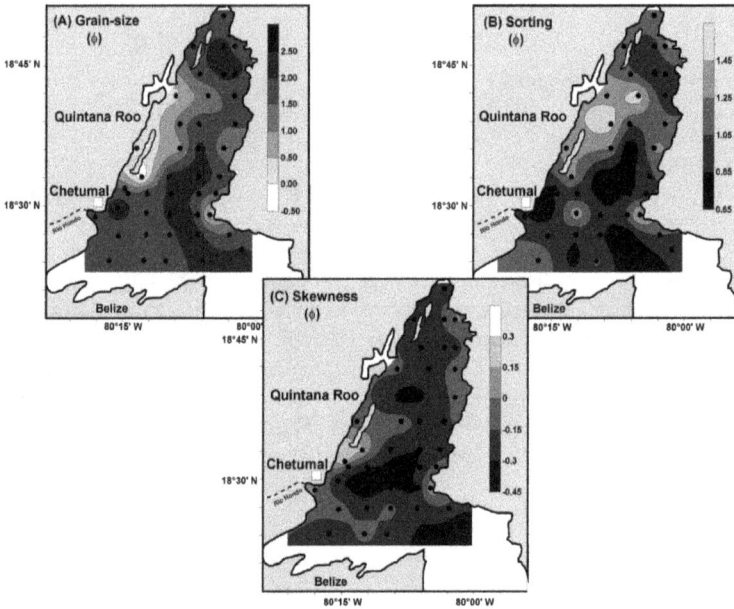

Figure 3. Spatial distribution of textural characteristics (A) Grain size (phi), (B) Sorting (phi), and (C) Skewness (phi) of sediments in Bahia Chetumal.

4.1.2. Bahia Magdalena

The grain-size distribution is relatively homogeneous in the bay with an interval in grain size from 2.5 to 3.5 ϕ with the exception of a textural gradient (increase in grain size) measured toward the mouth of the bay and Isla Margarita (Fig. 4A). In general, the sediments are well-sorted. Sorting is poor toward where the grain size increases (mouth of the bay and Isla Margarita (Fig. 4B). The skewness characteristics were systematically negative throughout the bay with certain trends toward positive values in the west, east, and toward the mouth of the bay and Isla Margarita (Fig. 4C).

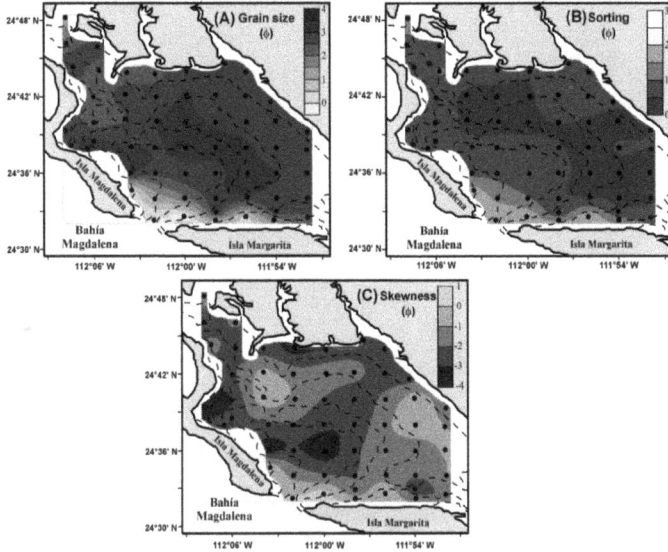

Figure 4. Spatial distribution of textural characteristics (A) Grain size (phi), (B) Sorting (phi), and (C) Skewness (phi) of sediments

4.2. Principal component analysis

4.2.1. Bahia Chetumal

The principal component analysis used for the textural data of sediment (grain size, sorting, and skewness) is shown in Figure 5A-C. The first eigenvalue, the grain size, and sorting explain 67% of the variance (Fig. 5A, B) and the second eigenvalue, skewness explains 28% of the variance (Fig. 5C). These two eigenvalues explain 95% of the variability of the textural characteristics. Only 5% of the variability is caused by other factors. The grain size and sorting showed a negative correlation. Thus, fine sediments are related to well-sorted clastic material and coarse-grained sediments are poorly sorted, whereas the skewness coefficient has a poor correlation with those textural characteristics.

4.2.2. Bahia Magdalena

The multivariate statistical analysis of principal components used for the Bahia Magdalena textural data indicated that two factors explain 98% of the total variability (Fig. 5D-F). Factor 1 is constituted by the grain size and sorting, and explains 74% of the variability (Fig. 5D, E), and factor 2, with 24% of the variability, corresponds to the skewness of the sediment (Fig. 5F). The grain size is inversely correlated with sorting, i.e. fine-grained sediments are better sorted with an skewness towards negative values. Only 2% of the variability is caused by other factors.

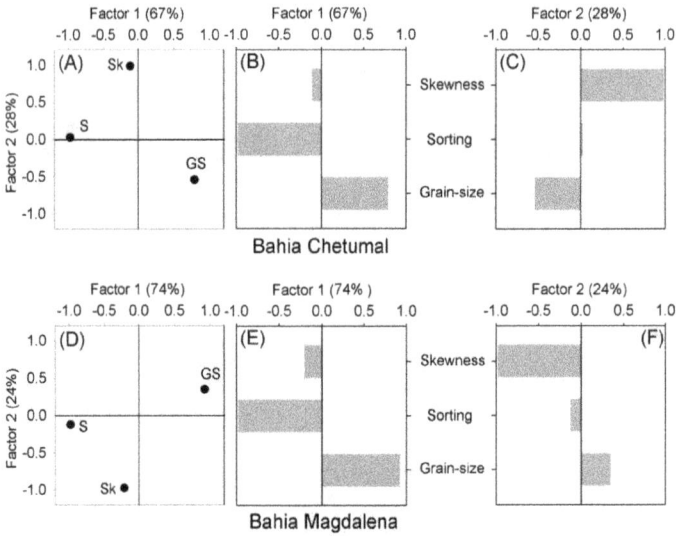

Figure 5. Graphic representation of the main factors and their respective loads obtained from the principal component analysis of sediment textural-data. GS: Grain-size, S: Sorting, and Sk: Skewness.

4.3. Net sediment transport

4.3.1. Bahia Chetumal

The vectors for the residual sediment transport are shown in Figure 6. In the north of the bay, the transport vectors show a SW direction. In the central region of the bay, transport vectors have a S-SE direction, except in the central-eastern margin, where the vector has a SW transport (Fig. 6). In the southern zone, the residual transport vectors showed a preferential direction to the S-SE. Stations 28, 29A, and 34 near the mouth of the Rio Hondo have a net sediment transport to the E-NE. In general, the surface sediment transport in Bahia Chetumal during sampling was from the head to the mouth of the bay (SW to SE direction). The smaller magnitudes of the transport vectors are located in the central part of the bay (Fig. 6) and this coincides with those areas of the bay in which the grain size tends to finer sizes.

4.3.2. Bahia Magdalena

The vectors of the residual-sediment transport described a cyclonic gyre in the central part and deep bay, whereas in the southeastern they describe an anticyclonic gyre (Fig. 7). In the northwestern margin of the bay, the residual-transport vectors showed a southeasterly direction (Fig. 7). On the coastal margin of Isla Margarita, the residual transport vectors denoted a pathway of sediment particles in a southwesterly direction (Fig. 7). The sediment transport to the mouth of the bay was not defined because only one sample was collected.

5. Discussion

5.1. Frequency distribution of magnitude vector

The vector analysis of the textural characteristics is a qualitative property that indicates the relative magnitude and the predominant direction of each vector, where two neighboring stations are exchanging material, without needing to provide quantitative data on the amount of material exchanged. The frequency distribution of the magnitude of the vectors was in the range of 0.4 to 0.7 (90% of the vector quantities) obtained in Bahia Chetumal. For Bahia Magdalena, the frequency distribution of the magnitude of the vector was heterogeneous, with only 34% of the vector magnitude in the range of 0.4 to 0.5 and the remaining 66% varied between magnitudes 0.1 to 0.4 and 0.6 to 1.0. Other studies have shown that 80% of vectors were characterized by vector magnitudes between 0.4 and 0.5 in the North Sea (Gao and Collins, 1994) and 0.9-1.0 for the northern Gulf of California (Carriquiry et al., 2001). This may suggest that hydrodynamic conditions in Bahia Magdalena are less homogeneous than those of Bahia Chetumal, the North Sea, and the northern Gulf of California, and therefore reflect the sedimentological environmental gradients. This suggests a higher variability in the exchange of materials between nearby stations (Carriquiry and Sanchez, 1999).

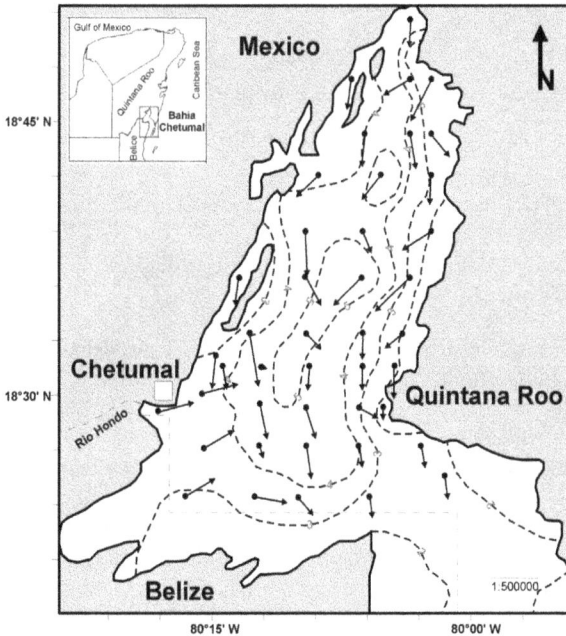

Figure 6. Dispersion of sediments in Bahia Chetumal inferred by the transport vectors. The vectors describe the main trajectory of the sedimentary material.

5.2. Transport and dispersion of sediments

The grain-size trend analysis has been widely used in marine and coastal environments to establish the net transport of sediments. In these studies, the net transport and dispersion of sediment were validated by comparing the residual vector transport with ocean currents (Carriquiry and Sanchez, 1999; Van Wesenbeck and Lanckneus, 2000; Carriquiry et al., 2001; Liu et al., 2002; Jia et al., 2002; Duman et al., 2004, 2006; Friend et al., 2006; Lucio et al., 2006; Duc et al., 2007). Thus, the GSTA is a useful tool to infer the movement of sediment particles in places where environmental hydrodynamics are poorly understood.

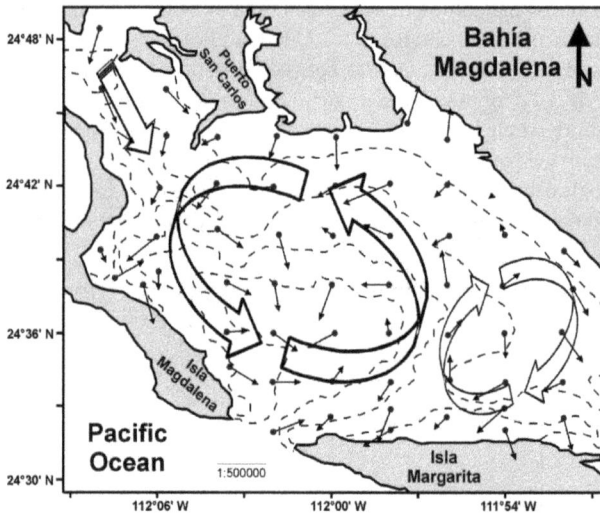

Figure 7. Dispersion of sediments in Bahia Magdalena inferred by the transport vectors. The circular arrows describe the main trajectory of the sedimentary material and residual tidal circulation.

The residual transport vectors allowed us to define that the net sediment transport is S-SW and S-SE inside Bahia Chetumal with a convergence towards the central part of bay consistent with the contours of the grain size and sorting. Hydrographic information is limited for Bahia Chetumal. The only earlier data come from Morales Vela et al. (1996) who measured the surface and bottom currents over a year inside the bay. The average surface current was 12 cm s^{-1} with a direction 183° ± 88°. On the bottom, the currents were slightly lower at 9 cm s^{-1} and a dominant direction of 182° ± 81°. Although the directions of surface currents varied between 137° and 240° and the bottom was 136° and 236°, both streams were in the S-SW and S-SE direction. Bahia Chetumal has a microtidal regime and tidal currents are expected to be weak. Therefore, the wind remains as a candidate to mix the water column. The prevalent wind direction was E–SE in the 70% of the total observed data. The role of prevailing winds and wind events is related to the orientation of the estuary. In Bahia Chetumal, the role of wind forcing in the circulation needs to be studied in detail (Carrillo et al., 2009). The persistence of relatively mixed conditions caused by strong wind events and the response of the plume to the direction of the wind deserve further study. However, the dispersion of sediment (S-SW and S-SE) is consistent with the direction of surface and bottom currents and winds reported for Bahia Chetumal (Morales Vela et al., 1996; Carrillo et al., 2009). The spatial distribution of fine to coarse sand suggests that the hydrodynamic conditions are sufficiently intensive to limit the deposition of silt and clay in the central and deep bay (Morales Vela et al., 1996; Sanchez et al., 2008). In Carrillo et al. (2009) it has been suggested that the magnitude of the surface and bottom currents are relatively low, consistent with Lankford (1977) who placed the Bahia Chetumal as a low energy environment. However, the currents (Morales Vela et al., 1996) inside the bay are able to resuspend (Shepard and Keller, 1978) and transport (Komar, 1977) sediment of 3 ϕ to 5 ϕ in the S-SW and S-SE directions (Sanchez et al., 2008).

The net sediment transport described a cyclonic gyre in the central part and deep portion and an anticyclonic gyre in the southeastern part of Bahia Magdalena. Studies of the circulation inside the bay are limited but useful to validate the sedimentary transport inferred from textural trends. In Sánchez Montante et al. (2007) the residual currents that result from the rectification of the forced circulation with the M2 tidal component were calculated. The results of residual tidal currents pointed to the presence of a cyclonic circulation in the central region of the bay and an anticyclonic circulation in the southeastern region. An experiment using the release of 58 particles inside the bay demonstrated the presence of a cyclonic gyre (Sánchez Montante et al., 2007), which corresponds to the residual tidal circulation and sediment transport and dispersion in the interior of the bay (Sanchez et al., 2010). An experiment with particles seeded in the bay shows they tend to remain inside and concentrate in areas that corresponded with the actual distribution of fish stocks. The areas of particle concentration were located along the coastal margin of the bay and the coasts of Islas Margarita and Magdalena. The sediment transport vectors defined convergence zones, which coincide with those areas in which the particles tend to be concentrated (Morales Zárate et al., 2006).

5.3. Variability of textural characteristics

The textural trends of the sediment have been extensively used to infer the possible paths of clastic material in different coastal environments. The combination of textural characteristics defines the existence of several examples for the inference of sediment transport and in all these, the sorting was better in the direction of the current (Sunamura and Horikawa, 1971; Mc Laren and Bowles, 1985; Gao and Collins, 1991, 1992; LeRoux, 1994a,b; Carriquiry and Sánchez, 1999; Carriquiry et al., 2001; Poizot et al., 2008; Sanchez et al., 2008, 2009, 2010; Sanchez and Carriquiry, 2012). The principal component analysis used for the textural characteristics of this study indicated that the spatial trends of grain size and sorting explain 67% and 74% of the variability, whereas the skewness explains 28% and 24% of the variability. The correlation analysis of grain size vs. sorting was significant ($F_{1,101,\alpha=0.05}=430$, $P<0.0000$) with a coefficient of determination ($R^2=0.88$). For grain size vs. skewness, the coefficient of determination ($R^2 = 0.28$) is consistent with the explained variability by the principal component analysis (22%). The results of Bahia Chetumal and Magdalena contrast with the variability in spatial trends of grain size and skewness that explain 95% of the variability in the Yellow Sea (Cheng et al., 2004).

In areas influenced by discharges from rivers, such as the Yellow Sea, the flocculation of the particles is one of the important factors affecting sorting and, to a lesser degree, grain size of the particles (Kranck and Milligan, 1991). This allows us to define the conditions of deposition on the basis of the analysis of the grain size (Fox et al., 2004). The difficulty of establishing a trend in the sediment texture is because flocculation can lead to a preference of the deposition of fine particles in the settling of selective sites of sediment that are related to the hydrodynamic conditions. The difference between the principal component analyses can be caused by processes derived from the flocculation of sediment, which allowed the establishment of poorly sorted material in the Yellow Sea. The discharge of ephemeral streams into Bahia Magdalena and the low discharge of the Rio Hondo into Bahia Chetumal promote well-sorted sediment deposition.

5.4. Implication of grain-size trend analysis

In general, the spatial trends of textural characteristics corroborate that the well-sorted fine-grained material is distributed over the central part of the bay and agreed with the preferential accumulation of organic matter, metals, and hydrocarbons in Bahia Chetumal (Ortiz Hernández and Sáenz Morales, 1999; García Ríos and Gold Bouchot, 2003; Álvarez Legorreta et al., 2005). This contrasts with Bahia Magdalena, which has no pollution problems, at least from potentially toxic elements (Shumilin et al., 2005) and organic matter (Sanchez, 2010, 2011).

The results obtained in the present and previous studies are of great significance for the region generally and particularly for the use and sustainable exploitation of natural resources, and for the definition of possible areas of environmental protection. By using the hypothesis that the contaminants preferentially associated with the fine particles of sediment, i.e. silts and clays, would follow the path of the sediment transport, the textural-trend analysis helps to identify the relationship between the discharge of pollutants and their sources, and helps to predict the transport and fate of contaminated sediments in marine environments. The development of Bahia Chetumal and Magdalena, caused by an increase of anthropogenic activities, undoubtedly will contribute to deterioration of these environments. The integration of research studies, including biological, chemical, and sedimentological, can be the basis for proposing monitoring programs, especially in areas where there are concerns about possible sources of pollution or are subject to a possible environmental impact, e.g. in the spawning and larval rearing areas of marine species of economic importance and the habitat and feeding sites of marine mammals such as the manatee and whales.

6. Conclusions

The grain size of surface sediments shows that particles are finer and better sorted with a skewness towards coarse particles in the central region and head of both bays. The principal component analysis indicated that the correlation between grain size and sorting was significant and explained 67% and 74% of the variability of the textural trend of grain size. The remaining 28% and 24% is explained by the skewness. In Bahia Chetumal, the net transport of sediment suggests that clastic material and particles (inorganic and organic) of anthropogenic or natural origin have a transport in the S-SW and S-SE directions, except near the mouth of the Rio Hondo where the net transport of sediment is preferentially E-NE. This indicates that there can be a greater influence on the dispersion of sediments and particulate pollutants that are discharged directly into the river, through the drainings of agricultural fields and urban waste. The convergence of the transport vectors in the central part of the bay is consistent with previous studies that described the spatial distribution of metals and hydrocarbons.

The net sediment transport described a cyclonic gyre in the central part and deep portion and an anticyclonic gyre in the southeastern part of Bahia Magdalena. These results agree with the residual tidal currents that showed the presence of a cyclonic circulation in the central region of the bay and an anticyclonic circulation in the southeastern region. The release of 58 particles inside the bay demonstrated the presence of a cyclonic gyre that corresponds to the residual tidal circulation and sediment transport and dispersion in the interior of the bay. The textural-trend analysis is a technique that can be useful for predicting the transport and fate of polluting sediments. The results of our research and the preliminary studies of geochemistry could be the basis to suggest new research and the monitoring of water quality and sediment, with special attention of areas where pollution problems exist or may be subjected to an eventual environmental impact.

Acknowledgements

We thank an anonymous reviewer for their valuable suggestions to this work. Financial support SIP 20110143 and 20120689, Instituto Politécnico Nacional, México. Thanks to Dr. Ellis Glazier for editing this English-language text.

Author details

Alberto Sanchez[1] and Concepción Ortiz Hernández[2]

*Address all correspondence to: alsanchezg@ipn.mx

1 Centro Interdisciplinario de Ciencias Marinas, Instituto Politécnico Nacional, La Paz, Baja California Sur, México

2 El Colegio de la Frontera Sur, Unidad Chetumal, Chetumal, Quintana Roo, México

References

[1] Acosta Ruíz J, Lara Lara J. Resultados fisicoquímicos en un estudio de variación diurna en el área central de Bahía Magdalena, B.C.S. Ciencias Marinas 1978; 5(1) 37-45.

[2] Álvarez Borrego S, Galindo Bect L, Chee Barragan A. Características hidroquímicas de Bahía Magdalena, B.C.S. Ciencias Marinas 1975; 2(1) 94-110.

[3] Álvarez Legorreta T, Sáenz Morales R. Hidrocarburos aromáticos policíclicos en sedimentos de la Bahía de Chetumal. In: Botello AV, Rendón von Osten J, Gold Bouchot G, Agraz Hernández C. (eds.) Golfo de México, Contaminación e Impacto Ambiental: Diagnóstico y Tendencias. Mexico: Universidad Autónoma de Campeche, Universidad Nacional Autónoma de México, Instituto Nacional de Ecología; 2005. p299-310.

[4] Buenfil Rojas M, Flores Cuevas N. Determinación de metales pesados (As, Cd, Hg y Pb) presentes en el Río Hondo, Quintana Roo. In: VI Congreso Internacional y XII Nacional de Ciencias Ambientales: Memorias en extenso, 6-8 June 2007, Chihuahua, Chihuahua. Mexico: Universidad Autónoma de Chihuahua, Academia Nacional de Ciencias Ambientales; 2007.

[5] Carriquiry JD, Sánchez A. Sedimentation in the Colorado River delta and Upper Gulf of California after nearly a century of discharge loss. Marine Geology 1999; 158(1-4) 125-145.

[6] Carriquiry JD, Sánchez A, Camacho-Ibar VF. Sedimentation in the northern Gulf of California after the elimination of Colorado River discharge. Sedimentary Geology 2001; 144(1-2) 37-62.

[7] Carrillo L, Palacios Hernández E, Yescas M, Ramírez Manguilar AM. Spatial and seasonal patterns of salinity in a large and shallow tropical estuary of the western caribbean. Estuaries and Coasts 2009; 32(5) 906-916.

[8] Cheng P, Gao S, Bokuniewicz H. Net sediment transport patterns over the Bohai Strait based on grain size trend analysis. Estuarine, Coastal and Shelf Science 2004; 60(2) 203-212.

[9] Davis J., editor. Statistics and data analysis in geology. New York: Wiley; 1986.

[10] De Jesús Navarrete A, Oliva Rivera JJ, Valencia Beltrán V, Quintero López N. Distribución de sedimentos en la Bahía de Chetumal, Quintana Roo, México. Hidrobiológica 2000; 10(1) 61-67.

[11] Díaz López C., Carrión Jiménez J.M., González Bucio J.L. Estudio de la contaminación por Hg, Pb, Cd y Zn en la Bahía de Chetumal, Quintana Roo, México. Revista Sociedad Química del Perú 2006; 72(1) 19-31.

[12] Duc DM, Nhuan MT, Ngoi CV, Nghi T, Tien DM, van Weering TjCE, van den Bergh GD. Sediment distribution and transport at the nearshore zone of the Red River delta, northern Vietnam. Journal Asian Earth Science 2007; 29(1) 558-565.

[13] Duman M, Avci M, Duman S, Demirkurt E, Duzbastilar MK. Surficial sediment distribution and net sediment transport pattern in Izmir Bay, western Turkey. Continental Shelf Research 2004; 24(9) 965-981.

[14]   Duman M, Duman S, Lyons TW, Avci M, Izdar E, Demirkurt E. Geochemistry and sedimentology of shelf and upper slope sediments of the south-central Black Sea. Marine Geology 2006; 227(1-2) 51-65.

[15]   Folk RL, editor. Petrology of Sedimentary Rock. Austin, TX: Hemphill Publishing Company; 1974.

[16]   Fox JM, Hill PS, Milligan TG, Boldrin A. Flocculation and sedimentation on the Pô River Delta. Marine Geology 2004; 203(1-2) 95-107.

[17]   Friedman GM. Distinction between dune, beach, and river sands from their textural characteristics. Journal of Sedimentary Petrology 1961; 31(4) 514-529.

[18]   Friend PL, Velegrakis AF, Weatherston PD, Collins MB. Sediment transport pathways in a dredged ria system, southwest England. Estuarine, Coastal and Shelf Science 2006; 67(3) 491-502.

[19]   Funes Rodríguez R, Gómez Gutiérrez J, Palomares García R. editors. Estudios ecológicos en Bahía Magdalena. Mexico: Centro Interdisciplinario de Ciencias Marinas. Instituto Politécnico Nacional, México; 2007.

[20]   Gallegos A, Czitrom S. Aspectos de la oceanografía física regional del Mar Caribe. In: Lavin MF. (ed) Contribuciones a la oceanografía física en México. Mexico: Unión Geofísica Mexicana; 1997. p225-242.

[21]   Gao S, Collins MB. A critique of the Mc Laren method for defining sediment transport paths: discussion. Journal of Sedimentary Petrology 1991; 61(1) 143-146.

[22]   Gao S, Collins MB. Net sediments transport patterns from grain size trends, based upon defi nition of 'transport vectors'. Sedimentary Geology 1992; 81(1-2) 47-60.

[23]   Gao S, Collins MB. Analysis of grain size trends, for defining sediment transport pathways in marine environments. Journal of Coastal Research 1994; 10(1) 70-78.

[24]   García Ríos V.A., Gold Bouchot G. Trace metals in sediments from Bahía de Chetumal, Mexico. Bulletin of Environmental Contamination and Toxicology 2003; 70(6) 1228-1234.

[25]   Ingram RL. Sieve analysis. In: Carver RE. (ed.) Procedures in Sedimentary Petrology. New York: Wiley Interscience; 1971. p49-68.

[26]   Jia JJ, Gao S, Xue YC. Sediment dynamic processes of the Yuehu inlet system, Shandong Peninsula, China. Estuarine, Coastal and Shelf Science 2003; 57(5-6) 783-801.

[27]   Kjerfve B. Tides of the Caribbean Sea. Journal of Geophysical Research 1981; 86(C5) 4243-4247.

[28]   Komar DP. Selective longshore transport rates of different grain size fractions within a beach. Journal of Sedimentary Petrology 1977; 47(4) 1444-1453.

[29] Kranck K, Milligan TG. Grain size in oceanography. In: Syvitski JPM. (ed.) Principles, methods and application of particle size analysis. New York: Cambridge University Press; 1991. p332-345.

[30] Lankford RR. Coastal lagoons of Mexico: Their origin and classification. In: Wiley M. (ed.) Estuarine Processes. New York: Academic; 1977. p182-215.

[31] LeRoux JP. An alternative approach to the identification of the end sediment transport paths based on grain size trends. Sedimentary Geology 1994a; 94(1-2) 97-107.

[32] LeRoux JP. A spreadsheet template for determining sediment transport vectors from grain size parameters: Computer and Geoscience 1994b; 20(3) 433-440.

[33] Liu JT, Liu K, Huang JC. The effect of a submarine canyon on the river sediment dispersal and inner shelf sediment movement in southern Taiwan. Marine Geology 2002; 181(4) 357-386.

[34] Lucio PS, Bodevan EC, Dupont HS, Ribeiro LV. Directional kriging: a proposal to determine sediment transport. Journal of Coastal Research 2006; 22(6) 1340-1348.

[35] Mc Laren P., Bowles D. The effects of sediment transport on grain size distributions. Journal of Sedimentary Petrology 1985; 55() 457-470.

[36] Mooers CK, Maul GA. Intra-Americas sea circulation. In: Robinson A.R., Brink K.H. (ed) The Sea. New York: Wiley; 1998. p183–208.

[37] Morales Vela B, Olivera Gómez D, Ramírez García P. editors. Conservación de los manatíes en la región del Caribe de México y Belice. Mexico: El Colegio de la Frontera Sur (ECOSUR)-Consejo Nacional de Ciencia y Tecnología (CONACyT); 1996.

[38] Morales Zárate MV, Aretxabaleta AL, Werner FE, Lluch Cota SE. Modelación de la circulación invernal y la retención de partículas en el sistema lagunar Bahía Magdalena-Almejas (Baja California Sur, México). Ciencias Marinas 2006; 32(4) 631-647.

[39] Obeso Nieblas M., Gaviño Rodríguez J.H., Jiménez Illescas A Modelación de la marea en el sistema lagunar Bahía Magdalena-Almejas, B.C.S., México. Oceánides 1999; 4(1) 79-98.

[40] Ortiz Hernández MC, Sáenz Morales R. Effects of organic material and distribution of fecal coliforms in Chetumal Bay, Quintana Roo, Mexico. Environmental Monitoring and Assessment 1999; 55(3) 423-434.

[41] Poizot E, Méar Y, Biscara L. Sediment Trend Analysis through the variation of granulometric parameters. A review of theories and applications. Earth-Science Reviews 2008, 86(1-4) 15-41.

[42] Sanchez A. editor. La Productividad Marina y su relación con los componentes biogénicos en el complejo lagunar de Bahía Magdalena-Almejas, Baja California Sur: Fase II. Mexico: Centro Interdisciplinario de Ciencias Marinas. Instituto Politécnico Nacional; 2010. p1-8.

[43] Sanchez A. editor. Distribución y composición de la materia orgánica en sedimentos del complejo lagunar de Bahía Magdalena-Almejas, Baja California Sur, México: Fase I. Mexico: Centro Interdisciplinario de Ciencias Marinas. Instituto Politécnico Nacional; 2011. p1-7.

[44] Sánchez A, Carriquiry J. Sediment transport patterns in Todos Santos Bay, Baja California, Mexico, inferred from grain-size trends. In: Mannig A. (ed) Sediment transport in Aquatic environments. Croatia: Intech; 2012. p1-18.

[45] Sánchez A, Alvarez Legorreta T, Sáenz Morales R, Ortiz Hernández MC, López Ortiz BE, Aguiñiga S. Distribución de parámetros texturales de los sedimentos superficiales en la Bahía de Chetumal: Implicaciones en la inferencia de transporte. Revista de la Sociedad Geológica Mexicana 2008; 25(3) 523-532.

[46] Sánchez A, Carriquiry J, Barrera J, López Ortiz BE. Comparación de modelos de transporte de sedimento en la Bahía Todos Santos, Baja California, México. Boletin de la Sociedad Geologica Mexicana 2009; 61(1) 13-24.

[47] Sánchez A, Shumilin E, López Ortiz BE, Aguíñiga S, Sánchez Vargas L, Romero Guadarrama A, Rodriguez Meza D. Sediment transport in Bahía Magdalena, inferred of grain-size trend analysis. Journal Latinoamerican of Aquatic Science 2010; 38(2) 167-177.

[48] SARH-CNA. Sinopsis geohidrológica del Estado de Quintana Roo. México: Secretaría de Recursos Hidraúlicos. 1987.

[49] Sánchez Montante O, Zaitsev O, Saldívar Reyes M. Condiciones hidrofísicas en el sistema lagunar Bahía Magdalena-Almejas. In: Funes-Rodríguez R, Gómez Gutiérrez J, Palomares García R. (eds.) Estudios ecológicos en Bahía Magdalena. Mexico: Centro Interdisciplinario de Ciencias Marinas. Instituto Politécnico Nacional; 2007. p1-28.

[50] Shepard FP, Keller GH. Currents and sedimentary processes in submarine canyons off the northeast United States. In: Stanley DJ, Kelling GK. (eds.) Sedimentation in Submarine Canyons, Fans and Trenches. Stroudsburg, Pennsylvania: Dowden, Hutchinson & Ross Inc.; 1978. p15-32.

[51] Shumilin E, Rodríguez Meza GD, Sapozhnikov D, Lutsarev S, Murrillo de Nava J. Arsenic concentrations in the surface sediments of the Magdalena-Almejas Lagoon complex, Baja California Peninsula, Mexico. Bulletin Environmental Contamination and Toxicology 2005; 74(3) 493-500.

[52] Sunamura T, Horikawa K. Predominant direction of littoral transport along Kujyukuri Beach, Japan. Coastal Engineering in Japan 1971; 14 107-117.

[53] Van Wesenbeck V, Lanckneus J. Residual sediment transport paths on a tidal sand bank: a comparison between the modified Mc Laren model and bedform analysis. Journal of Sedimentary Research 2000; 70 (3) 470-477.

[54] Zaitsev O, Cervantes Duarte R, Sánchez Montante O, Gallegos García A. Coastal upwelling activity on the Pacific shelf of the Baja California Peninsula. Journal Oceanography 2003; 59(4) 489-502.

Chapter 3

Sediment Transport Modeling Using GIS in Bagmati Basin, Nepal

Rabin Bhattarai

Additional information is available at the end of the chapter

1. Introduction

Soils play an important role in the maintenance of global food supplies, an ever increasingly important role as total population expands. The first 'Global Assessment of Human-Induced Soil degradation' (GLASOD) was published in 1990 and estimated that 1.97 billion hectares, equivalent to an area of 15% of total land cover, suffered degradation from the mid 1940's up to 1990. The more recent GLASOD (Global Assessment of Soil Degradation) survey has indicated more than 10^9 ha of the land surface of the world are currently experiencing serious soil degradation as a result of water erosion. For total suspended sediment yield from the land to the oceans, values closer to 15-20 x 10^9 tons year $^{-1}$ have been most frequently cited, whereas average global specific total sediment load is approximately 140-188 tons km^{-2} year^{-1} [23]. On a global scale, the loss of 75 billion tons of soil costs the world about US $400 billion/year (at US$3/ton of soil for nutrients and US$2/ton of soil for water), or approximately US$70/person/year [13].

Erosion prediction is the most widely used and most effective tool for soil conservation planning and design. Because it is impossible to monitor the influence of every farm and ranch management practice in all ecosystems under all weather conditions, erosion predictions are used to rank alternative practices with regard to their likely impact on erosion. These erosion predictions are thus an essential part of soil conservation programs. Assessment of soil erosion as to how fast soil is being eroded is helpful in planning conservation work. Estimates of the rate of soil loss may then be compared with what is considered acceptable and the effects of different conservation strategies can be determined. Modeling can be an effective method of predicting soil loss under a wide range of conditions as it can provide a quantitative and consistent approach to estimating soil erosion and sediment transport. Using remote sensing and GIS to parameterize such models allows them to be applied over local, regional and global scales.

Two main types of model: empirically based and process based are available for predicting soil erosion and sediment transport. Empirically based technology means regression or lumped mathematical models, which were developed using the experimental data of plot studies on erosion by water. Zingg [29] and Musgrave [18] equations are examples of initial steps towards the empirical soil erosion models. Universal Soil Loss Equation, USLE [26], later revised as Revised USLE or RUSLE [20] is one such model developed in the USA with more than 10,000 plot years of research data and experience of soil scientists. It is the most widely used model for soil erosion estimation because of the simplicity. It is based on the set of mathematical equations that estimate average annual soil loss from inter-rill and rill erosion. In addition, the equation combines interrelated physical and management parameters such as soil type, rainfall pattern, and topography that influence the rate of erosion. Erosion Productivity Impact calculator (EPIC) model [25], which was developed to assess the effect of soil erosion on soil productivity, also uses USLE and Modified USLE (MUSLE) model [24] to simulate erosion process. Chemical, Runoff and Erosion from Agricultural Management Systems (CREAMS) model [11], Agriculture Non-point Source Pollution model (AGNPS) model [28], and Soil and Water Assessment Tool (SWAT) model [1] are the examples of hybrid models which are based on USLE/MUSLE/RUSLE for the erosion estimation but use the sediment transport approach on the basis of continuity equation for sediment yield estimation.

Physically or process based models are intended to represent the essential mechanisms controlling erosion and sediment transport process. These models are the synthesis of individual component that affect the erosion and transport process. Aerial Non-point Source Watershed Environmental Response Simulation (ANSWERS) model [2], Kinematic Runoff and Erosion model (KINEROS) model [27], European Soil Erosion model (EUROSEM) model [17], and Water Erosion Prediction Program (WEPP) [19] are examples of process based models. Although physically based models try to emulate the physical processes involved in soil erosion and sediment transport, the weakness of these models is numerous parameters they need for calibration and also suffer from the problem of equifinality [3].

The overall aim of the study is the modeling of soil erosion and transport processes in distributed manner so that erosion, deposition and sediment yield can be computed and verified with the observations in data limited conditions. To achieve this objective, an empirical model was framed within Geographic Information System (GIS) to predict soil erosion in distributed manner. Then, the sediment delivery approach is used to predict sediment yield in this study. For the empirical approach, the revised form of the USLE model, RUSLE, is used to predict erosion potential on a cell-by-cell basis in conjunction with SEDD model to determine the catchment sediment yield by using the concept of sediment delivery ratio [7].

2. Methodology

A very popular empirical model, known as USLE is used to estimate soil erosion in this study. Then, sediment delivery approach is used to estimate the sediment yield which a part of eroded sediment that appears at watershed outlet. Empirical methods such as the USLE have been found to produce realistic estimates of surface erosion (and also sediment yield) over areas of small size [26, 10 4]. Sediment delivery distributed (SEDD) model couples USLE with a spatial disaggretion criterion of sediment delivery processes. The revised form of USLE, commonly known as RUSLE, is expressed as:

$$A = R * K * L * S * C * P \tag{1}$$

Where, A = average annual soil loss predicted (ton ha^{-1}), R = rainfall runoff erosivity factor (MJ mm ha^{-1} hr $^{-1}$), K = soil erodibility factor (ton ha hr MJ^{-1} ha^{-1} mm^{-1}), L = slope length factor, S = slope steepness factor, C = cover management factor and P = support practice factor.

The value of RUSLE factors are computed using the following methods as described in the Agricultural Handbook 703 [20].

$$R = \frac{1}{n} \sum_{i=1}^{n} \left(\sum_{j=1}^{m} E_j (I_{30})_j \right) \tag{2}$$

Where, n = total number of years, m = total number of rainfall storms in i th year, I_{30} = maximum 30 minutes intensity (mm hr $^{-1}$), E_j = total kinetic energy (MJ ha^{-1}) of j th storm of i th year and it is given as:

$$E_j = \sum_{i=1}^{p} e_k * d_k \tag{3}$$

Where, p = total number of divisions of j th storm of i th year, d_k = rainfall depth of k th division of the storm (mm), e_k = kinetic energy (MJ ha^{-1} mm^{-1}) of k th division of the storm and is given as [20]:

$$e_k = 0.29(1 - 0.72e^{(-0.05i_k)}) \tag{4}$$

Where, i_k = intensity of rainfall of k th division of the storm (mm hr^{-1})

If λ is the horizontal projection of the slope length (in meter), then L factor is given as:

$$L = \left(\frac{\lambda}{22.1} \right)^m \tag{5}$$

Where, λ = contributing slope length (in meter), m = variable slope length exponent.

The slope-length exponent 'm' is related to the ratio β of rill erosion (caused by flow) to interrill erosion (principally caused by raindrop impact) by the following equation:

$$m = \beta / (1 + \beta) \tag{6}$$

For moderately susceptible soil in both rill and inter-rill erosion, McCool et al. (1989) suggested the equation:

$$\beta = \frac{(Sin\theta / 0.0896)}{3.0(Sin\theta)^{0.8} + 0.56} \tag{7}$$

Where, θ = slope angle (degrees).

The slope steepness factor S is evaluated from the following equations (McCool et al., 1987):

$$S = 10.8 \sin \theta + 0.03 \ \ for \ s \ \ 9\%$$
$$S = 16.8 \sin \theta - 0.50 \ for \ s \geq 9\% \tag{8}$$

Where, s = slope in percentage.

C and P factors are assigned to different grid according to land cover while K factor is estimated using the soil data.

In a catchment, not all eroded soil reaches the catchment outlet but a part of the soil eroded in an overland region gets deposited within the catchment. The values of ratio of sediment yield to total surface erosion, which is termed as sediment delivery ratio (D_R), for an area are found to be affected by catchment physiography, sediment sources, transport system, texture of eroded material, land cover etc. [23]. However, variables such as catchment area, land slope and land cover have been mainly used as parameters in empirical equations for D_R [9, 12].

Ferro & Minacapilli [5] and Ferro [1997] hypothesized that D_R in grid cells is a strong function of the travel time of overland flow within the cell. The travel time is strongly dependent on the topographic and land cover characteristics of an area and therefore its relationship with D_R is justified. Based on their studies on probability distribution of travel time, the following relationship was assumed herein for a grid cell lying in an overland region of a catchment:

$$D_R = \exp(-\gamma t_i) \tag{9}$$

Where, t_i = travel time (hr) of overland flow from the ith overland grid to the nearest channel grid down the drainage path and γ = coefficient considered as constant for a given catchment.

The travel time for grids located in a flow path to the nearest channel can be estimated if the lengths and velocities for the flow paths are known. The direction of flow from one cell to a neighboring cell is often ascertained by using an eight direction pour point algorithm in grid-based GIS analysis. Once the pour point algorithm identifies the flow direction in each cell, a cell-to-cell flow path is determined to the nearest stream channel and thus to the catchment outlet. If the flow path from cell i to the nearest channel cell traverses m cells and the flow length of the i th cell is l_i (which can be equal to the length of a square side or to a diagonal depending on the direction of flow in the i th cell) and the velocity of flow in cell i is v_i, the travel time t_i from cell i to the nearest channel can be estimated by summing up the time through each of the m cells located in that flow path:

$$t_i = \sum_{i=1}^{m} \frac{l_i}{v_i} \tag{10}$$

In this study, the method of determination of the overland flow velocity proposed by the US Soil Conservation Service was chosen due to its simplicity and the availability of the information required (SCS, 1975). The flow velocity (v_i) is considered to be a function of the land surface slope and the land cover characteristics:

$$v_i = a_i * S_i^b \tag{11}$$

Where, b = a numerical constant equal to 0.5 [22, 5], S_i = slope of the i th cell and a_i = a coefficient related to land use [8]. Introducing equations (10) and (11) into equation (9) gives

$$D_R = \exp(-\gamma \sum_{i=1}^{m} \frac{l_i}{a_i S_i^{0.5}}) \tag{12}$$

It should be noted that $l_i / S_i^{0.5}$ is the definition of travel time used by Ferro & Minacapilli [5]. Values of the coefficient a_i for different land uses were adopted from [8] and are presented in Table 4.

If S_E is the amount of soil erosion produced within the i th cell of the catchment estimated using equation (1), then the sediment yield for the catchment, S_y, was obtained as follows:

$$S_y = \sum_{i=1}^{n} D_R * S_E \tag{13}$$

Where, n = the total number of cells over the catchment and the term D_R = the fraction of S_E that ultimately reaches the nearest channel. Since the D_R of a cell is hypothesized as a function of travel time to the nearest channel, it implies that the gross erosion in that cell multiplied by the D_R value of the cell becomes the sediment yield contribution of that cell to the nearest stream channel. The D_R values for the cells marked as channel cells are assumed to be unity.

3. Study Area

The study area selected for this study is Bagmati Basin, Nepal. The basin is chosen because of its bio-climatic diversity due to elevation differences from valley floors to mountain summits, and related land use changes having influence on soil erosion, which is considered typical for the Middle Mountains of Nepal. Bagmati is the draining river from the Kathmandu city which is the capital of Nepal. The Bagmati basin covers an area of 3,500 km² in total and drains out of Nepal across the Indian State Bihar to reach the Ganges. The watershed with the elevation ranging from 57 m to 2,913 m is situated at latitude of 26° 30′ to 28°N and longitude 85° to 86°E. The watershed can be divided into three main areas: the upper, middle and the lower Bagmati watershed areas (BWA). The Upper Bagmati Watershed Area covers the whole of the Kathmandu valley including its source at Shivapuri. From the Chovar gorge, the river flows into the Middle Bagmati watershed Area across the Mahabharat and Siwalik ranges. The catchment area of upper and middle Bagmati basin is about 2,800 km². The terrain of the upper and middle BWA is rugged and comprised of several steep mountains except Kathmandu valley. The area of upper and middle Bagmati basin draining to Karmaiya is considered in the study on the basis of data availability.

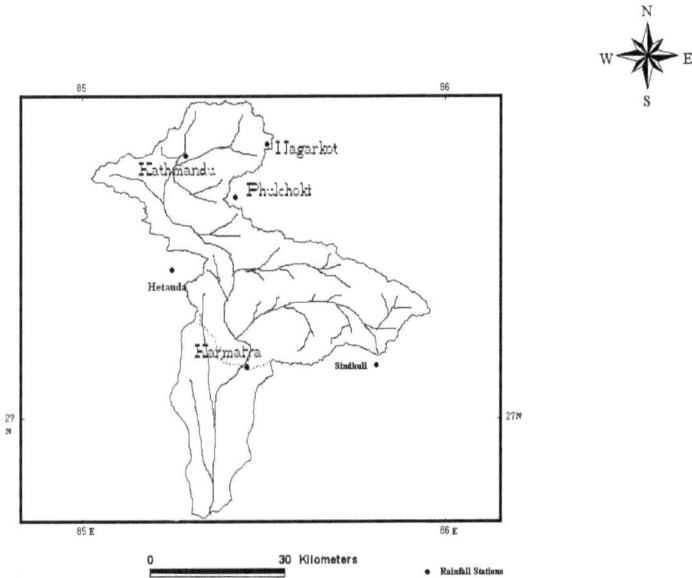

Figure 1. Bagmati basin, Nepal

The climate of the Bagmati watershed can be subdivided into three altitude/climate zones. These are: (a) Subtropical sub humid zone below 1,000 m: the southern most parts of the Bagmati watershed area including the Siwaliks region lie in this zone, (b) Warm temperate humid zone between 1,000-2,000 m: a large part (more than 60%) of the BWA lies in warm temperate humid zone between 1,000 – 2,000m altitudes and (c) Cool temperate humid zone between 2,000-3,000 m: only a small portion (about 5%) of the Bagmati watershed falls above

2,000 m. The annual average rainfall in the watershed is about 1,800 mm and it produces 1,400 mm of runoff per year on average, which accounts for about 75% of annual average rainfall. In the basin, steep slope in mountainous area and land use change are the major factors of soil erosion, which is considered typical for the Middle Mountains of Nepal. Total population in the catchment is about 1.5 millions. Figure 1 shows the map of the catchment along with streams and tributaries.

S N	Data type	Stations	Location		Duration	Remarks
			Lat. (N)	Long.(E)		
1	Rainfall	Daman	27° 36'	85° 05'	1987-97	Daily
		Hetauda	27° 25'	85° 03'	1987-97	Daily
		Godavari	27° 35'	85° 05'	1987-97	Daily
		Airport	27° 42'	85° 22'	1990-97	Daily
					1993-97	Hourly
		Nagarkot	27° 42'	85° 31'	1990-97	Daily
		Sindhuligadhi	27° 17'	85° 58'	1990-97	Daily
		Karmaiya	27° 07'	85° 28'	1990-97	Daily
2	Sediment	Karmaiya	27° 07'	85° 28'	1990-91,93,95-97	Daily

Table 1. Description of hydrologic data set

S N	Description	Scale or grid resolution	Source	Remarks
1	DEM	90 m (SRTM DEM)	USGS	Raster
2	Landuse	1:25,000	BIWMP	Vector
3	Soil	1:25000	BIWMP	Vector
4	Watershed boundary	1:25,000	BIWMP	Vector
5	River network	1:25000	BIWMP	Vector

Table 2. Description of spatial data set

Hydrologic data (rainfall, evaporation, suspended sediment concentration) for the basin are obtained from Department of Hydrology and meteorology (DHM). Digital Elevation Model (DEM) data, in 90 m resolution, was obtained from obtained from United States Geological Survey (USGS) (available at: http://srtm.usgs.gov). STRM DEM provides comprehensive and consistent global coverage of topographically derived data sets, including streams, drainage basins and ancillary layers. Other spatial data set such as: soil, land use, basin boundary, river network are obtained from Bagmati Integrated Watershed Management

Programme (BIWMP). The details of hydrologic data are provided in Table 1 while Table 2 contains the details about spatial data set.

As observed in the DEM of the watershed (Figure 2), the elevation varies significantly from as low as 137 m to as high as 2913 m from mean sea level. Lower part of the watershed is relatively flat compared to the upper and middle part. Kathmandu, the capital of Nepal lies in the upper part of the watershed. One third of the watershed is relatively flat as 34% of the watershed area has slope in the range of 0 - 10%. About 50% of the area has mild slope ranging from 10 - 30%. Remaining 15% watershed contains high slope with slope value more than 30%.

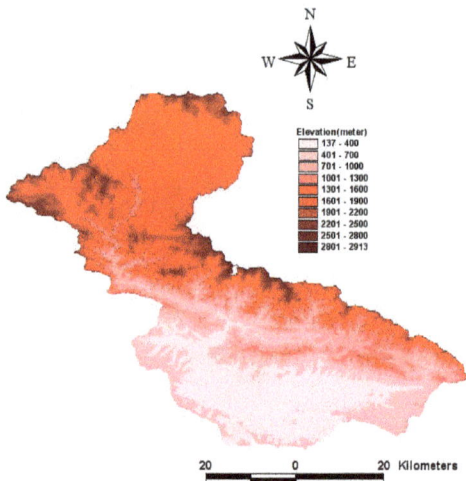

Figure 2. Digital Elevation Model (DEM) of Bagmati basin (SRTM DEM, 90 m resolution)

The land use in the watershed is observed to be mixed type. Cultivated land is major land use pattern in the upper part of the watershed while in middle and lower part of the watershed, forest area is seen to be dominant land use type. Majority of built-up area falls on the upper part of watershed, which represents Kathmandu. The land use pattern in the watershed is presented in Figure 4. 5. More than half of the watershed area (58%) is covered by forest. Cultivated land accounts for 38% of the area of the watershed while nearly 4% of the land in the watershed is barren. The land use distribution in the watershed is presented in Figure 3. The most extensive soils in the area are Dystrochrepts, Hapludalfs and Haplumbrepts, which occupy most of the hilly and mountaineous land. The texture of these soils is sandy/loamy in nature that varies from sandy clay to loam. The Dystrochrepts are also the most important soils in the inner Terai valleys. Soil type Rhododtalfs is commonly found in the gently undulating slopes and restricted to scattered, quasi-subtropical areas in the lower Hiamlayas. These soils are prone to severe soil erosion. The soil in the south face on the low altitude Mahabharat range is Dystrochrepts and Hapludalfs. These soils are mostly cultivated. The Haplaquepts are the dominant soils in the Terai plain as well as on paddy fields in hilly areas and elsewhere. Major soil types in the mountainous lands are Haplumbrepts and Dystrochrepts. Loamy soil texture is dominant in the watershed as demonstrated in Figure 4.

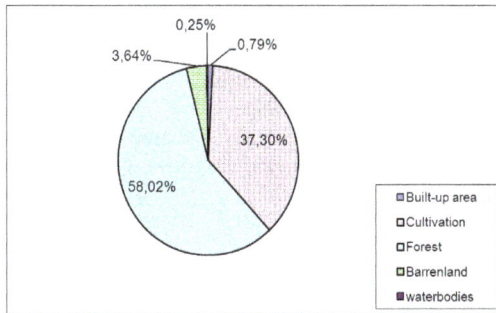

Figure 3. Land use distribution in Bagmati basin

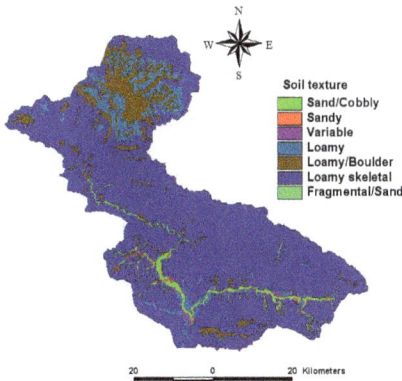

Figure 4. Soil map of Bagmati basin

Data Preparation And Simulation

Revised Universal Soil Loss Equation is one of the simplified models, which predicts soil erosion from hillslopes. The factors such as rainfall runoff erosivity factor (R) associated with the model represent the effects of climatic parameters in soil erosion while soil erodibility factor (K) represents the nature of the soil, its characteristics and influence in soil erosion. Topography and land use practices are other major factors incorporated in the model to account their effects in soil erosion.

Out of seven rainfall stations in Bagmati basin, one station measures hourly rainfall while remaining six other stations measures daily rainfall. So, rainfall data from these seven stations are analyzed to find the correlation in the rainfall pattern. The analysis of daily, monthly and annual rainfall trends of these stations showed that the trend was similar for all these stations. This helped to in disaggregating daily rainfall data into hourly data for the remaining six stations. For the basin, rainfall erosivity index "R" value was computed for monthly basis R value for was computed using equations (2), (3) and (4) since sediment yield information was available on monthly basis. Soil erodibility (K) factor values were assigned on grid by grid basis on the basis of soil texture [21] of the basin and assigned K values are presented in Table 3. Topographical parameters (L, S) were extracted from 90 m resolution SRTM digital elevation model (DEM) obtained from USGS (http://srtm.usgs.gov). Equation (5) was used for L factor calculation while S factor was computed using equation (8) for each cell. Similarly, C value, which depends on land use, was obtained from different literature [21], [16]. The C values used assigned for different land use in the basin are tabulated in Table 4. In case of P factor, the value is taken 0.5 for agricultural land and for rest of the land use; P value is assigned to be 1.

Textural Class	Organic matter content (%)		
	0.5	2	4
Fine sand	0.0211	0.0184	0.0132
Very fine sand	0.0553	0.0474	0.0369
Loamy sand	0.0158	0.0132	0.0105
Loamy very fine sand	0.0580	0.0501	0.0395
Sandy loam	0.0356	0.0316	0.0250
Very fine sandy loam	0.0619	0.0540	0.0435
Silt loam	0.0632	0.0553	0.0435
Clay loam	0.0369	0.0329	0.0277
Silty clay loam	0.0487	0.0422	0.0343
Silty clay	0.0329	0.0303	0.0250

Table 3. Soil Erodibility factor by soil texture in SI unit (ton ha hr MJ^{-1} ha^{-1} mm^{-1})

S N	Land Use	C value basis	C Value	a value
1	Cultivation	Crops, disturbed land	0.4000	1.55
2	Water body	Depositional sinks	0.0001	3.08
3	Forest	Forest	0.0020	0.76
14	Barren land	Fallow	1.0000	3.08
15	Built-up area	Paved, occasional construction	0.0005	6.19

Table 4. Cover management factor (C) on the basis of land use

Results And Discussion

"Once RUSLE parameters for Bagmati basin was computed following the procedure out-lined alobe, sediment delivery ratio (SDR) map for Bagmati basin computed using Equation (12)". The SDR map for the basin is presented in Figure 5 below. It is observed that flat areas around the south and north parts of the watershed has low sediment delivery ratios while the hilly areas within the watershed had higher values for sediment delivery ratio. This find-ing is consistent with the fact that steep areas are supposed to have higher sediment deliv-ery ratio compared to flat areas. In terms of watershed management perspective, the areas with higher values of SDR should be given higher priority compared to areas with lower SDR values for implementation of erosion control measures in this watershed.

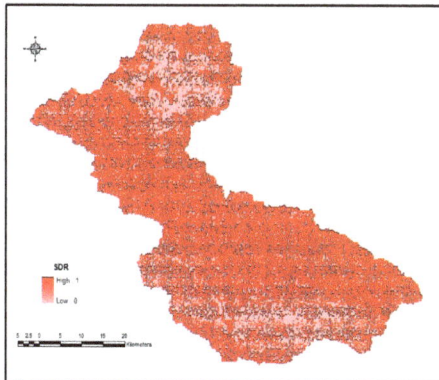

Figure 5. Sediment Delivery Ratio (SDR) map for Bagmati Basin

The sediment yield data are available for only few months of the year for Bagmati basin. So, it was not possible to analyze the long term sediment yield value and thus, monthly computation is carried out. Soil erosion map and SDR map was used to compute the sediment yield value at the watershed outlet. The Observed monthly sediment yield was compared with the computed as seen in Figure 6 below. The simulated result using this approach is fairly consistent with the observed data although this methodology slightly overpredicted sediment yield for the most of the observed months. There can be several reasons which can lead to overestimation of sediment yield values. For example, only one rainfall station had hourly measurement while remaining stations recorded daily values. It was assumed that the rainfall pattern over the watershed was similar. If rainfall data with finer temporal resolution were available for all the stations, the computed of R value would have been more reliable.

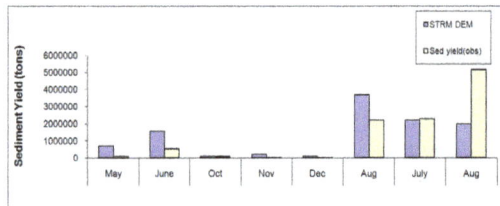

Figure 6. Comparison of simulated and observed sediment yield

The comparison of observed and computed sediment yield also indicate that great care is required in the selection of input values for the rainfall (R) and soil erodibility (K) factors. The USLE model was developed from the data suing the experiments that were carried out on a standard plot of 22.1 m length of uniform 9% slope. So, USLE-based performance can expected to be better for finer (for example 30 m) DEM resolution. Earlier studies have demonstrated that DEM resolutions can affect the outcome of RUSLE based simulations and better agreement can be obtained using fine DEM resolution [4]. Similarly, RUSLE results may be improved if more detailed soil, land use/cover data are available.

The model prediction may have been improved if γ coefficient was calibrated using the measured sediment yield values at mean annual scale for SDR computation. During SDR calculation, the sensitivity analysis of the parameter γ showed that the computed Sy was not very sensitive to γ in equation (12). The variation of γ value by 15 times (from 0.1 to 1.5) changed the Sy value only 10%. Since large variation in γ affected Sy insignificantly during sensitivity analysis, γ value was taken as 1 in the computation for simplicity. The sensitivity analysis has supported the findings of Jain & Kothyari [10] where they had reported that Sy was not very sensitive to γ in their study.

Conclusion

Soil erosion is a natural process. Modeling a natural process using mathematical simulation involves use of complex relationships. The number of factors associated with such complex process imposes their effect in various degrees. It is, thus, essential to consider only those factors, which are likely to have dominant effects in the process while carrying out mathematical simulation. This simplifies the process and is acceptable in most cases. Universal Soil Loss Equation (USLE) (and its revised form, RUSLE) is one of such simulation model, which predicts soil erosion from hillslopes. The factors such as rainfall runoff erosivity factor (R) associated with the model represent the effects of climatic parameters in soil erosion while soil erodibility factor (K) represents the nature of the soil, its characteristics and influence in soil erosion. Topography and land use practices are other major factors incorporated in the model to account their effects in soil erosion.

This study is an attempt to estimate soil erosion and sediment yield at Bagmati River basin using existing conceptual methods and GIS. This methodology can be used for the identification of sediment source areas and prediction of sediment yield at a catchment scale with available optimum data sets. ArcGIS was used for discretizing the catchment into grid cells of different resolutions. Grid cell slope, drainage direction and catchment boundary were generated from DEM using pour point method. The DEM was further analyzed to classify the grid cells into overland flow and channel region by using channel initiation threshold area approach. After preparing different USLE parameter layers, the gross surface erosion map was computed. The sediment delivery ratio of overland flow cell was assumed to be a function of the travel time of overland flow from given cell to the nearest downstream channel cell. For channel cells, the sediment delivery ratio was assumed to be unity. The computed and observed values were observed to have some discrepancy for monthly sediment yield. The variation is resulted by the few assumptions made during the analysis. In the study, computation of soil erodibility value (K) was based on soil texture only. Similarly, constant cover management factor (C) values were used instead of time varying because of the lack of series of land-use map for different years. Use of finer resolution DEM can also improve the estimation of slope length (L) and Slope steepness (S) factor. Improved results can be expected if these enhancements are incorporated. The proposed modeling framework is simple and can be a useful tool in conservation planning with reasonable reliability at data scarce areas.

Acknowledgements

The author would like to thank Department of Hydrology and Meteorology, Nepal and Bagmati Integrated Watershed Management Programme for providing the data used in this study. The present work benefited from the input of Dr. Dushmanta Dutta (CSIRO, Australia), who provided valuable comments and assistance to the undertaking of the research summarised here.

Author details

Rabin Bhattarai

Department of Agricultural and Biological Engineering, University of Illinois at Urbana-Champaign, Urbana, IL 61801, USA

References

[1]   Arnold, J. G., Srinivasan, R., Muttiah, R. S., & Williams, J. R. (1998). Large area hydrologic modeling and assessment part I: model development. *Journal of American Water Resources Association*, 34(1), 73-89.

[2]   Beasley, D. B., Huggins, L. F., & Monke, E. J. (1980). ANSWERS: A model for watershed planning. *Trans. ASAE*, 23(4), 938-944.

[3]   Beven, K., & Freer, J. (2001). Equifinality, data assimilation, and uncertainty estimation in mechanistic modelling of complex environmental systems using the GLUE methodology. *Journal of Hydrology*, 249(1-4), 11-29.

[4]   Bhattarai, R., & Dutta, D. (2007). Estimation of soil erosion and sediment yield using GIS at catchment scale. *Water Resources Management*, 21(10), 1635-1647.

[5]   Ferro, V., & Minacapilli, M. (1995). Sediment delivery processes at basin scale. *Hydrol. Sci. J.*, 40(6), 703-717.

[6]   Ferro, V. (1997). Further remarks on a distributed approach to sediment delivery. *Hydrol. Sci. J.*, 42(5), 633-647.

[7]   Ferro, V., & Porto, P. (2000). Sediment delivery distributed (SEDD) model. *J. Hydrol. Engng ASCE*, 5(4), 411-422.

[8]   Haan, C. T., Barfield, B. J., & Hayes, J. C. (1994). Design Hydrology and Sedimentology for Small Catchments. *Academic Press, New York*.

[9]   Hadley, R. F., Lal, R., Onstad, C. A., Walling, D. E., & Yair, A. (1985). Recent Developments in Erosion and Sediment Yield Study. *UNESCO (IHP) Publication, Paris, France*.

[10] Jain, M. K., & Kothyari, U. C. (2000). Estimation of soil erosion and sediment yield using GIS. *Hydrol. Sci. J.*, 45(5), 771-786.

[11] Kinsel, W. G. (1980). CREAMS: A Field Scale Model for Chemicals, Runoff, and Erosion From Agricultural ManagementSystems. *U.S. Department of Agriculture, Conservation Report*, 26, 640.

[12] Kothyari, U. C., & Jain, S. K. (1997). Sediment yield estimation using GIS. *Hydrol. Sci. J.*, 42(6), 833-843.

[13] Lal, R. (1990). Soil Erosion in the tropics, principles & management. *New York. McGraw-Hill Inc.*

[14] McCool, D. K., Brown, L. C., Foster, G. R., Mutchler, C. K., & Meyer, D. (1987). Revised slope steepness factor for the universal soil loss equation. *Trans. ASAE*, 30, 1387-1396.

[15] McCool, D. K., Foster, G. R., Mutchler, C. K., & Meyer, L. D. (1989). Revised slope length factor for the Universal Soil Loss Equation. *Trans. ASAE*, 32, 1571-1576.

[16] Morgan, R. P. C. (1995). Soil erosion & conservation. *Longman, UK*.

[17] Morgan, R. P. C., Quinton, J. N., Smith, R. E., Govers, G., Poesen, J. W. A., Auerswald, K., Chisci, G., Torri, D., & Styczen, M. E. (1998). The European Soil Erosion

Model (EUROSEM): A dynamic approach for predicting sediment transport from fields and small catchments. *Earth Surface Processes and Landforms,* 23, 527-544.

[18] Musgrave, G. W. (1947). The quantitative evaluation of factors in water erosion- a first approximation. *Journal of Soil and water conservation,* 2(3), 133-138.

[19] Nearing, M. A., Foster, G. R., Lane, L. J., & Flinkener, S. C. (1989). A process based soil erosion model for USDA water erosion prediction project technology. Trans. ASCE , 32(5), 1587-1593.

[20] Renard, K. G., Foster, G. R., Weesies, G. A., Mc Cool, D. K., & Yoder, D. C. (1996). Predicting Soil Erosion by Water: A guide to conservation planning with the Revised Universal Soil Loss Equation. *Agricultural Handbook 703. Agricultural Research Services. US Department of Agriculture.*

[21] Schwab, Glenn. O., Frevert, Richard. K., Edminster, Talcott. W., & Barnes, Kenneth. K. (1981). Soil and water conservation engineering. *John Willey & Sons, New York, USA.*

[22] Soil Conservation Service. (1975). Urban hydrology for small watersheds. *USDA Technical release* [55].

[23] Walling, D. E. (1988). Erosion and sediment yield research-some recent perspectives. *J. Hydrol.,* 100, 113-141.

[24] Williams, J. R., & Berndt, H. D. (1977). Sediment yield prediction based on watershed hydrology. *Trans. ASAE,* 20(6), 1100-1104.

[25] Williams, J. R., Jones, C. A., & Dyke, P. T. (1984). The EPIC model and its application, Proceeding of ICRISAT-IBSNAT-SYSS Symposium on Minimum data sets for Agro-technology Transfer. *March 1983, Hyderabad. India.*

[26] Wischmeier, W. H., & Smith, D. D. (1978). Predicting rainfall erosion losses-a guide to conservation planning. *US Department of Agriculture, Agricultural Handbook 537.*

[27] Woolhiser, D. A., Smith, R. E., & Goodrich, D. C. (1990). KINEROS, A kinematic run-off and erosion model: documentation and user manual. *ARS-77, USDA-Agricultural Research Service.*

[28] Young, R., Onstad, C. A., Bosch, D. D., & Anderson, W. P. (1987). AGNPS: Agricultural Non-Point Source Pollution Model: awatershed analysis tool. USDA-Agricultural Research Service. *ConservationResearch Report 35, U.S. Department of Agriculture, Washington, D.C.*

[29] Zingg, R. W. (1940). Degree and length of land slope as it affects soil loss in runoff. *Agriculture Engineering, ,* 21(2), 59-64.

Chapter 4

Sediment Transport Dynamics in Ports, Estuaries and Other Coastal Environments

X. H. Wang and F. P. Andutta

Additional information is available at the end of the chapter

1. Introduction

Given ever expanding global trade, the international economy is linked to the well-being of major coastal infrastructures such as waterways and ports. Coastal areas comprise about 69% of the major cities of the world; therefore the understanding of how coastal aquatic environments are evolving due to sediment transport is important. This manuscript discusses topics from both modelling and observation of sediment transport, erosion and siltation in estuarine environments, coastal zones, ports, and harbour areas. It emphasises particular cases of water and sediment dynamics in the high energy system of the Po River Estuary (Italy), the Adriatic Sea, the Mokpo Coastal Zone (South Korea), the Yangtze Estuary and the Shanghai Port, the Yellow Sea (near China), and Darwin Harbour (Northern Australia). These systems are under the influence of strong sediment resuspension/deposition and transport that are driven by different mechanisms such as surface waves, tides, winds, and density driven currents.

The development of cities around ports is often associated with the expansion of port activities such as oil, coal, and gas exportation. Such development results in multiple environmental pressures, such as dredging to facilitate the navigation of larger ships, land reclamation, and changes in the sediment and nutrient run-off to catchment areas caused by human activities [1]. The increase in mud concentrations in coastal waters is a worldwide ecological issue. In addition, marine sediment may carry nutrients and pollutants from land sources. An understanding of sediment transport leads to a better comprehension of pollution control, and thus helps to preserve the marine ecosystem and further establish an integrated coastal management system [e.g., 2-3]. [4] observed that many historical sandy coasts have been replaced by muddy coasts, and is considered permanent degradation. Additionally, [5] reported that recreational and maritime activities may be adversely impacted by processes of sediment resuspension and deposition. It was shown by [6] that increased sediment concentration in the Adriatic Sea has affected the growth of phytoplankton at the subsurface, because sunlight penetration is considerably reduced.

Before proceeding with the key issues about the transport of sediment in the previously mentioned systems, a brief and summarized overview of the main characteristics and dynamics of sediment transport is provided to contribute to the understanding of this chapter. In general, sediment particles considered in transport of sediment cycle, consist of non-cohesive and cohesive sediment types (Fig. 1a). (a) Sediments of particle size $d_{50} <$ 4 μm, mud or clay, are classified as a cohesive sediments. In contrast, (b) particle size $d_{50} >$ 64 μm may be weakly cohesive; however, these particles are included in the non-cohesive group, and range from mud through to sand [7]. The dynamics of sediment transport rely upon water circulation, salinity concentration, biological interaction, and sediment type. Cohesive sediments, such as clay and small-particle mud, are often transported in the water column, as these sediments are easily suspended by water currents. Alternatively, non-cohesive sediments, such as sand, are usually transported along the bottom by the processes of saltation, rolling, and sliding. Many numerical models include these processes and are based on empirical experiments, often performed in laboratories. These experiments provide estimates of the bed load transport according to particle size, bottom stress, and a threshold stress for initial bed movement [7].

The interaction between sediments is also an important feature pertinent to the transport of sediment. The interaction among cohesive sediments (mud) is different from that of non-cohesive sediments (sand). Cohesive sediments may aggregate, forming flocs of typical sizes of 100-200 μm. This aggregation process is called flocculation, and is caused by chemical or biological interaction. Flocculation is important for increasing the settling velocity; flocculated sediment particles settle faster on the bottom. "*Chemical flocculation*" is started by salinity ions that attach to the small mud particles, causing electronic forces between these particles, which start aggregating and thus forming a larger mud floc. In contrast, "*Biological flocculation*" is caused by bacteria and plankton, which produce exopolymer (i.e. a transparent mucus) that acts as glue between mud particles. This mucus results in the formation of extremely large flocs (~1000 μm in size), known as snow flocs [7].

The concentration of sediment near the surface may affect the formation of snow flocs, because sunlight penetration in the water column is decreased due to increased suspended-sediment concentration (SSC) and thereupon the reduced light penetration inhibits the production of plankton. In high turbidity waters, i.e. SSC > 0.5 g l^{-1}, marine snow is scarce; however, in less turbid water, i.e. SSC < 0.1 g l^{-1}, marine snow is common. Also, algae mats formation may influence the degree of erosion, because they decrease the propensity of sediment resuspension. In contrast, the influence of animal burrows may facilitate erosion [7]. Because of the different types of sediments and the flocculation process, the profile of the vertical distribution of SSC varies considerably. This vertical profile may indicate a well-mixed distribution, a smooth increase in sediment concentration with depth, or a depth-increase concentration with a step shape, called lutocline (Fig. 1b). The lutocline inhibits vertical mixing and thus conserves a nepheloid layer (i.e. bottom layer of high sediment concentration).

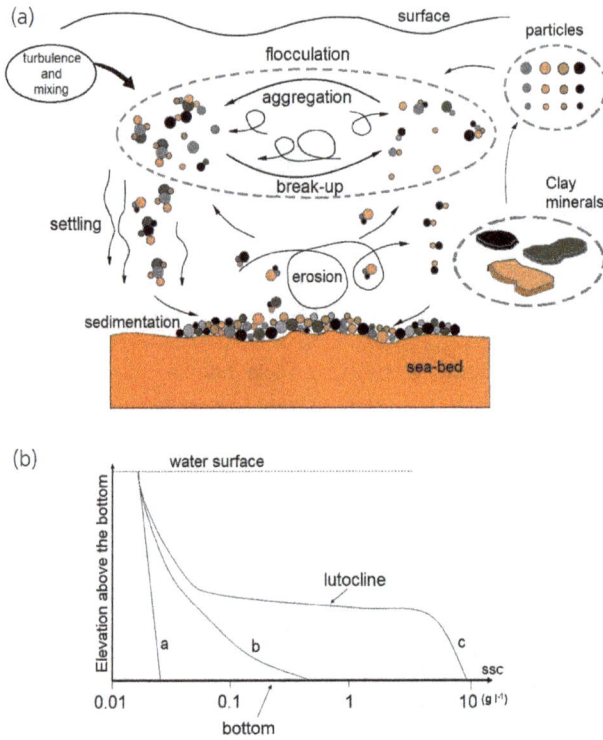

Figure 1. (a) Cycle of deposition and resuspension of cohesive sediment involved in particle aggregation and breakup. (b) Three typical vertical profiles of suspended sediment concentration in estuaries, where a, b and c denote a nearly well-mixed, partially-mixed and step shape profiles, respectively [source: 7-8].

This chapter gives an overview of four important suspended sediment transport processes that occur in ports, estuaries and other coastal environments. The following topics are investigated, based upon research on sediment dynamics at the University of New South Wales, Australia:

- The importance of including wave-currents when modelling sediment transport, showing the effect of waves generated during Bora events on SSC and net sediment flux in the Northern Adriatic Sea;

- The effect of increased SSC, combined with increased irradiance factor (Fc) of photosynthetically active radiation (PAR), on phytoplankton blooms (PB), with analysis of the PB event that occurred between January and April 2001 in the Mokpo Coastal Zone (Korea);

- The effect of coastal constructions on sediment transport, with analysis of the effect of dikes on the Yangtze River Delta, and problems with silting in the navigation channel of Shanghai Port (China);

- Tidal circulation modelling, specifically the role of mangrove and tidal flat areas in caus-ing tidal asymmetry, and the effect on the transport of suspended sediment in Darwin Harbour (Australia).

2. Description of study sites

2.1. The Po River and the Adriatic Sea

The Po River (~12.5° E and ~45° N) is 680 km in length, and is located in the northern area of the Adriatic Sea. It provides up to 50% of the total fresh water discharge into the Adriatic Sea (Fig. 2). The annual mean river flow is ~ 46 km^3/year, with the maximum river discharge events typ-ically occurring during the spring and a few times in autumn. The climate is temperate, with average temperatures of over 10° C in summer, and over 0° C in winter, and a runoff of 250-750 mm/year. Strong northeasterly winds prevail in winter, known as Bora events (typical wind speed ~30 m s^{-1}). These winds are usually ~10 ° C cooler than the water in the Adriatic Sea. In contrast, the southeasterly winds, which are often less intense and occur during summer and autumn, are known as Scirocco. The Bora and Scirocco wind conditions result in downwelling and upwelling events, respectively, in the western Adriatic coast [9-11].

The Adriatic Sea is a semi-enclosed sea, being one of the arms of the Mediterranean Sea. The Adriatic is connected to the eastern part of the Ionian Sea through the Ontranto Strait (Fig. 2). This sea is approximately 800 km long and 200 km wide. Depths vary from less than 200 m in the northern area, up to 1320 m in the southern area – with such depths covering an entire ~120 km wide expanse (i.e. South Adriatic Pit, [12]), and reduce to less than 800 m at the 70 km wide Ontranto Strait [10]. The eastern coast of the Adriatic Sea comprises numer-ous islands varying in main diameter from a few tens of meters up to tens of kilometres, and this coastline has many zones of high steepness. In contrast, the western coast has isobaths running parallel to the coastline and a smoother slope compared to the eastern coast (Fig. 2). The Adriatic Sea receives the runoff of 28 rivers, mostly located along the coast of Italy. The main river inflow to the Adriatic is from the Po River; however, the rivers Tagliamento, Piave, Brenta, and Adige together contribute a runoff of ~15.2 km^3/year, which is nearly one third of the Po's total runoff. The remaining 23 rivers in the Adriatic provide an average runoff of ~8 km^3/year [11].

The general circulation in the Adriatic Sea has been studied using field data and numerical simulations by [1, 10, 13-16], and is observed to be a cyclonic (anti-clockwise) circulation that is highly variable with the seasons [10, 13-17]. The annual water temperature excursion ex-ceeds 15° C. [1] observed the intense boundary current on the western side, the Western Adriatic Coastal Current (WACC), which is both thermohaline and wind driven. The WACC reaches maximum velocities during winter, under the influence of strong northeast-erly wind stress, i.e. Bora events [18]. The thermohaline component of the WACC is mainly forced by river discharge from the Po River, and thus reaches maximum intensity during spring and autumn [9, 19]. The position of the WACC is deflected from the inshore areas in winter, towards the shelf slope during the summer by an opposite wind-driven current due to Scirocco events. Similar processes showing boundary currents being pushed offshore by opposing wind-driven currents have also been observed at other shelves, such as: the Great Barrier Reef, the continental shelf north of the Monterey Bay, and New Jersey shelf [20-22].

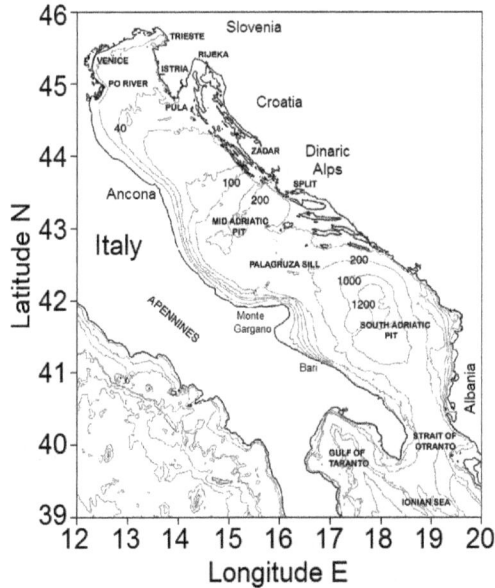

Figure 2. The Adriatic Sea location and a synthesized description of the main circulation.

Thermal balance in the Adriatic Sea is complex and influenced by river discharge (e.g. the Po River), surface heat flux by the wind (Bora and Scirocco events), and heat flux through the Otranto Strait. Water masses of the Northern Adriatic are renewed each year when the colder and denser water mass sinks and moves along the seabed to the deep basin of the Adriatic [1]. This northern Adriatic water mass forms a *"denser cascade water"*, which, for the Adriatic Sea, is caused by temperature gradients, while for many aquatic systems, located in Tropical and Sub-tropical areas, this is often caused by hypersaline waters [e.g. 23-30]. During the spring and summer, however, the water mass in the northern Adriatic is warmed up and forms a well-defined thermocline. Furthermore, the water discharge from the Po River is an important controlling factor to the baroclinic currents in the basin of the Adriatic Sea [9]. The thermal balance within the Adriatic Sea is also maintained by the net heat inflow through the Otranto Strait from the Ionian Sea [1].

The annual load of sediment from the Po River is 10-15 x 10^6 tons/year. The sediment in the Northern Adriatic Sea is mainly formed by sand with grain size varying from 50 to 2000 μm, and silt with grain size between 2 and 50 μm. The smaller sediment particles, i.e. clay, are also observed, however, they do not provide the major contribution of fine sediment in the northern area [1]. This chapter concerns the sediment transport in the Adriactic Sea of two classes: sediment particles larger and smaller than 50 μm grain size. [31] suggested that fine sediments such as silt and clay are mostly supplied from the Northern Adriatic Sea Rivers (e.g. Po River). Sediment is supplied into the sea and later dispersed through local circulation. Because the general circulation of the Northern Adriatic Sea is cyclonic, with the presence of the WACC on the western coast, there is a possibility that the sediment input from the rivers in the Northern Adriatic is transported southward by the coastal current. Therefore, the bottom sediment distribution would be predominantly sorted by the grain size according to their respective settling velocities.

2.2. The Youngsan River and the Mokpo Coastal Zone (Korea)

The Youngsan River Estuary (YRE) is located in the Mokpo coastal zone (MKZ), in the southwestern area of South-Korea (Fig. 3). The annual mean river flow is ~1.5 km^3/year, and the sediment load to the Yellow Sea is 0.7 x 10^6 tons/year. The climate is temperate, with average temperatures typically between 1.7° C and 4.4° C during winter and between 21.4° C and 26.1° C in summer. Maximum rainfall generally occurs during summer, accounting for 50 to 70% of the annual precipitation. Annual runoff is 250-750 mm. [11, 32-33]. The Mokpo area is located at the southeastern boundary of the Yellow Sea, and the YRE is connected to the Yellow Sea through four narrow inlets (i.e. ~1-3 km wide).

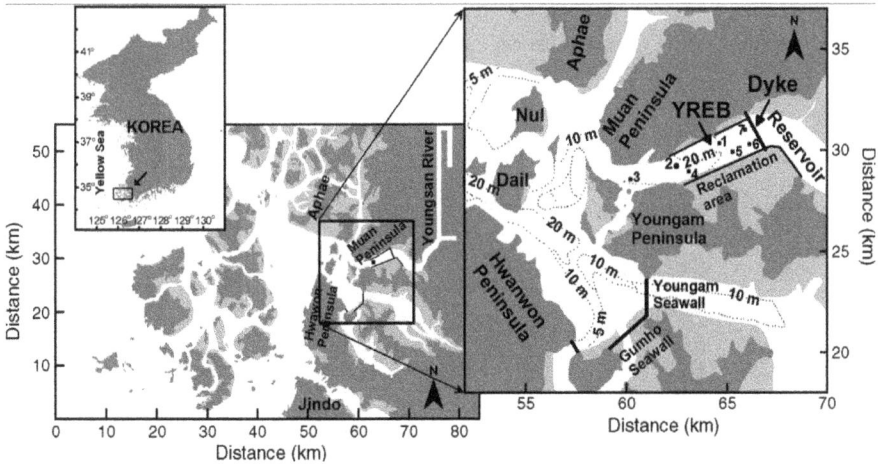

Figure 3. The domain of the hydrodynamic-sediment transport model at the southwest coast of Korea. The inset shows the location of the 1-D ecosystem model (●) in the Youngsan River Estuarine Bay (YREB) [source: 33].

Tidal features in the YRE are mixed, but predominantly semidiurnal according to the criteria of A. Courtier of 1938 [34], with the tidal form number [$N_f=(K_1+O_1)/(M_2+S_2)=0.28$]. Although there is the presence of many island and tidal flat areas, the tidal currents of the YRE are ebb dominant. Ebb/flood dominance is characterized by a shortened ebbing/flooding period, resulting in stronger ebb/flood currents, respectively. In addition, the flooding periods are nearly twice as large as the ebbing periods [35-37]. The ebb dominance is likely to be caused by important features such as the many scattered islands, combined with the extensive tidal flats [35]. Moreover, [38] observed that ebb dominance is likely to appear in regions of abundant tidal flats.

To add complexity to such tidal asymmetry problems (e.g. flood and ebb dominance), the MCZ has three important sea structures: the dike built in 1981, the Youngam seawall built in 1991, and the Geumho seawall built in 1994. Since the construction of these structures, changes in the tidal characteristics such as the increased amplitudes have been observed [37, 39-40].

This chapter section aims to show that in order to properly predict the variability in phytoplankton mass production in the turbid waters of the MCZ, it is important to use a 3D sediment transport model, coupled with the ecosystem model. This solves the variable vertical dynamics of sediment resuspension and mixing [33].

2.3. Yangtze River and the Shanghai Port in the East China Sea

The Yangtze River or Changjiang River (Fig. 4) is the third longest river in the world (6300 km), and the fourth in terms of both water flow (~900 km³/year) and sediment discharge (470-490 x 10⁶ tons/year), with the transport of a dissolved load of 180 x 10⁶ tons/year [11, 41-44]. The climate of this area is temperate, with temperatures of over 10° C in summer and 0° C winter, and an average runoff of 250-750 mm/year, the maximum river discharge occurring in summer [11]. The Yangtze is a mesotidal estuary according to the criteria of A. Courtier of 1938 [34], with a mean tidal range of 2.7 m [43].

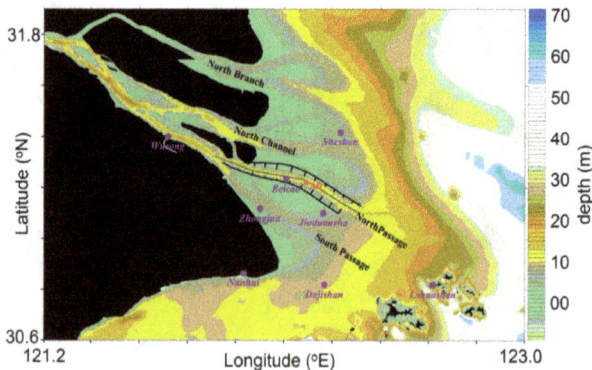

Figure 4. The Yangtze River or Changjiang River in China, and the indication of the navigation channel used for the trades of Shanghai Port [source: 44].

The sediment of the Yangtze River Estuary (YRE) mainly consists of small sediment particles of less than 63 μm (over 95%). The system is dominated by small sediment particles that lead to a highly turbid environment, and therefore the near bottom SSC can reach or exceed 4 g/l [44-47].

The Yangtze connects to the coastal zone through four inlets, namely North Branch, North Channel, North Passage, and South Passage. The main physical mechanics driving the transport of suspended sediment (TSS) varies between the four inlets: (a) in the South Passage TSS is mainly driven by tidal distortion, (b) in the North Passage TSS is dominated by gravitational circulation and tidal distortion, (c) in the North Channel TSS is dominated by gravitational circulation, and (d) for the North Branch the main mechanisms are not well described [44, 48-49]. For the North Passage other mechanisms are also suggested to contribute to the TSS and formation of the estuarine turbidity maximum zone (ETM), which include advective transport and turbulence suppression by salinity or suspended sediment induced stratification [50]. [51] performed a large analysis of the temporal and spatial variation of fluid mud, and flocculated settling. However, the joint contribution of the different TSS driving mechanisms with geometry is quite complex and requires further investigation.

The TSS in the Yangtze River Estuary (YRE) has been studied for many years [e.g. 41, 48-49, 52-59]. However, since the completion of the Deep Navigation Channel in 2011, important changes to the local hydrodynamics, and thus to the transport of sediment, are expected. In addition, there is the effect of the fluvial sediment trap by the Three Gorges Dam, which caused a significant decrease to fluvial sediment load [59-60]. Although the reduction of fluvial sediment has been reported, the silting problem attracted attention because the estimate deposition of sediment in the navigation channel was over 100% of the original yearly average predicted value, i.e. 30 million m^3 [62]. Recently, [44, 63] have reported that the greater siltation within the delta of the YRE is mostly influenced by the redistribution of local sediment through processes such as erosion and deposition within the delta area.

On Yangtze Estuary is the Shanghai Port, the world's busiest container port, which is extremely important to the economy of China. During 2010 and 2011, this port handled nearly 30 million container units per year. To facilitate local navigation, the Deepwater Navigation Channel (DNC) was built, 92 km in length and 12.5m deep. Although the channel comprises two dikes of nearly 50 km each, as well as 19 groins built to increase speed along the DNC, silting is still an issue, and dredging maintenance is greater than originally predicted [44, 55, 64-67].

A 1-DV model was applied to study the fine suspended sediment distribution at the South Channel-North Passage of the YRE [68]. Then, a 2D vertical integrated model was used to simulate, and subsequently to investigate the characteristics of tidal flow and suspended sediment concentration at this channel [69]. From these studies it was observed that new features had formed after the finalization of the shipping channel; however, the model used did not include the baroclinic component, which is an important factor in the transport of sediment.

2.4. Darwin Harbour (Australia)

Darwin Harbour (DH) is a shallow estuary, with a typical depth of less than 20 m and a maximum depth of up to ~40m. The harbour is situated in the Northern Territory (NT) of Australia, and connects to the Timor Sea. The land surrounding DH is occupied by the cities of Palmerston and Darwin (the latter is the capital of NT). DH is defined as the water body south of a line from Charles Point (west point) to Gunn Point (east point), and comprises the Port Darwin, Shoal Bay and the catchments of the West Arm, Middle Arm and East Arm

[Darwin Harbour Advisory Committee 2003]. DH forms two adjacent embayments. The western embayment receives the freshwater inputs predominantly from the Elizabeth River (flowing into the East Arm), the Darwin River, Blackmore River and Berry Creek (flowing into the Middle Arm), while the eastern embayment receives freshwater input from the Howard River [70]. DH area comprises numerous tidal flats and mangroves, with nearly 5% of the whole mangrove area in the Northern Territory, i.e. ~274 km². [71-72]

Darwin Harbour is forced by semi-diurnal tides, and is classified as a macro-tidal estuary (tidal form number Nf = 0.32). The maximum observed tidal range is 7.8 m, with mean spring and neap tidal ranges of 5.5 m and 1.9 m, respectively [11, 73-76].

Evaporation usually exceeds rainfall throughout the year, except during the wet season. From February to October, the evaporation rate ranges from 170 mm to 270 mm, respectively, with an average annual evaporation rate of ~2650 mm. The fresh-water input into DH is negligible in the dry season, and evaporation exceeds river discharge. Therefore, in the dry season salinity concentrations in the harbour may become at least 0.8 psu higher than the adjacent coastal waters [77].

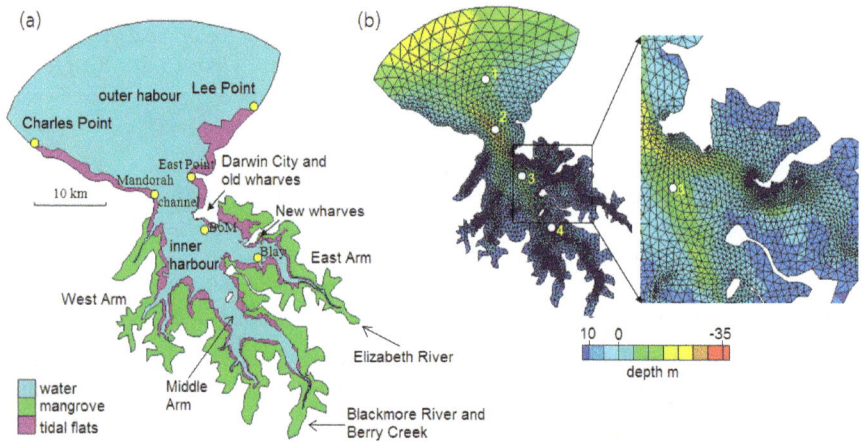

Figure 5. (a) Model domain of Darwin Harbour with indication of the harbour areas and data available to calibrate and validate the model (yellow dots). (b) Unstructured numerical mesh used for the simulations, where colour corresponds to depth in meters, and numbered points indicate location of sampling stations used to analyse tidal asymmetry along the harbour [source: 76].

The climate of this region is tropical savannah, with average monthly temperatures of over 20° C throughout the year. DH is located in a subarid/humid area with a typical rainfall of 1500-1600 mm/year (rainfall of 2500 mm in exceptionally wet years). Runoff typically varies between 100 and 750 mm/year, with maximum runoff usually occurring between October and April [11]. Although DH is of great economic importance to the NT, most of the current knowledge about the main driving forces for the local hydrodynamics is due to efforts by [75-76, 78-80]. To add complexity to the understanding of the hydro-dynamical and morphological changes in DH, the combined effect from the headlands, rivers, and embayments create a complicated bathymetry that leads to the formation of many tidal jets within narrow channels, eddies etc.

[76] Conducted some research at the western embayment of DH (Fig. 5a), and provided a calibrated and validated model to study the hydrodynamics in the harbour (Fig. 5b). From this study, the role that the mangrove and tidal flat areas play on the tidal asymmetry could be verified. It was thus confirmed that a decrease in area of the tidal flat and mangroves would lead to increased tidal asymmetry of flood dominance, and, because of this, result in the net sediment transport to the inner harbour area.

3. Methods

This chapter addresses the different study regions, followed by independent research and numerical modelling. As such, we have provided the methodology in separate sub-sections. Each of the following sub-sections summarizes the field work conducted, the calibration, and validation of the model for the four study sites.

3.1. Setting up of the numerical model for the Northern Adriatic Sea

For the Adriatic Sea, a sediment transport model similar to that of [81] was used, with im-provements made by incorporating the effect of wave current [1, 82-84]. The Adriatic Inter-mediate Model was based on the Princeton Ocean Model (POM) [85]; with the horizontal resolution of 5 km applied to a structured mesh. The model had 21 vertical layers that used the sigma coordinate, with a high vertical resolution was used near the surface and bottom. The simulations had the time steps of 7 and 700 seconds for the external and internal modes. The 2.5 turbulent closure method of Mellor-Yamada was used, and the diffusivity coefficient for SSC was assumed to be equal to that of heat and salt, and viscosity according to [86].

The flocculation of fine suspended sediment is mostly observed near the Po River mouth, and in areas before reaching the ocean [87]. Because of that, flocculation or aggregation processes were neglected, and thus all sediment behaves as a non-cohesive type and moving as a Newtonian fluid. For the fine sediment, i.e. silt and fine sand (20 < d < 60 μm), resuspen-sion was caused by turbulence. Inertia of sediment particles was also neglected, and their vertical velocity parameterized by a small settling velocity (w_s). For more information about the settling velocity, sediment source in the Adriatic Sea, and all the physical and numerical parameters used in the model, please refer to [1, 82-84].

The tides are known to be relatively week in the Northern Adriatic Sea; however, [83] in-cluded the tides to observe the tidal current effect on sediment transport. For the bottom stress two expressions were applied, an expression that considered the wave orbital velocity on the bottom, and the other expression that neglected this effect. The third version of the SWAN model was used to simulate the waves. The model was used in the stationary mode to compute the wave fields under the forcing of 6 hour interval.

The suspended sediment concentration was assumed not to affect water density. It is important to note that this last assumption is only valid for low concentrations of SSC, such as those lower than 1 g/l [e.g., 88-90]. The conservation of SSM in the water column was applied and the fluid considered incompressible. [2] showed that the Adriatic Sea is supplied by a riverine sediment input that is ~ 1.67 Mt/month, with the Po River contributing nearly 70% of that. The other rivers along the Adriatic coast had the equal contribution, which represented the remaining 30 % of the sediment input.

The sediment dynamics in the Northern Adriatic Sea is induced by riverine sources or resuspended sediment from the seabed. Simulations were conducted to quantify the different mechanisms responsible for the transport of suspended sediment [1], and the simulations examining in details the wave-current interaction [84].

The numerical simulations by [1] were: (a_1) simulation forced only by the Po River plume, (b_1) simulation forced by the Po River plume and wind stress. (c_1, d_1 and f_1) simulation forced by the Po River plume, wind stress, and additional wave forcing. For the simulation assuming wind effect, the assumed wind conditions were the Bora and Scirocco, which are typical wind conditions of the region [e.g., 9, 14-15, 19, 91] and summarized in [Table 2, in 1]. A homogeneous field with initial temperature of 12° C, and salinity 38 psu were assumed in the model. These are representative of ambient winter conditions without stratification. Simulations for a 30 day period were made, assuming continuous discharge from the Po River.

In contrast, the numerical simulations by [84] were: (a_2) simulation forced without waves, tides, and SSC effect on water density; (b_2) simulation forced by waves, but without tides and SSC effect on water density; (c_2) similar to b_2, except with waves assumed to be aligned with bottom currents; (d_2) simulation forced with waves and tides, but without SSC effect on water density; (e_2) simulation forced with waves and SSC effect on water density, but without tides. River runoff was assumed to be continuous from 1 January 1999 to 31 January 2001. The initial conditions were obtained from climatological simulation of the Adriatic Sea circulation in [17], and the sediment model was coupled with the hydrodynamical model from 1 December 2000.

3.2. Setting up of the numerical model for the Mokpo Coastal Zone

The simulation for MCZ consists of a 3D hydrodynamical model coupled with the sediment transport model, and a 1-D biogeochemical model [33]. The Princeton Ocean Model (POM) was chosen [85]. This model used the 2.5 turbulence closure scheme [92], and included the effect of sediment concentration on water density, and the stability function on the drag bottom coefficient [93]. The biological 1-D Modular Ecosystem Model (MEM) is based on the European Regional Sea Ecosystem Model [94]. This model constrains the physical and geophysical environmental conditions such as photosynthetically active radiation, temperature, and salinity. It also includes the trophic interactions between biological functional groups [95-96].

The simulations were run from January to April 2001, and the vertical salinity and temperature data used to calibrate/validate the hydrodynamical model were obtained by Mokpo National University at seven stations in the Youngsan River Estuary. The period of simulation partially covers the winter to spring seasons, and the distribution of the 7 sampling stations covers areas near the river mouth and upstream regions.

Hourly data from the hydrodynamic and sediment model were provided to the biogeochemical model. Specifically, temperature was used to compute the metabolic response to the biota, salinity was used for oxygen saturation concentration, the vertical diffusion coefficient was used for the biogeochemical-state variables, and the combination of sea surface elevation with suspended sediment concentration was used to estimate light penetration in the water column. The river discharge from the Youngsan Reservoir was also included. The water depth at the Youngsan River Estuarine Bay was assumed to be 21 m. The horizontal grid resolution of the model was 1km, with 18 vertical sigma layers. The open boundary was forced with the four main tidal components, i.e. M_2, S_2, K_1 and O_1. Nodal corrections and astronomical arguments were included to predict tides during the period of simulation.

The initial concentration values of pelagic biogeochemical are listed in (Table 1). The phytoplankton population, biomass content of carbon, nitrogen, phosphorus, and silicon for each phytoplankton group were obtained from [97]. Model sensitivity tests were performed in 8 different simulations, by assuming different parameters for light attenuation and vertical mixing rates, see Table 4 in [33]. A complete description of the whole setting of the model, the modelling experiments, and additional numerical and physical parameters is provided in [33].

	Nitrate	Phosphate	Ammonia	Silicate	DO
Surface	2.25	0.26	1.15	6.02	333
Bottom	1.36	1.20	1.20	7.36	340

Table 1. Initial condition assumed for concentrations of pelagic biogeochemicals (unit in mmol m^{-3}) [source: 33].

3.3. Setting up of the numerical model for the Yangtze Estuary

To study the hydrodynamics and transport of sediment, the 3D Princeton Ocean Model (POM) was used. This model uses a structured mesh and resolves the equations for momentum, temperature, and salinity using the finite differences method. The vertical coordinate is sigma [85, 98-99], and the turbulent closure method is described in [92, 100], while to compute the vertical mixing processes [101] was used. To compute the horizontal diffusion of momentum, the Smagorinsky diffusion scheme [86] was used. The complete description of the model is shown by [99]. The wetting and drying scheme for the domain is implemented in the model, with a minimum water depth established to avoid negative values [102-103].

To calibrate and validate the model in order to study the transport of sediment in the Deep Navigation Channel DNC of the YRE, field data measured in 2009 were used. The data were collected after the construction of the two dikes and 19 groins; however, the water depth was about 10.5 m at that time [43-44]. The equation used in the model, the initial conditions for the hydrodynamics, and initial sediment distribution are all described in [44]. The physical and numerical parameters are summarized in table 2.

Parameter	Description	Value
W_{s50}	free settling velocity	$-1.715\times10^{-5}(ms^{-1})$
m_1	empirical settling coefficient	-0.014
n_1	empirical settling coefficient	2.20
m_2	empirical settling coefficient	2.89
n_2	empirical settling coefficient	2.80
C_0	Flocculate empirical coefficient	$0.20 \ (kgm^{-3})$
a	empirical coefficient	10.0
β	empirical coefficient	0.5
E_0	empirical erosion coefficient	$2.0\times10^{-5}(kgm^{-2}s^{-1})$
τ_c	critical shear stress for erosion or deposition	$0.05(kgm^{-1}s^{-2})$

Table 2. Parameters used in the sediment transport model [source: 44].

Tidal harmonic components from 8 sites were used to verify the model, and the root mean square error (RMSE). The tidal components used in the model were observed to represent nearly 95% of the tidal oscillation (i.e. M_2, S_2, K_1 and O_1). Tidal currents were used to verify the water speed, and a good agreement was achieved. Salinity measurements were used to verify the proper simulation of mixing in the YRE, and aside from the periods of highly vertical stratification during ebb currents, the model properly simulated the temporal variation in salinity at the sampling sites. The final validation of the model was to verify the proper simulation of the transport of SSC, and in general the model could reproduce the physical mechanism driving the transport of sediment well. In summary, aside from drawbacks such as over-mixing of salinity during the neap tide due to the 2.5 Mellor-Yamada turbulence closure scheme, the model was calibrated and verified, and thus was still a valuable tool to study and understand the influence the navigation channel has on the transport of sediment within YRE.

3.4. Setting up of the numerical model for the Darwin Harbour

To simulate the hydrodynamics and transport of sediment for Darwin Harbour, the unstructured numerical model FVCOM was applied [104]. The mesh was formed by 9,666 horizontal grid cells, and 20 vertical layers using sigma coordinate. The horizontal resolution varied between ~ 20 to ~3,300 m, with the higher resolution areas in the inner harbour and lower resolution in the outer harbour [76].

To force the model at the external open boundary, tidal forcing was used in the coastal area between Charles Point and Lee Point. The tidal components were obtained from TPXO7.2 global model. The semi-diurnal components used to force the model were (M_2, S_2, N_2 and K_2), while the diurnal components were (K_1, O_1, P_1 and Q_1). Three shallow-water components (M_4, MS_4, MN_4) and two extra tidal components of low frequency were also used, i.e.

M_f and M_m. For the internal boundary, e.g. upstream river zones, there are three sources of fresh water in the domain (Elizabeth River, Blackmore River and Berry Creek); however, the simulation was for the dry season and thus river discharge was negligible [76]. In the dry season, the small presence of density-driven currents is often confined upstream of Darwin Harbour, and they are often less than 3% of the maximum tidal current intensity. At the surface, the wind is an important mechanism to cause sediment resuspension, by wind-driven currents and waves [105]. The macro-tides in Darwin Harbour (typical tidal oscillation between 3.7 and 7.8 m), however, dominate the transport of sediment with tidal currents of up to ~ 3m s^{-1} [106]. The additional effects from wind, river discharge and the heat flux at the free-surface boundary were negligible, allowing the simulation to be forced by tides alone.

The bottom drag coefficient (C_d) was set to be a function of the water depth (see Eq. 2 in 76). The mangrove area was treated differently, because the influence of roots and trees significantly increase the friction and thus reduce water speed [107]. From empirical experiments C_d was observed to vary between 1 and 10, and its value relies upon tidal conditions, mangrove species, and patchiness of mangrove distribution. Therefore, the main value for C_d was set to 5. The remaining numerical and physical parameters, such as the viscosity and diffusion coefficient, are all described in more detail in [76].

For the initial conditions, constant values for salinity (33 psu) and temperature (25°C) were used. These are characteristic values during the dry season, and, with the zero river input, result in a barotropic model. The simulation started on 20th of June 2006 (00:00:00), with a one second time step, and duration of 31 days. Six different simulations were analysed, different sizes of tidal flats and mangrove areas were assumed, and one simulation excluding the presence of tidal flats and mangrove areas. These simulations provide an understanding of the independent effects of tidal flats and mangroves in the tidal asymmetry of Darwin Harbour. There were three main numerical experiments, namely: (Exp. 1) where tidal flats and mangrove areas were considered, (Exp. 2) where mangrove areas were removed from the domain, and (Exp. 3) where both mangrove and tidal flat areas were removed from the domain.

4. Results and discussion

4.1. Sediment transport in the Adriatic Sea

The key results from [84] are summarized as follows:

The Bora wind generated barotropic southward longshore currents that connected to the partially buoyancy driven WACC. This resulted in surface water currents of up to 1.3 m s^{-1} near the Po River mouth, and maximum bottom currents of 0.3 m s^{-1} near Ortona. These general features were all in concordance with [108]. The smooth wind conditions resulted in a reduced interior vorticity, which is caused by the orographic incisions around the Dinaric Alps [109]; however, the good representation of the WACC in the Nothern Adriatic Shelf combined with the wave-currents provided a realistic physical representation of sediment transport during the Bora event. The Bora wind caused higher wave heights on the western coast than on the eastern coast. The wave direction was mainly aligned with the wind direction in the Adriatic Sea; however, the direction was mainly perpendicular when approaching the western coast because of wave refraction.

For low and moderate wind conditions, the modelled waves showed good agreement with observed waves. The measurements used to verify the model were obtained at the buoys at Ancona and Ortona. During strong wind conditions, such as Bora events, the model showed good results compared to observations of the waves at Ortona, while for Ancona the wave response was underestimated by 50%. This was caused by the low horizontal scale resolution of 40 km ECMWF wind fields. Due to the complex orography, the model is incapable of resolving the fine wind variability [91, 110].

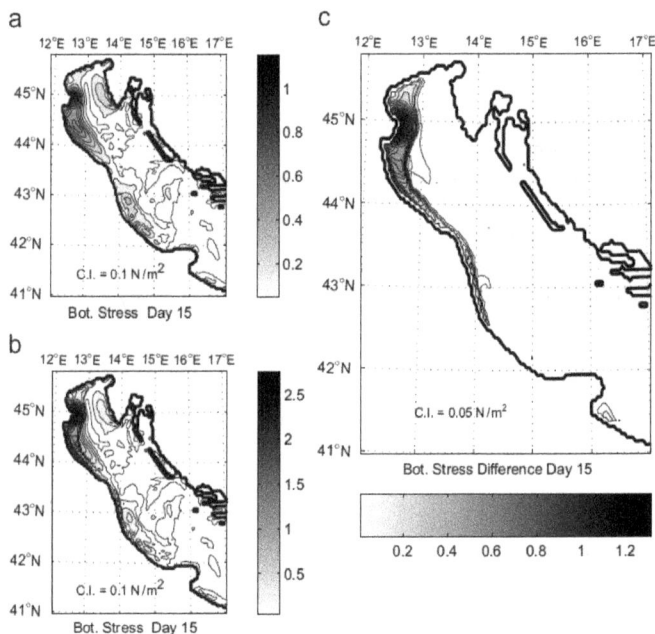

Figure 6. The bottom stress on 15 January 2001, predicted by Experiment 1 (a), Experiment 2 (b), and their difference (c) [source: 84].

Waves and currents have been shown to affect sediment resuspension in the Bottom Boundary Layer (BBL). Recent field studies conducted near the Po River delta were used to analyse the effect of wave-current interaction [e.g., 111-113]. The simulation without the wave effect (experiment 1) showed the bottom current reaching ~ 0.34 m s^{-1} during the Bora event. The Bora event caused the bottom stress to increase from 0.01 N m^{-2} to 0.66 N m^{-2} (Fig. 6a).

This increased bottom stress results in considerable erosion, and a subsequent increase in concentration of both fine and course sediment near the bottom. During periods without Bora winds, the resuspension was weak, and the fine sediment from the Po River discharge dominated the SSC in the water column. For experiment 2, in which the effect of wave-current interaction was considered (Fig. 6b), the bottom stress reached a maximum value of 2.2 N m^{-2}, and the concentration of fine and coarse sediments increased by 80%. (Fig. 7b). The wave-current interaction increased the bottom drag coefficient from 0.0048 to values of up to 0.015 (not shown).

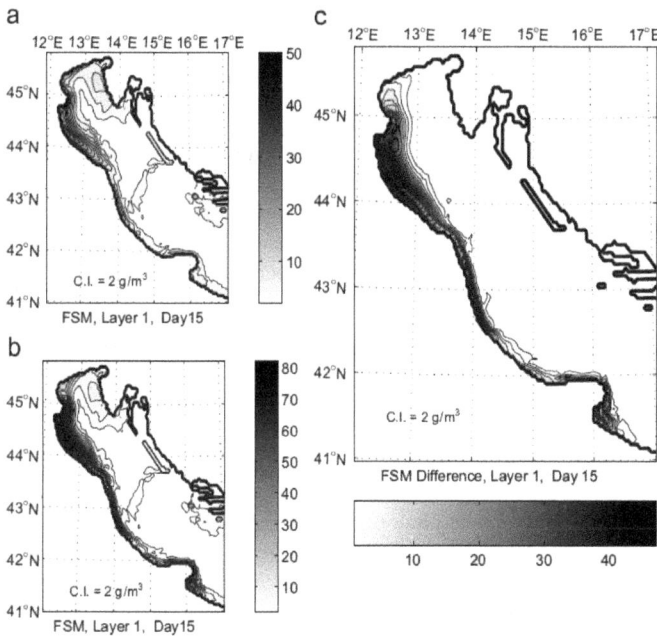

Figure 7. The surface (layer 1) fine sediment concentration on 15 January 2001, predicted by experiment 1 (a), Experiment 2 (b), and their difference (c) [source: 84].

In experiment 1, a high bottom stress was predicted in the north-east shelf along the Italian coast. Therefore, high SSC was obtained (i.e. 28 g m^{-3}) near the Po River delta, with similar concentration of coarse and fine sediments, which were very well vertically mixed. In experiment 2 the distribution patterns were very similar to those of experiment 1. The bottom stress and sediment concentration increased in magnitudes to 1.3 N m^{-2} and 50 g m^{-3}, respectively. This increase was verified along the coast (~ 200 km long), southwards from the Po River nearshore zone. It was observed that the prediction of the fine suspended sediment concentration showed good agreement with [112]. [113] has also obtained SSC in the range of 70-100 g m^{-3}.

The sediment flux was analysed in experiment 1, using results from a cross-sectional area. During the Bora event on 15th of January, an upwelling net sediment flux was observed, despite the downwelling currents along the Italian coast. The horizontal fields near the surface and bottom were also analysed, and a net sediment flux with maximum values of 20 g m^{-2} s^{-1}, was observed southwards near the Italian coast. The predictions were in agreement with estimates by [113] in a different Bora event. In experiment 1, the southward flux of 10.5 t s^{-1} (fine sediment) and 9.3 t s^{-1} (coarse sediment) was calculated at a cross-section area (N) near the Po River delta in the Adriatic Sea. In contrast, for experiment 2, the flux of fine and coarse sediment increased to 25.6 t s^{-1} and 24.1 t s^{-1}, respectively.

Experiment 3 was conducted to verify the effect from the wave and current aligned to the sediment transport. The net sediment flux of fine and coarse sediment was over predicted by 8% and 9%, respectively, from experiment 2. These small differences are evidence that the wave propagation direction had little effect on sediment flux. This was shown for strong wave conditions by [114].

To observe the effect of the tides, experiment 4 was conducted. It considered the same conditions as experiment 2, with the additional influence of tidal currents. The four main semi-diurnal tidal components were used (i.e. M_2, K_2, N_2, S_2), and the three main diurnal components (i.e. K_1, O_1, P_1). The sediment flux at the cross-sectional area (N) was observed to be reduced by 1%, compared with results from experiment 2.

The final experiment was similar to experiment 2. It included, however, the additional influence of sediment concentration on water density, using a simple bulk relation from [83]. Because of the small SSC, the effect on sediment flux was negligible, with slight changes of less than 1%.

4.2. Factors driving the phytoplankton bloom in the Mokpo coastal zone

The results showed little change in phytoplankton biomass throughout the entire water column during January. In contrast, in February the concentration of phytoplankton biomass increased at the surface. In the first two weeks of March this biomass increased significantly, and reached maximum values at the surface after two weeks (i.e. mid March). The model simulated the timing of the phytoplankton blooms well; however, the maximum biomass obtained from the theoretical results was 2-3 times lower than observed values. Some possible explanations for this underestimation are the short period of simulation (4 months), and the use of a 1-D biogeochemical model instead of a 3D model. Moreover, when the sluice gates located upstream are opened, the system temporally becomes a salt-wedge estuary. This increased fresh water input may cause the sinking riverine phytoplankton detritus that flows along the bottom. Additionally, there is a likely effect from the resuspended phytoplankton and/or detritus that were not incorporated in the model [33].

A good correlation was found between the variation in phytoplankton biomass and diatom concentrations. Despite the observed increase in radiance Q (W m^{-2}), the diatom bloom finished in April. Diatom concentration decreases mostly because of lack of dissolved phosphate (P) and silicate (Si) in the euphotic zone. This depletion is caused by phytoplankton uptake combined with stratification-induced limitation of nutrient supply near the bottom. It has been shown that phytoplankton blooms develop due to a decrease in vertical mixing rates. Therefore, the increased vertical mixing that occurs from early January to mid-February inhibits the permanence of phytoplankton cells in the euphotic area [33]. Vertical mixing was increased due to the colder wind decreasing surface water temperatures, which then caused convective overturn in the water column.

The features of phytoplankton biomass (PB) variation were examined for light attenuation coefficients (Fc) of 0.43, 0.46 and 0.49 (Fig. 8). The PB over the euphotic zone increased slowly. All the blooms started at the same time, reaching nearly the same maximum of PB, i.e. 35 mg m^{-2}. However, the rates of increase were different, with maximum growth rate verified to occur for Fc = 0.49. Moreover, variation between neap and spring tides has also been shown to affect the PB. Water turbidity is increased due to larger sediment resuspension caused by the strong spring tidal current, and thus the PAR attenuation may affect phytoplankton production. The model results showed that the absence of the effect from SSC reduces PAR attenuation, and therefore changes phytoplankton production (not shown). Vertical mixing also affects the PB, because phytoplankton cells are taken to deeper layers where light attenuation is higher [33]. [115] proposed that blooms happen when phytoplankton growth exceeds the rate of vertical movement. Therefore, a decrease in vertical mixing results in phytoplankton blooms in coastal and oceanic waters [e.g. 116-118].

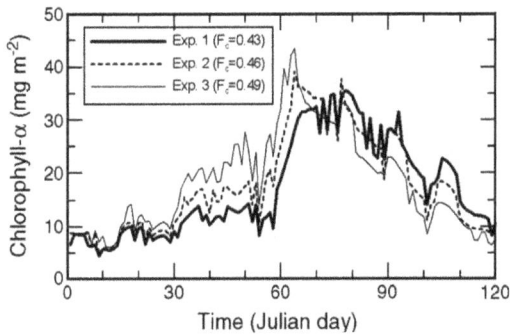

Figure 8. Temporal variation of the vertical integrated chlorophyll-a concentration in the euphotic zone. The numerical experiment for different values of Fc and the simulation start time at January 2001 [source: 33].

4.3. The effect of dikes on sediment transport in the Yangtze River Estuary

Tidal harmonic components were calculated using measurements from 8 water level stations, and later compared with results of the model. The root mean square error obtained for the 4 main components was under 10%, and the shift of tidal phases less than 10 degrees. In general, the model showed reasonable agreement compared with measurements of tidal oscillation, water current, and temporal variation of salinity and SSC. Details of the model calibration and verification are shown in [44]. The results of the SSC calculated from the model are shown in

(Fig. 9a). Differences between simulations and observations of the speed increase of the surface flood currents, during the rising sea level, were verified. Observations showed more steepness during flood tides, and, for a short period of time, measured flood current was approximately 20% higher than the model result. In contrast, water currents near the bottom consistently showed good agreement between modelled and observed values at any given time. Even with the minor differences found, the model was properly calibrated, and thus shown to be a valuable tool for studying sediment transport mechanisms in the YRE [44].

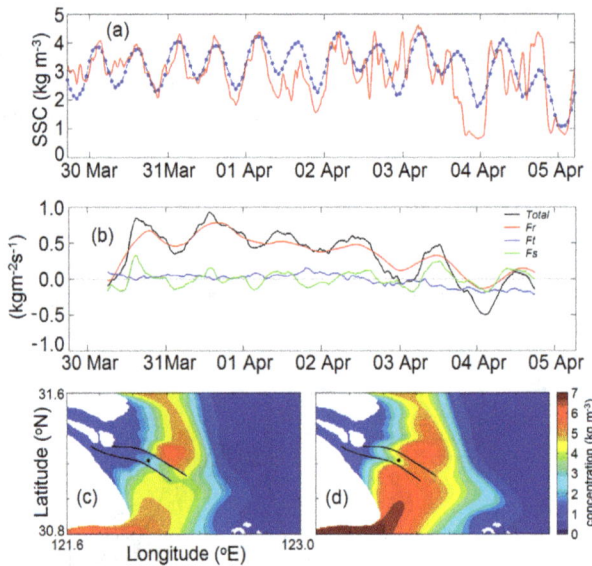

Figure 9. (a) Measured (red line) and model simulated (blue dotted line) suspended sediment concentration at 62 cm above the seabed. (b) The tidally averaged residual suspended sediment flux and its three components, i.e. residual flow (*Fr*), tidal pumping (*Ft*), and shear dispersion (*Fs*), where the positive value indicates ebb transport of SSC. (c) and (d) show the tidally averaged suspended sediment concentration after and before the DNC construction, respectively. The location of the dikes is indicated in both figures (c and d), and the location of the sampling station in the DNC (black dot).

Figure 9b shows field data results of the sediment transport near the bottom. During the spring tides between 30th of March and 2nd of April, there was more suspended sediment present at the onset of the ebb tide than at the onset of the flood tide. This resulted in a residual and tidal pumping transport of sediment downstream, where residual transport dominated the total transport of suspended sediment (TSS). Commonly, high residual TSS causes a downstream shear dispersion TSS, whereas low residual transport causes an upstream shear dispersion transport. In contrast, during the neap tide observed from the 2nd of April onwards, the bottom horizontal water velocities were much smaller than those observed during spring tides. Because of this, sediment resuspension decreased considerably, and therefore the residual transport of SSC was reduced. In addition, tidal pumping was observed to transport SSC mainly landward. The residual transport and shear dispersion were observed to alternate SSC transport between upstream and downstream directions.

The results showed that the maximum SSC is generated in the sand bar area (Figs. 9c,d). The simulation in which the effect of the two dikes was applied (Fig. 9c) showed a discontinuity in the high SSC caused by the DNC, and the maximum SSC was observed adjacent to the north dike (e.g. SSC ~ 6 kg m^{-3}). In contrast, the simulation without the presence of dikes showed high SSC within the entire sand bar area near the estuarine mouth. The SSC decreasing upstream indicated a low local resuspension of sediment and a low sediment input from river discharge. In addition, the Three Gorges Dam caused a significant decrease in sediment load to the Yangtze River [60-61]. This local sediment resuspension is therefore evidence that the silting occurring within the DNC is caused by redistribution of sediment between the shoals on each side of the dikes.

Comparison between the simulations with and without the dikes, found enhanced ebb dominance to occur with the inclusion of the dikes (not shown). These changes have previously been observed and reported [63], by comparing field measurements obtained before and after construction of the dikes. After construction, the traditional sediment transport path across the estuarine mouth was blocked; however, some sediment may still be transported around the dike edges, and a small amount may enter the channel. This sediment intruding into the DNC is likely to cause siltation near the seaward side of the DNC.

4.4. Tidal asymmetry in Darwin Harbour

The amplitude of the M_2 tidal component was observed to gradually increase from the open boundary (i.e. outer harbour) to nearly 1.7 m in the arms, and decrease from the arms to nearly 1.0 and 0 meters in the tidal flat and mangrove areas, respectively. The decrease in amplitude for the tidal flat and mangrove areas was caused by large energy dissipation due to bottom friction. The phase shifted in areas of the outer harbour and harbour arms, and this shift was caused by reduced wetting-drying areas. The maximum horizontal velocity is about 3.0 m s^{-1} in the Middle Arm (see Fig. 5). The variation of M_2 tidal current ellipses were observed at many vertical layers and locations of the domain. In the outer harbour the M_2 currents were up to 0.3 m s^{-1}, and up to 0.6 m s^{-1} in the channel. The horizontal currents were observed to decrease by almost 30 % from the surface to the bottom [76].

Comparison of the three numerical experiments from simulations that neglected the tidal flat or mangrove areas, showed an increase in the tidal amplitude for the inner and outer harbour. The maximum increase was obtained when both tidal flats and mangroves areas were not considered in the domain (0.02 m), while there was an increase of 0.01 m for the simulation neglecting only the mangrove areas. Although the variation of tidal amplitude was observed to be quite small, for the M_4 tidal component, there was an increase in amplification of almost 50% when the mangrove areas were removed, and almost 75% when both the tidal flats and mangrove areas were removed. The phase of the M_4 tidal component was observed to advance almost 20 degrees when the mangrove areas and tidal flats were not included in the domain [76].

To verify the effect of mangrove and tidal flat areas on tidal asymmetry, the tidal asymmetry skewness parameter γ was calculated. This allows identification of the major factors controlling tidal asymmetry. For γ, it was verified that the main tidal components controlling tidal asymmetry were M_2 and M_4, and so the expression to calculate γ was:

$$\gamma_{M_2/M_4} = \frac{\frac{3}{2}a_{M_2}^2 a_{M_4}^2 \sin\left(2\phi_{M_2} - \phi_{M_4}\right)}{\left[\frac{1}{2}\left(a_{M_2}^2 + 4a_{M_4}^2\right)\right]^{\frac{3}{2}}} \tag{1}$$

where a and ϕ are respectively the amplitudes, and phases of the astronomical tides M_2 and M_4.

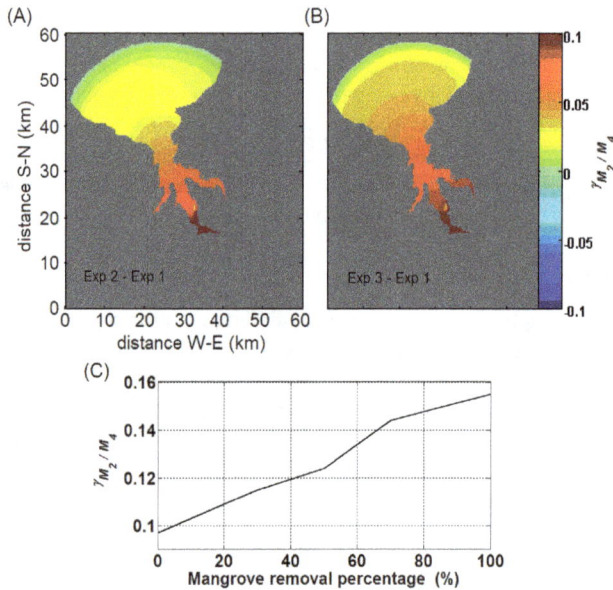

Figure 10. Predicted changes of the coefficient (γ), with results obtained from the difference between experiment 2 and 1 (A), and 3 and 1 (B). (C) Parameter γ calculated for different percentages of mangrove area removed in the East Arm near Station Blay [source: 76].

Figure 10 shows that the removal of mangroves and tidal flat areas resulted in an increased tidal asymmetry. In turn, Darwin Harbour would have more flood dominance in tidal currents (Figs. 10A,B). The maximum increase in asymmetry was observed in experiment 3. The experiments in which different percentages of the mangrove areas were removed are shown in figure 10C. The relation between removed mangrove area and increased tidal asymmetry skewness factor was observed to be almost linear, and increased by 0.1 if 100% of the mangrove areas were removed. These results demonstrate that the tidal flats and mangrove zones function as sponge zones for dampening tidal asymmetry [76].

5. Conclusions

This chapter provided information about four important suspended sediment transport processes. It showed how wave current interaction increases SSC and therefore affects the net transport of sediment, and demonstrated the importance of applying a coupled hydro-dynamical and biogeochemical numerical model to better simulate phytoplankton blooms. In addition, the effect of coastal construction on sediment transport was described, along with the role of tidal flats and mangrove areas in causing tidal asymmetry. Specific conclusions for each case study are as follows:

5.1. Effect of wave current interaction on sediment transport in coastal zones

The general features of the model during the Bora event in January 2001 predicted the WACC, and the large wave fields generated by the strong winds agreed with observations on the western Adriatic shelf. The bottom boundary layer was resolved, and the wave-current interaction was able to be implemented in the model. Moreover, the tidal effect on sediment transport was also examined in the Northern Adriatic Sea.

The model showed that the Bora event occurring from 13–17 January 2001 caused waves with height of 2 m, and period of 5 s. The wave-current near the bottom layer resulted in strong sediment resuspension. The wave effect, combined with the longshore coastal currents and the turbulent driven vertical flux, caused a large southward sediment flux. The flux was estimated using data from a cross-section area of the Adriatic Sea, near the Po River. The experiment including the wave-current interaction showed larger southward sediment flux than the experiment with waves neglected. The sediment flux was clearly maintained by the strong vertical mixing and the bottom sediment resuspension. The results also showed that the southward flux was more confined to the northern area of the Adriatic Sea, and sediment plumes were confined to the western Adriatic shelf north of Ancona.

Simulation results have shown that in areas between the Po River and Ancona, sedimentation and erosion rates doubled due to the combined motion of wave–current interaction in the BBL during the Bora event. Bora events typically occur 9 times during winter seasons [119], and thus the influence of waves is important in the long-term sediment transport at this coastal area. The annual sedimentation rate near the Po River mouths was found to be 2-6 cm yr^{-1} [120]. The sedimentation rate predicated during the Bora event of 13-17 January 2001 represented nearly 10% of the annual rate. This prediction agrees with observations if we assume the average of ~ 10 Bora events per year. In addition, from experiment 2, it was verified that the high concentration of suspended sediment along the western coast, north of Ancona, was locally driven by waves. Moreover, the wave-current interaction with the BBL enhanced the bottom stress and caused increased sediment resuspension. From the third experiment it was observed at the cross-sectional area (N), that wave direction and tides had little effect on sediment resuspension in the northeastern coast of the Adriatic Sea.

5.2. Effect of SSC on phytoplankton bloom

Simulation of phytoplankton biomass dynamics was obtained during the winter and spring seasons. The model was run using irradiance forcings and physical oceanographic forcings such as SSC, temperature, salinity, tides, and vertical diffusivity. The results from winter revealed that cooling of the surface waters induced vertical mixing, and thus inhibited the growth of phytoplankton biomass. In contrast, periods of neap tides combined with increased freshwater discharge decreased vertical mixing, and in turn triggered PB spring bloom. In the tidally turbid coastal waters, such as the MCZ, the small neap tidal currents were not capable of increasing SSC. As a result, more light was observed though the water column, favouring increases in phytoplankton biomass [33].

Results revealed that prediction of phytoplankton blooms are very sensitive to the light attenuation factor (Fc), vertical diffusion (Kv), and suspended sediment concentration (SSC). Moreover, simulations using depth-averaged diffusivity and monthly averaged vertical diffusivity, may hinder the primary production processes. Finally, the simulations revealed the importance of coupling the 3D hydro-sediment model with the ecosystem model, which leads to more realistic estimates of phytoplankton biomass oscillation [33].

5.3. Influence of coastal construction on sediment transport

The simulations were well calibrated and validated using field data measured in the DNC in Shanghai Port on March 2009. Combining the field measurements with the extrapolated results for the whole domain of the YRE, we have verified that the system is highly stratified during neap tides; however, close to well mixed during spring tides. Calculation of the Richardson number revealed the dominance of fresh water inflow over vertical mixing during neap tides; while, in spring tides, vertical mixing overcomes the buoyancy of the fresh water river inflow [44].

The simulations revealed that the transport of sediment into the DNC comes from open areas, rather than from upstream areas as is commonly expected. Therefore, the input from the Yangtze River is not the major source of sediment, causing siltation, in this navigation channel. The dikes were observed to inhibit the majority of the alongshore sediment transport near the delta zone; however some of the sediment transported towards the dikes veers south, and as it approaches the edges of the dikes a small amount may enter the channel. It was verified that, in the North Channel, the estuarine maximum turbidity zone was generated predominantly by gravitational circulation, whereas, in the South Passage, it was generated mainly by tidal distortion effects. For the DNC there was no maximum turbidity generated locally. Some turbid waters in adjacent areas were observed to be transported upstream in this channel. In summary, the simulation results revealed the magnitude of the maximum turbidity zone in the North Passage, and showed that most of the sediment depositing into the DNC is caused by sediment redistribution from adjacent zones [44].

5.4. Tidal asymmetry modulated by tidal flats and mangrove areas

The results obtained from the simulations, combined with field observations, showed that the circulation in DH is mainly driven by tides during the dry season, with negligible wind and river influence. DH is a semidiurnal system, dominated primarily by the M_2 tidal component, and secondarily by the S_2 component. The maximum velocities observed were in the Middle Arm, with peaks of over 3.0 m s^{-1} at the surface layer. The Middle Arm also had the maximum and minimum vertical shear of the horizontal velocities, which would therefore result in some of the maximum vertical mixing zones of DH.

The sensitivity test evidenced that tidal flats and mangrove areas play an important role in modulating the amplitude and phase of some tidal components. In particular, the amplitude of the M_4 component was observed to increase by almost 75%, and the phase was verified to advance/delay a few degrees in the outer/inner harbour respectively. In addition, the amplitude and phase of the M_2 component were observed to change in the inner harbour, with an amplitude increase of 1% and phase advance of 4 degrees.

The parameter gamma was calculated, showing variation in the tidal elevation skewness. Gamma was observed to increase by 100% in the absence of mangrove areas, and by 120% in the absence of both tidal flats and mangrove areas. Furthermore, the increase in the elevation skewness correlated almost linearly with the decrease in mangrove area. It was therefore suggested that a similar effect would be found with a decrease in tidal flat areas. The findings of this study show how important flooding estuarine areas, e.g. tidal flats and mangroves, are in modulating tidal asymmetry. These findings could be further used to verify similar effect in other estuaries, bays, and harbour areas. As tidal asymmetry strongly affects sediment transport in the estuaries, care must be taken in terms of the reclamation of the mangrove areas and tidal flats around the harbour watershed.

Acknowledgements

X. H. Wang and F. P. Andutta were supported by a 2011 Australian Research Council/ Linkage Project – LP110100652. This work was also supported by the National Computational Infrastructure National Facility at the Australian National University. This is a publication of the Sino-Australian Research Centre for Coastal Management, paper number 10.

Author details

X. H. Wang[*] and F. P. Andutta

*Address all correspondence to: hua.wang@adfa.edu.au

School of Physical, Environmental and Mathematical Sciences, University of New South Wales at Australian Defence Force Academy UNSW-ADFA, Australia

References

[1] Wang, X. H., & Pinardi, N. (2002). Modeling the dynamics of sediment transport and resuspension in the northern Adriatic Sea. *Journal of Geophysical Research*, 107(C12), 3225, doi:10.1029/2001JC001303.

[2] Frascari, F., Frignani, M., Guerzoni, S., & Ravaioli, M. (1988). Sediments and pollution in the Northern Adriatic Sea. *Annals of the New York Academy Sciences*, 534, 1000-1020.

[3] Giordani, P., Hammond, D. E., Berelson, W. M., Montanari, G., Poletti, R., Milandri, A., Frignani, M., Langone, L., Ravailoi, M., Rovatti, G., & Rabbi, E. (1992). Benthic fluxes and nutrient budgets for sediments in the Northern Adriatic Sea: Burial and recycling efficiencies. *Science of the Total Environment, Supplement*, 251-275.

[4] Wolanski, E. (2007). *Estuarine Ecohydrology*, Elsevier, Amsterdam, 157.

[5] Warren, R., & Johnsen, J. (1993, June 14-16, 1993). Cohesive sediment modelling for coastal lagoons. Kuala Lumper, Malaysia. *paper presented at International Colloquium and Exposition on Computer Applications in Coastal and Offshore Engineering (ICE-CA COE'93)*.

[6] Vichi, M., Pinardi, N., Zavatarelli, M., Matteucci, G., Marcaccio, M., Bergamini, M. C., & Frascari, F. (1998). One-dimensional ecosystem model tests in the Po Prodelta area (Northern Adriatic Sea). *Environmental Modelling & Software*, 13, 471-481.

[7] Wolanski, E., Andutta, F. P., & Delhez, E. (2012). Estuarine hydrology. *In: Encyclopedia of Lakes and Reservoirs. Lars Bengtsson, Reginald Herschy, Rhodes Fairbridge (eds.)*, Springer, In press.

[8] Maggi, F. (2005). Flocculation dynamics of cohesive sediment. *PhD-Thesis*, Delft University of Technology, Delft, 136.

[9] Kourafalou, V. H. (1999). Process studies on the Po River plume, North Adriatic Sea. *Journal of Geophysical Research*, 104, 29963-29985.

[10] Wang, X. H. (2005). A numerical study of sediment transport in a coastal embayment during a winter storm. *Ocean Modelling*, 10, 253-271.

[11] Milliman, J. D., & Farnsworth, K. L. (2011). *River Discharge to the Coastal Ocean: A Global Synthesis*, Cambridge University Press, 392.

[12] Vilibić, I. (2003). An analysis of dense water production on the North Adriatic shelf. *Estuarine, Coastal and Shelf Science*, 56, 697-707.

[13] Malanotte-Rizzoli, P., & Bergamasco, A. (1983). The dynamics of the coastal regions of the Northern Adriatic Sea. *Journal of Physical Oceanography*, 13, 1105-1130.

[14] Artegiani, A., Bregant, D., Paschini, E., Pinardi, N., Raicich, F., & Russo, A. (1997a). The Adriatic Sea general circulation. Part I: air-sea interactions and water mass structure. *Journal of Physical Oceanography*, 27, 1492-1514.

[15] Artegiani, A., Bregant, D., Paschini, E., Pinardi, N., Raicich, F., & Russo, A. (1997b). The Adriatic Sea general circulation. Part II: Baroclinic circulation structure. *Journal of Physical Oceanography*, 27, 1515-1532.

[16] Zavatarelli, Z., Pinardi, N., Kourafalou, V. H., & Maggiore, A. (2002). Diagnostic and prognostic model studies of the Adriatic Sea general circulation: Seasonal variability. *Journal of Geophysical Research*, 107, 3004, doi:10.1029/2000JC000210.

[17] Zavatarelli, M., & Pinardi, N. (2003). The Adriatic Sea modelling system: a nested approach. *Annales Geophysicae*, 21, 345-364.

[18] Poulain, P. M. (2001). Adriatic Sea surface circulation as derived from drifter data between 1990 and 1999. *Journal of Marine System*, 29, 3-32.

[19] Raicich, F. (1996). On the fresh water balance of the Adriatic Sea. *Journal of Marine System*, 9, 305-319.

[20] Haley, J. P. J., & Lermusiaux, P. F. J. (2010). Multiscale two-way embedding schemes for free-surface primitive equations in the "Multidisciplinary simulation, estimation and assimilation system". *Ocean Dynamics*, 60, 1497-1537.

[21] Ramp, S. R., Lermusiaux, P. F. J., Shulman, I., Chao, Y., Wolf, R. E., & Bahr, F. L. (2011). Oceanographic and atmospheric conditions on the continental shelf north of the Monterey Bay during August 2006. *Dynamics of Atmospheres and Oceans*, 52, 192-223.

[22] Andutta, F. P., Ridd, P. V., & Wolanski, E. (submitted). *Age and the flushing time in the Great Barrier Reef coastal waters*.

[23] Nunes, Vaz. R. A., Lennon, G. W., & Bowers, D. G. (1990). Physical behaviour of a large, negative or inverse estuary. *Continental Shelf Research*, 10, 277-304.

[24] Diop, E. S., Soumare, A., Diallo, N., & Guisse, A. (1997). Recent change of mangroves of the Saloum river estuary, Senegal. *Mangroves and Salt Marshes*, 1, 163-172.

[25] Largier, J. L., Hollibaugh, J. T., & Smith, S. V. (1997). Seasonally hypersaline estuaries in Mediterranean-climate regions. *Estuarine, Coastal and Shelf Science*, 45, 789-797.

[26] Lavin, M. F., Godinez, V. M., & Alvarez, L. G. (1998). Inverse-estuarine features of the upper gulf of California. *Estuarine, Coastal and Shelf Science*, 47, 769-795.

[27] Webster, I. T. (2010). The hydrodynamics and salinity regime of a coastal lagoon- The Coorong Australia- Seasonal to multi-decadal timescales. *Estuarine Coastal and Shelf Science*, 90, 264-274.

[28] Andutta, F. P., Ridd, P. V., & Wolanski, E. (2011). Dynamics of hypersaline coastal waters in the Great Barrier Reef. *Estuarine, Coastal and Shelf Sciences*, 94, 299-305.

[29] Winant, C. D., & de Velasco, G. G. (2012). Dynamics of hypersaline estuaries: Laguna San Ignacio, Mexico. Chapter 12 in, 2, *In: Uncles R J, Monismith S G (Ed.), Water and Fine Sediment Circulation. In: Wolanski E, McLusky D (Series Ed.), The Treatise on estuarine and coastal Science*, Elsevier.

[30] Nunes, Vaz. R. A. (2012). The salinity response of an inverse estuary to climate change & desalination. *Estuarine, Coastal and Shelf Sciences*, 98, 49-59.

[31] Brambati, A., Bregant, D., Lenardon, G., & Stolfa, D. (1973). Transport and sedimentation in the Adriatic Sea. *Museo Friulano di Storia Nat.*, Udine, Italy, Publ. 20, 60.

[32] Byun, D. S., Wang, X. H., & Holloway, P. E. (2004). Tidal characteristic adjustment due to dyke and seawall construction in the Mokpo Coastal Zone, Korea. *Estuarine, Coastal and Shelf Science*, 59, 185-196.

[33] Byun, D. S., Wang, X. H., Zavantarelli, M., & Cho, Y. K. (2007). Effects of resuspended sediments and vertical mixing on phytoplankton spring bloom dynamics in a tidal estuarine embayment. *Journal of Marine Systems*, 67, 102-118.

[34] Defant, A. (1960). *Physical Oceanography*, Pergamon Press, New York, 2, 598.

[35] Kang, J. W. (1999). Changes in tidal characteristics as a result of the construction of sea-dike/sea-walls in Mokpo coastal zone in Korea. *Estuarine, Coastal and Shelf Science*, 48, 429-438.

[36] Kang, J. W., & Jun, K. S. (2003). Flood and ebb dominance in estuaries in Korea. *Estuarine, Coastal and Shelf Science*, 56, 187-196.

[37] Kang, J. W., Moon, S. R., & Lee, K. H. (2009). Analyzing sea level rise and tide characteristics change driven by coastal construction at Mokpo Coastal Zone in Korea. *Ocean Engineering*, 36, 415.

[38] Speer, P. E., & Aubrey, D. G. (1985). A study of nonlinear tidal propagation in shallow inlet/estuarine systems. Part II: theory. *Estuarine, Coastal and Shelf Science*, 21, 207-224.

[39] Lee, S. W. (1994). Change in tidal height at Mokpo Harbour due to the construction of Youngsan River Estuary Dyke. *Korean harbour hydraulics*, 18, 27e37, in Korean.

[40] Jeong, M. S., Jeong, D. D., Shin, S. H., & Lee, J. W. (1997). Tidal changes in the harbor due to the development of Mokpo Coastal Zone. *Journal of Korean Harbour*, 11, 1e8, in Korean.

[41] Milliman, J. D., Shen, H., Yang, Z., & Meades, R. H. (1985). Transport and deposition of river sediment in the Changjiang estuary and adjacent continental shelf. *Continental Shelf Research*, 4(1/2), 37-45.

[42] Milliman, J. D., & Syvitski, J. P. M. (1992). Geomorphic/tectonic control of sediment discharge to the ocean: the importance of small mountainous rivers. *Journal of Geology*, 100(5), 525-544.

[43] Cao, Z., Wang, X. H., Guan, W., Hamilton, L. J., Chen, Q., & Zhu, D. (submitted). Observations of nepheloid layers in the Yangtze estuary, China. *Marine Technology Society Journal.*

[44] Song, D., & Wang, X. H. (inpreparation). *Observation and Modelling Study of the Turbidity Maximum in the Deepwater Navigation Channel in the Yangtze River Estuary.*

[45] He, Q., Li, J., Li, Y., Jin, X., & , Y. (2001). Field measurements of bottom boundary layer processes and sediment resuspension in the Changjiang estuary. *Science in China (Series B)*, 44(Supp.), 80-86.

[46] Shi, J. Z., Zhang, S. Y., & Hamilton, L. J. (2006). Bottom fine sediment boundary layer and transport processes at the mouth of the Changjiang Estuary, China. *Journal of Hydrology*, 327(1/2), 276-288.

[47] Liu, G., Zhu, J., Wang, Y., Wu, H., & Wu, J. (2011). Tripod measured residual currents and sediment flux Impacts on the silting of the Deepwater Navigation Channel in the Changjiang Estuary. *Estuarine, Coastal and Shelf Science*, 93(3), 192-201.

[48] Li, J., & Zhang, C. (1998). Sediment resuspension and implications for turbidity maximum in the Changjiang Estuary. *Marine Geology*, 148(3-4), 117-124.

[49] Shi, Z., & Kirby, R. (2003). Observations of fine suspended sediment processes in the turbidity maximum at the North Passage of the Chanjiang estuary, China. *Journal of Coastal Research*, 19(3), 529-540.

[50] Shi, Z. (2004). Behaviour of fine suspended sediment at the North passage of the Changjiang Estuary, China. *Journal of Hydrology*, 293(1-4), 180-190.

[51] Shen, H., & Pan, D. (2001). *Turbidity maximum in the Changjiang Estuary*, China Ocean Press, Beijing, in Chinese with English introduction, 194.

[52] Yang, Z., Milliman, J. D., & Fitzgerald, M. G. (1982). Transfer of Water and Sediment from the Yangtze River to the East China Sea, June 1980. *Canadian Journal of Fisheries and Aquatic Sciences*, 40(S1), 72-82.

[53] Milliman, J. D., Hsueh, Y., Hu, D., Pashinski, D. J., Shen, H., Yang, Z., & Hacker, P. (1984). Tidal phase control of sediment discharge from the Yangtze River. *Estuarine, Coastal and Shelf Science*, 19(1), 119-128.

[54] Beardsley, R. C., Limeburner, R., Yu, H., & Cannon, G. A. (1985). Discharge of the Changjiang (Yangtze River) into the East China Sea. *Continental Shelf Research*, 4(1-2), 57-76.

[55] Su, J., & Wang, K. (1986). The suspended sediment balance in Changjiang Estuary. *Estuarine Coastal and Shelf Science* , 23(1), 81-98.

[56] Shen, H., Li, J., Zhu, H., Han, M., & Zhou, F. (1993). Transport of the suspended sediment in the Changjiang Estuary. *International Journal of Sediment Research*, 7, 45-63.

[57] Shen, J., Shen, H., Pan, D., & Xiao, C. (1995). Analysis of transport mechanism of wa-
      ter and suspended sediment in the turbidity maximum of the Changjiang Estuary.
      Acta Geographica Sinica, 50(5), 411-420, in Chinese with English abstract.

[58] Hamilton, L. J., Shi, Z., & Zhang, S. Y. (1998). Acoustic backscatter measurements of
      estuarine suspended cohesive sediment concentration profiles. *Journal of Coastal Re-
      search*, 14(4), 1213-1224.

[59] Gao, J., Yang, Y., Wang, Y., Pan, S., & Zhang, R. (2008). Sediment dynamics of turbid-
      ity maximum in Changjiang River mouth in dry season. *Frontiers of Earth Science in
      China*, 2(3), 249-261.

[60] Chen, X., & Zong, Y. (1998). Coastal erosion along the Changjiang deltaic shoreline,
      China: History and prospective. *Estuarine Coastal and Shelf Science*, 46(5), 733-742.

[61] Yang, Z., Wang, H., Saito, Y., Milliman, J. D., Xu, K., Qiao, S., & Shi, G. (2006). Dam
      impacts on the Changjiang (Yangtze) River sediment discharge to the sea: The past
      55 years and after the Three Gorges Dam. *Water Resource Research*, 42, W04407, doi:
      10.1029/2005WR003970.

[62] Wu, H., & Guo, W. (2004). The physical model study on suspend sediment back silt-
      ing in navigation channel of Yangtze River Estuary. *Shanghai Estuarine and Coastal
      Science Research Centre*, in Chinese.

[63] Tan, Z., Fan, Q., Zheng, W., & Zhu, J. (2011). Analysis of reasons for the siltation in
      North Passage of Yangtze Estuary. *Port & Waterway Engineering* [1], 29-40, in Chinese
      with English abstract.

[64] He, S., & Sun, J. (1996). Characteristics of suspended sediment transport in the tur-
      bidity maximum of the Changjiang River Estuary. *Oceanologia et Limnologia Sinica*,
      27(1), 60-66, in Chinese with English abstract.

[65] Chen, J., Li, D., Chen, B., Hu, F., Zhu, H., & Liu, C. (1999). The processes of dynamic
      sedimentation in the Changjiang Estuary. *Journal of Sea Research*, 41(1-2), 129-140.

[66] Wu, J., Wang, Y., & Cheng, H. (2009). Bedforms and bed material transport pathways
      in the Changjiang (Yangtze) Estuary. *Geomorphology*, 175-184.

[67] Ge, J., Ding, P., & Chen, C. (2010). Impacts of Deep Waterway Project on local circula-
      tions and salinity in the Changjiang Estuary, China. Shanghai, China. *paper presented
      at the 32nd International Conference on Coastal Engineering*.

[68] Shi, J. Z. (2010). Tidal resuspension and transport processes of fine sediment within
      the river plume in the partially-mixed Changjiang River estuary, China. *A personal
      perspective Geomorphology*, 121(3-4), 133-151.

[69] Shi, J. Z., Zhou, H. Q., Liu, H., & Zhang, Y. G. (2010). Two-dimensional horizontal
      modeling of fine-sediment transport at the South Channel-North Passage of the par-
      tially mixed Changjiang River estuary. *China Environment Earth Sciences*, 61(8),
      1691-1702.

[70] Wilson, D., Padovan, A., & Townsend, S. (2004). The Water Quality of Spring and Neap Tidal Cycles in the Middle Arm of Darwin Harbour during the Dry Season. *Water Monitoring Branch, Natural Resource Management Division*, Department of Infrastructure, Planning and Environment, Darwin.

[71] Brocklehurst, P., & Edmeades, B. (1996). The Mangrove Communities of Darwin Harbour. *Tech Memo.* [96].

[72] Tien, A. T. (2006). Influence of deep aquifer springs on dry season stream water quality in Darwin Rural Area. *Water Monitoring Branch Natural Resource Management Division Report*, 6/2006D, 54.

[73] Michie, M. G. (1987). Distribution of foraminifera in a macrotidal tropical estuary: Port Darwin, Northern Territory of Australia. *Australian Journal of Marine and Freshwater Research*, 38, 249-259.

[74] Woodroffe, C. D., Bardsley, K. N., Ward, P. J., & Hanley, J. R. (1988). Production of mangrove litter in a macrotidal embayment, Darwin Harbour, N.T., Australia. *Estuarine, Coastal and Shelf Science*, 26(6), 581-598.

[75] Li, L., Wang, X. H., Sidhu, H., & Williams, D. (2011). Modelling of three-dimensional tidal dynamics in Darwin Harbour, Australia. *in Proceedings of the 15th Biennial Computational Techniques and Applications Conference (CTAC-2010)*, 52, C103-C123.

[76] Li, L., Wang, X. H., Williams, D., Sidhu, H., & Song, D. (2012). Numerical study of the effects of mangrove areas and tidal flats on tides: A case of study of Darwin Harbour, Australia. *Journal of Geophysical Research*, In press.

[77] Padovan, A. V. (1997, December). The water quality of Darwin Harbour: October 1990 - November 1991. *Water Quality Branch, Water Resources Division, Department of Lands, Planning and Environment, NT Government, Report* [34/1997D].

[78] Williams, D., Wolanski, E., & Spagnol, S. (2006). Hydrodynamics of Darwin Harbour. *In: Wolanski E, editor. The Environment in Asia Pacific Harbours*, Springer, Printed in The Netherlands, 461-476.

[79] Williams, D. (2009). Part 1: hydrodynamics and sediment transport. *In Dredging of sand from Darwin Harbour, hydrographic and marine life, Australian Institute of Marine Science, Arafura Timor Research Facility*, Brinkin, Northern Territory, 1-33.

[80] Ribbe, J., & Holloway, P. E. (2001). A model of suspended sediment transport by internal tides. *Continental Shelf Research*, 21, 395-422.

[81] Wang, X. H. (2001). A numerical study of sediment transport in a coastal embayment during a winter storm. *Journal of Coastal Research*, 34, 414-427.

[82] Wang, X. H. (2002). Tide-induced sediment resuspension and the bottom boundary layer in an idealized estuary with a muddy bed. *Journal of Physical Oceanography*, 32, 3113-3131.

[83] Wang, X. H., Pinardi, N., & Malacic, V. (2007). Sediment transport and resuspension due to combined motion of wave and current in the northern Adriatic Sea during a Bora event in January 2001: A numerical modelling study. *Continental Shelf Research*, 27, 613-633.

[84] Blumberg, A. F., & Mellor, G. L. (1987). A description of a three-dimensional coastal ocean circulation model. *in Three dimensional Coastal Ocean Models, edited by N. S. Heaps*, American Geophysical Union, Washington D. C., 1-16.

[85] Smagorinsky, J. (1963). General circulation experiments with the primitive equations, I: the basic experiment. *Monthly Weather Review*, 91, 99-164.

[86] Nittrouer, C., Miserocchi, S., & Trincardi, F. (2004). The PASTA project: investigation of Po and Apenine sediment transport and accumulation. *Oceanography*, 17(4), 46-57.

[87] Adams, C. E., Jr , , & Weatherly, G. L. (1981). Some effects of suspended sediment stratification on an oceanic bottom boundary layer. *Journal of Geophysical Research*, 86, 4161-4172.

[88] Trowbridge, J. H., & Kineke, G. C. (1994). Structure and dynamics of fluid muds over the Amazon continental shelf. *Journal of Geophysical Research*, 99, 865-874.

[89] Fohrmann, H., Backhaus, J. O., Laume, F., & Rumohr, J. (1998). Sediments in bottom-arrested gravity plumes: numerical case studies. *Journal of Physical Oceanography*, 28, 2250-2274.

[90] Cavaleri, L., & Bertotti, L. (1997). In search of the correct wind and wave fields in a minor basin. *Monthly Weather Review*, 125, 1964-1975.

[91] Mellor, G. L., & Yamada, T. (1982). Development of a turbulence closure model for geophysical fluid problems. *Reviews of Geophysics and Space Physics*, 20(4), 851-875.

[92] Byun, D. S., & Wang, X. H. (2005). Numerical studies on the dynamics of tide and sediment transport in the western tip of the southwest coast, Korea. *Journal of Geophysical Research*, 110, C03011.

[93] Baretta, J. G., Ebenhöh, W., & Ruardij, P. (1995). The European regional seas ecosystem model, a complex marine ecosystem model. *Netherlands Journal of Sea Research*, 33, 233-246.

[94] Legendre, L., & Rassoulzadegan, F. (1995). Plankton and nutrient dynamics in marine waters. *Ophelia*, 41, 153-172.

[95] Vichi, M., Oddo, P., Zavatarelli, M., Coluccelli, A., Coppini, G., Celio, M., Fonda, Umani. S., & Pinardi, N. (2003). Calibration and validation of a one-dimensional complex marine biogeochemical fluxes model in different areas of the northern Adriatic Sea. *Annales Geophysicae*, 21(2), 413-437.

[96] Yoon, Y. H. (2000). On the distributional characteristics of phytoplankton community in Mokpo Coastal Waters, Southwestern Korea during low temperature season. *Journal of Institute for Basic Sciences*, Yosu Natinia University, 2, 71-82, in Korean.

[97]   Blumberg, A. F., & Mellor, G. L. (1983). Diagnostic and prognostic numerical circula-
       tion studies of the South Atlantic Bight. *Journal of Geophysical Research*, 88(C8),
       4579-4592.

[98]   Mellor, G. L. (2004). Users guide for a three-dimensional, primitive equation, numeri-
       cal ocean model. *Program in Atmospheric and Oceanic Sciences*, Princeton University,
       Princeton, NJ, 08544-0710.

[99]   Mellor, G. L., & Yamada, T. (1974). A hierarchy of turbulence closure models for
       planetary boundary layers. *Journal of the Atmospheric Sciences*, 31(7), 1791-1806.

[100] Mellor, G. L. (2001). One-dimensional, ocean surface layer modeling: a problem and
       a solution. *Journal of Physical Oceanography*, 31(3), 790-809.

[101] Oey, L. Y. (2005). A wetting and drying scheme for POM. *Ocean Modelling*, 9(2),
       133-150.

[102] Oey, L. Y. (2006). An OGCM with movable land-sea boundaries. *Ocean Modelling*,
       13(2), 176-195.

[103] Chen, C., Liu, H., & Beardsley, R. C. (2003). An Unstructured Grid, Finite-Volume,
       Three-Dimensional, Primitive Equations Ocean Model: Application to Coastal Ocean
       and Estuaries. *J. Atmos. Oceanic Technol.*, 20, 159-186.

[104] Wright, L. D., Boon, J. D., Xu, J. P., & Kim, S. C. (1992). The bottom boundary layer of
       the bay stem plains environment of lower Chesapeake Bay. *Estuarine, Coastal and
       Shelf Science*, 35(1), 17-36, doi: 10.1016/S0272-7714(05)80054-X.

[105] Mehta, A. J. (1988). Laboratory studies on cohesive sediment deposition and erosion.
       in Physical processes in estuaries, edited by J. Dronkers and W.V. Leussen, Springer Verlag,
       Berlin, 427-445.

[106] Mazda, Y., Magi, M., Kogo, M., & Hong, P. N. (1997). Mangroves as a coastal protec-
       tion from waves in the Tong King delta, Vietnam. *Mangroves and salt marshes*, 1(2),
       127-135, doi: 10.1023/a:1009928003700.

[107] Lee, C. M., Askari, F., Brook, J., Carniel, S., Cushman-Rosin, B., Dorman, C., Doyle, J.,
       Flament, P., Harris, C. K., Jones, B. H., Kuzmic, M., Martin, P., Ogston, A., Orlic, M.,
       Perkins, H., Poulain, P., Pullen, J., Russo, A., Sherwood, C., Signell, R. P., & Thaler,
       Detweiler. D. (2005). Northern Adriatic response to a wintertime Bora wind event.
       EOS Transactions, 86(16), 19.

[108] Pullen, J., Doyle, J. D., Hodur, R., Ogston, A., Book, J. W., Perkins, H., & Signell, R.
       (2003). Coupled ocean-atmosphere nested modeling of the Adriatic Sea during win-
       ter and spring 2001. *Journal of Geophysical Research*, 108, C103320.

[109] Signell, R. P., Carniel, S., Cavaleri, L., Chiggiato, J., Doyle, J., Pullen, J., & Sclavo, M.
       (2005). Assessment of operational wind models for the Adriatic Sea using wind and
       wave observations: a first analysis focused on the Venice lagoon region. *Journal of
       Marine System*, 53, 217-233.

[110] Matteucci, G., & Frascari, F. (1997). Fluxes of suspended materials in the north Adriatic Sea (Po Prodelta area). *Water, Air and Soil Pollution*, 99, 557-572.

[111] Boldrin, A., Langone, L., Miserocchi, S., Turchetto, M., & Acri, F. (2005). Po River plume on the Adriatic continental shelf: dispersion and sedimentation of dissolved and suspended matter during different river discharge rates. *Marine Geology*, 222-223, 135-158.

[112] Traykovski, P., Wiberg, P. L., & Geyer, W. R. (2006). Observations and modeling of wave-supported sediment gravity flows on the Po prodelta and comparison to prior observations from the Eel shelf. *Continental Shelf Research*, 111, C03S17, doi: 10.1029/2005JC003110.

[113] Grant, W. D., & Madsen, O. S. (1979). Combined wave and current interaction with a rough bottom. *Journal of Geophysical Research*, 84, 1797-1808.

[114] Huisman, J., Von Oostveen, P., & Weissing, F. J. (1999). Critical depth and critical turbulence: two different mechanism for the development of phytoplankton blooms. *Limnology and Oceanography*, 44, 1781-1787.

[115] Gran, H. H., & Braarud, T. (1935). A quantitative study of the phytoplankton in the Bay of Fundy and the Gulf of Marine. *Journal of the Biological Board of Canada*, 1, 279-467.

[116] Sverdrup, H. U. (1953). On conditions for vernal blooming of phytoplankton. *Journal du Conseil- Conseil International Pour L'exploration de La Mer*, 18, 287-295.

[117] Cloern, J. E. (1991). Tidal stirring and phytoplankton bloom dynamics in an estuary. *Journal of Marine Research*, 49, 203-221.

[118] Wang, X. H., Oddo, P., & Pinardi, N. (2006). On the bottom density plume on coastal zone off Gargano (Italy) in the southern Adriatic Sea and its inter-annual variability. *Journal of Geophysical Research*, 111, C03S17, doi:10.1029/2005JC003110.

[119] Palinkas, C. M., Nittrouer, C. A., Wheatcroft, R. A., & Langone, L. (2005). The use [7]Be to identify event and seasonal sedimentation near the Po River delta, Adriatic Sea. *Marine Geology*, 222-223, 95-112.

Sediment Dynamic Processes

Chapter 5

Flocculation Dynamics of Mud: Sand Mixed Suspensions

Andrew J. Manning, Jeremy R. Spearman,
Richard J.S. Whitehouse, Emma L. Pidduck,
John V. Baugh and Kate L. Spencer

Additional information is available at the end of the chapter

1. Introduction

Sediments present in muddy estuaries and tidal inlets are regarded as being predominantly cohesive. These muds are usually composed of both clay and silt minerals combined with organic matter (Winterwerp and van Kesteren, 2004), and with the exception of very low particle concentrations or extremely high energy flow conditions, muddy particles occur as a spectra of floc sizes (D) when entrained into suspension (Kranck and Milligan, 1992).

In reality, natural sediments tend to comprise a mixture of different particle sizes, non-cohesive sediment including fine sands and, because of the interaction between these different fractions, the mixture behaves in a different way than the constituent parts (Whitehouse et al., 2000). Uncles et al. (1998) found that the proportion of mud and sand in subtidal and intertidal sediments can vary both temporally and spatially (e.g. Uncles et al, 1998). Fig. 1 shows an example of mud and sand in close proximity in the Eden Estuary (east coast of Scotland).

Very little is quantitatively known about how mixtures of cohesive and non-cohesive sediments, of different ratios and concentrations, interact whilst in suspension in turbulent flows and the effect this has on the resultant flocs formed and their flocculation properties, in particular the settling velocity. This has important implications for sediment transport modelling. Drawing on key literature and new data, this chapter will provide an overview of mixed sediment flocculation dynamics and how they can influence sediment transport.

The first part of this chapter reviews the theoretical aspects relating to the flocculation of mud:sand mixtures. It commences with a brief review of flocculation processes (2), followed by an overview of segregation environments verses flocculating suspensions (3), and then the biological influences on mixed sediment flocculation are summarised (4). The second part of

Figure 1. Sand and muddy sediments in close proximity, Eden Estuary, Fife (east coast of Scotland).

the chapter (5-7) draws on the findings of recent empirical studies assessing mixed sediment floc behaviour. The laboratory experimental protocols and findings are reported with floc data in spectral and parameterised formats presented and discussed. The potential implications of mud:sand flocculation on sediment transport modelling are also discussed (8-9).

2. Flocculation factors

From a sediment transport perspective, knowledge of the settling rate of sediments in suspension is vital in determining depositional fluxes and sediment transport rates. Sand is a non-cohesive material and therefore does not flocculate in pure sand suspensions. The settling velocity (Ws) is generally proportional to the square of the particle size or diameter (D). Conversely, mud is strongly cohesive and flocculates forming small, compact microflocs as well as larger, more porous macroflocs (Eisma, 1986; Manning, 2001; Manning and Dyer, 2002a,b) – Fig. 2. Flocculation is a dynamically active process which readily reacts to changes in hydrodynamically generated turbulent shear stresses (τ) (e.g. Krone, 1962; Parker et al., 1972; McCave, 1984; van Leussen, 1994; Winterwerp, 1998; Manning, 2004a), suspended particulate matter (SPM) concentration, together with salinity, mineralogy and biological stickiness.

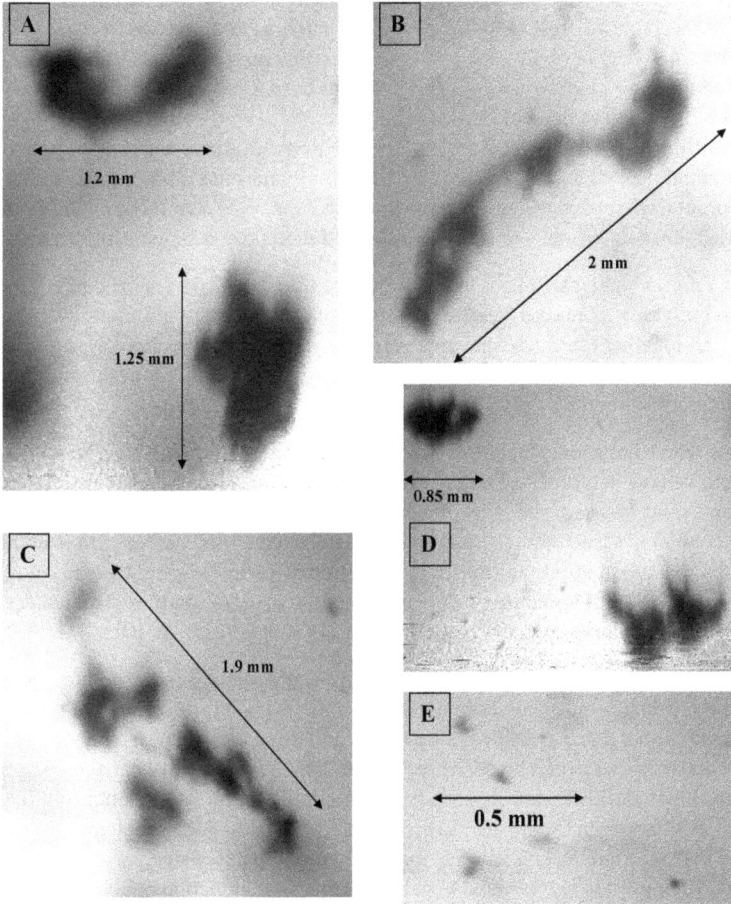

Figure 2. A selection of floc images from a predominantly muddy origin. A) A ragged cluster-type macrofloc (top) and a simple stringer composed of two macroflocs interlinked by organic fibres (bottom); B) a 'string of pearls' type macrofloc; C) a long interlinked stringer comprising two clustered macroflocs; D) ragged macroflocs settling; and E) a selection of small slow settling microflocs, some of which are probably the result of macrofloc fracturing and subsequent break-up during a turbulent event which exceeded the original macrofloc structural integrity threshold.

Flocculation can significantly alter the sediment transport patterns throughout an estuary, and floc properties can vary both in time and space. For example, Manning et al. (2006) showed that during spring tidal conditions in the Tamar Estuary (UK), macroflocs can typically reach 1-2 mm in diameter. These flocs demonstrate settling velocities up to 20 mm.s^{-1}, but their effective densities ϱ_e (i.e. the floc bulk density less the water density) are generally less than 50 kg.m^{-3}, which means they are prone to break-up when settling through a region of high turbulent shear.

There are, however, many estuarial environments where mud and fine sand co-exist as a single mixture (Mitchener et al., 1996) and this creates the potential for these two fractions to combine and exhibit some degree of interactive flocculation (Manning et al., 2007, 2009). The erosion and consolidation of mixtures of mud and sand has been thoroughly reviewed (Williamson, 1991; and Whitehouse et al, 2000), and there have been some studies that have examined mixed sediment settling (e.g. Dankers et al., 2007). However, very little investigation has been devoted to the potential flocculation that may occur when mud and sand mixtures are entrained into suspension, as it was not considered to be an important factor. This could be a valid assumption for a segregational environment, where the mud and sand do not combine into a single matrix.

When we refer to 'mixed sediment flocculation' in this chapter, we are primarily referring to suspension mixtures of mud (typically composed of clay minerals and fine silts up to 63 μm in diameter together with organic matter) and predominantly non-cohesive sediments (typically up to the size of fine sands, i.e. about 100-200 μm, as larger grains are unlikely to directly interact with mud).

Previous research has shown that a clay content of between 5 – 10% can cause natural sediment mixtures to behave in a cohesive manner (Dyer, 1986; Raudviki, 1998). Thus, different ratios of mud and sand can vary the level of cohesion, which will influence the resultant level of flocculation. Biological activity, more commonly associated with cohesive sediments, has been highlighted to play an important role in the cohesion of sediments (e.g. Paterson and Hager-they, 2001). However, it is extremely difficult to quantify such a complex sedimentary matrix in a fundamental manner, primarily as a result of a lack of verification data.

Of the various processes that occur during a tidal cycle, flocculation of the sediment is regarded as one of the primary mechanisms that can affect the deposition, erosion and consolidation rates. An individual floc may comprise up to 10^6 individual particulates. As flocs grow in size their effective densities generally decrease (Koglin, 1977; Tambo and Watanabe, 1979; Klimpel and Hogg, 1986) and their settling speeds rise due to a Stokes' Law relationship (Dyer and Manning, 1998) between D and Ws. Furthermore, low density flocs also demonstrate settling velocities that are significantly quicker than the individual cohesive particles (~ 1-5 μm in diameter). The cohesive nature of these particulates is a combination of both the electrostatic charging of the clay minerals as they pass through brackish to highly saline water, and various sticky biogenic coatings, such as mucopolysaccharides (e.g. Paterson 1989).

Van Leussen (1988) theoretically assessed the comparative influence of the three main collision mechanisms: Brownian motion, turbulent shear and differential settling, and deduced that turbulent shear stresses, principally those generated by velocity gradients present in an estuarine water column, were the dominant flocculation mechanism. This mechanism was deemed most effective for turbulent shear stresses ranging between 0.03-0.8 Pa. These stresses are representative of those typically experienced in the near bed region of many European macrotidal and mesotidal estuaries and hence estuaries are ideal environments for flocculation.

The energy for turbulent mixing is derived from the kinetic energy dissipated by the water flowing across the sediment bed. The frictional force exerted by the flow per unit area of the bed is the shear stress (turbulent shear stress during turbulent flow conditions). The efficiency with which the particles flocculate is a reflection of the stability of the suspension (van Leussen, 1994). A suspension is classified as unstable when it becomes fully flocculated, and is stable when all particles remain as individual entities.

As low to medium levels of turbulent shear stress can promote floc growth, high levels of turbulence that occur during a tidal cycle, can cause disruption to the flocculation process by instigating floc break-up, and eventually pull the constituent components of a floc apart. As turbulent activity increases, both turbulent pressure differences and turbulent shear stresses in the flow rise. If the floc structural integrity is less than the imposing turbulent induced forces, the floc will fracture. Also, aggregate break-up can occur as a result of high impact particle collisions during very turbulent events. Floc break-up by three-particle collisions tends to be the most effective (Burban et al., 1989). Hence, turbulent shear stress can impose a maximum floc size restriction on a floc population in tidal waters (McCave, 1984). Eisma (1986) observed a general agreement between the maximum floc size and the smallest turbulent eddies as categorised by Kolmogorov (1941a, b).

3. Segregation and flocculation

This section looks at how mud and sand can co-exist within an aquatic environment. Mud:sand sediment mixtures may behave either in a segregated way, or interact through flocculation. The phenomenon of mud:sand segregation considers the mud and sand to operate as two independent suspensions (van Ledden, 2002) and, as such, very little bonding occurs, and flocculation interactions between the cohesive and non-cohesive sediment fractions are non-existent. Mixed sediment experiments have shown that mud particles and sand grains which behave in a segregated manner, settle simultaneously but as independent fractions to form two well sorted layers at the bed/water interface (Ockenden and Delo, 1991; Migniot, 1968; Williamson and Ockenden, 1993).

Williamson (1991) reviewed a number of the characteristics of mud:sand mixtures in the natural environment (some of the key findings are summarised in this paragraph). The review investigated the distributions and characteristics of mud and sand mixtures based on a literature search and a review of relevant fieldwork data. Some of the features common to both mud and sand, such as: spatial distributions, vertical layering, bioturbation, depositional characteristics and flocculation, were described. The review suggested that muddier sediments were generally found in regions of lower dynamic activity and sandier sediments in higher energy regions. However, the local distributions could only be explained by local hydrody-namic analysis and these data were often lacking, which did not allow a complete picture to be obtained. Flocculation and the effects of salinity distributions were found to be important in governing the mud distributions, with a muddy reach often being found in the flocculation zone. The vertical profile of settled mud and sand was also investigated, with laminations of mud and sand often being found. The thickness of the layers in the laminated sediment profiles were typically sub-millimetre to a few millimetres. The process of bioturbation (i.e. the reworking of the bed sediments by living organisms) can potentially produce a mixing of bed sediment particles prior to resuspension (e.g. Nowell et al., 1981; Paterson et al., 1990; Widdows et al., 2004). Thus a bed which is initially deposited as a discretely segregated layering of mud and sand may be transformed into a quasi-homogeneous mixture.

Van Ledden (2003) states that mud and sand can be deposited as mixtures or in alternating layers in estuaries. An example of this is visible in the upper part of Fig 1. Additionally, biological activity such as bioturbation (i.e. the reworking of the bed sediments), can mix the sediment particles. As a result the mud content in many parts of an estuary may not be uniform, but can become segregated both vertically and horizontally – a phenomenon known as mud:sand segregation (van Ledden, 2003).

Mud:sand segregation can have a direct influence on the settling velocity of the sediments once entrained. For instance, the settling velocity of individual sand grains could be reduced as they pass through a layer of flocculating muddy sediments in close proximity to the sea bed. Van Ledden (2003) provides three examples which illustrate the importance of why a physical understanding of the distribution of mud and sand in estuarine systems is important:

- Large mud content variations at the bed surface indicate that both mud and sand contribute to bed level changes in estuaries and tidal inlets. These will affect the navigable depth and high water levels.

- Cohesive muddy sediments have the propensity to adsorb contaminants (Förstner and Wittmann, 1983). This, in turn, has a direct effect on water quality and related environmental issues (e.g. Uncles et al., 1998). The amount of segregation present on both temporal and spatial scales will provide an indication to the potential degree of pollution in bed sediments.

- The mud content in sediment beds is a crucial habitat parameter, which controls the distribution of flora and fauna in estuarine systems (e.g. Reid and Wood, 1976; Kennish, 1986; Widdows et al., 2004). Dyer et al. (2000), for example, showed that the sediment type and grain size are the best physical descriptors of floral and faunal assemblages in the upper zone of intertidal mudflats.

Van Wijngaarden (2002a, 2002b) examined the mud:sand content distributions in the upper 300 mm of the bed in the Haringvliet – Holland Diep (The Netherlands). Mud content varied from less than 15% at the mouths of most of the river branches feeding into the system, to nearly two thirds mud in the channels of the Holland Diep. Fast settling sand grains accumulated at the end of river branches whereas the slower-settling muddy suspensions were transported further downstream due to settling lag into the central part of the Holland Diep. The segregation is, to a large extent related to varying bed levels throughout the system and variations in the turbulent shear stresses (van Ledden, 2003), which influence erosion, deposition and transport.

There are also many locations where mud and sand co-exist as a mixture (Mitchener et al., 1996) and this creates the potential for these two fractions to combine within a flocculation matrix when re-entrained into suspension (Manning et al., 2007). When sand is added to a predominantly muddy matrix, Mitchener et al. (1996) found that this increased the binding potential between the clay particles, for example as found in the subtidal mud patches off Sellafield in the Irish Sea (Feates and Mitchener, 1998). Thus the physical effect of adding cohesive mud to a sandy environment can create increased bed stability, which can potentially lead to mixed sediment flocs forming when the eroded bed is entrained (Kamphuis and Hall, 1983; Alvarez-Hernandez, 1990; Williamson and Ockenden, 1993; Torfs, 1994; Mitchener et al., 1996; and Panagiotopoulus et al., 1997). Even where sand and mud are considered to be fairly well segregated at the bed, sand and mud can co-exist in suspended sediment transport. Spearman et al. (2011) describe an example in the outer Thames Estuary (UK), renowned for being a sandy area, where the flux of suspended sediment of mud and sand are of the same magnitude.

Therefore, in a segregated environment, both mud and sand are present acting in a completely independent manner. In a flocculating environment, the mud and sand particles are interacting to form flocs which demonstrate very different characteristics (e.g. D, Ws, ϱ_e) from their compositional base. The nature of the sedimentary regime is best determined by observational measurements rather than being able to be determined *a priori*. This can pose additional problems for the prediction and modelling of suspended sediment transport in mixed sediment estuarine environments and this will be considered in Section 9.

4. Role of biology in mud: Sand mixtures

Although not directly examined in the laboratory experiments which will be discussed later in this chapter, it is important to consider other effects of which a key one is due to biological factors influencing the grains in suspension. These factors work in addition to the primary chemico-physical ones to make mixed sediment flocculation possible. In predominantly muddy/silty environments, benthic microphytobenthos contribute up to half the total auto-trophic production in an estuarine system (Underwood and Kromkamp, 1999; Cahoon, 1999). Biostabilisation can increase particle cohesion, for example: epipelic diatoms (e.g. Paterson and Hagerthey, 2001) secrete extra-cellular polymeric substances (EPS; Tolhurst et al., 2002) as they move within the sediments. EPSs are regarded as highly effective stabilisers of muddy sediments (e.g. de Brouwer et al. 2005; Gerbersdorf et al. 2009; Grabowski et al., 2012).

The influence of biology on sand is reported to a much lesser extent in the literature, however sand grains that are exposed to long-term biological activity, may also develop a cohesive bio-coating which could increase the particle collision efficiency when they are entrained. Hickman and Round (1970) reported that sand particles can be joined by 'epipsammic' diatoms which attach to sand grains. Epipsammic macro-algal forms either adnate to the grain surface or attach to sand grains by their mucilage stalks. Epipsammic diatoms which are attached to sand grains, demonstrate strong adhesive properties to the grain surface (Harper and Harper, 1967). When fine sand and biology are combined into a single matrix, they can form "microbial mats" and the binding strength of these mats can be extremely high. Little (2000) states that because these types of algal threads are sticky with EPS, they can efficiently trap fine sand grains. These sticky bio-coatings can increase the collision efficiency (Edzwald and O'Melia,

1975) of particles when entrained into suspension, thus allowing fine sand grains to adhere with the clay fraction and form the cage-like structure around fine sand particles. Through microscopic photography, Wolanski (2007) observed the formation of large muddy flocs formed by mud creating a sticky membrane around large non-cohesive silt particles.

5. Experimental approaches

When investigating the role sand may play in the flocculation process, several important research questions need to be considered, including:

i.   How does the settling velocity of mixed sediment flocs vary in response to different mud:sand mixtures?

ii.   What effect does turbulence have on mixed sediment flocculation?

iii.   Do resuspended sand particles favour interacting with microflocs or macroflocs more, and enhance their settling dynamics?

iv.   If mixed sediment flocculation occurs, are sand grains directly incorporated into both microfloc and macrofloc fractions?

v.   Does flocculation have an effect on the distribution of the particle mass and the mass settling flux (MSF) of different suspended mud:sand mixtures?

In order to address aspects of the above questions, a series of new controlled laboratory environment research were initiated to quantitatively examine the flocculation and interaction between suspended sand and mud sediment mixtures. Other aspects of mud:sand behaviour have been assessed in laboratory environment measurements (e.g. Ockenden and Delo, 1988; Williamson and Ockenden, 1993; Torfs, 1994; Torfs et al., 1996; Dankers et al., 2007). During the new experiments, suspensions of mud and sand, of different total concentrations, were sheared at different rates in a mini-annular flume and the resultant floc properties observed. The new experimental runs primarily comprised pre-determined mud:sand mixtures complemented with some additional data from naturally occurring mud:sand sediment mixtures.

5.1. Annular flume simulations

This study utilised a mini-annular flume to create a consistent and repeatable turbulent environment (see Fig. 3A) (Manning and Whitehouse, 2009). The annular flume has an outer diameter of 1.2 m, a channel width of 0.1 m and a maximum depth of 0.15 m, along with a detachable motor driven rotating roof (10 mm thick) to create the flow for cohesive sediment experiments (e.g. Manning and Dyer, 1999). Maximum flow speeds of approximately 0.7 m.s^{-1} can be produced in the lower half of the water column, created by 10 mm deep paddles attached to the underside of the roof. A Nortek mini-ADV (Acoustic Doppler Velocimeter) probe was used to calibrate the flow in terms of velocity and turbulent kinetic energy (TKE) at a distance of 22 mm (the floc extraction height) above the flume channel base.

Figure 3. The mini-annular flume (A) and the LabSFLOC instrument set-up (B).

5.2. Floc property measurements

Representative floc populations were measured using the LabSFLOC version 1.0 – Laboratory Spectral Flocculation Characteristics – instrument (Manning, 2006). This utilises a high magnification Puffin (model UTC 341) monochrome all-magnetic Pasecon tube video camera (Manning and Dyer, 2002a), to observe particles settling in a Perspex settling column (see Fig. 3B), allowing for minimal disruption of the particles. The video camera, positioned 75 mm above the base of the column, views all particles in the centre of the column that pass within a 1 mm depth of field, 45 mm from the lens. The video camera has an annulus of six high intensity red 130 mW LED's (light emitting diodes) positioned around the camera lens, which results in the flocs being viewed as silhouettes and produces a clear image of their size and structure. Whilst other studies may refer to muddy and/or mud-sand mixture particles as aggregates, for simplicity this study will refer to all aggregated combinations of particles as flocs.

5.3. Flume experimental protocols

The flume was filled with 45 litres of saline water (salinity = 20 ±0.2), to a depth of 0.13 m. The mixed sediments (both pre-determined and natural) were introduced into the flume as slurries of known SPM (suspended particulate matter) concentrations. Gravimetric analysis of extracted water samples was used to monitor the ambient concentration during the flume runs and check they were within the required experimental tolerances. For each run, different rotation speeds were used to shear the sediment slurries at shear stresses (τ) ranging from 0.06-0.9 Pa ±5% (equivalent Kolmogorov microscale values are: 381 - 138 µm ; equivalent G-values, the root mean square of the gradient in the turbulent velocity fluctuations, are: 7.1 – 54.2 s^{-1}) at the floc sampling point. Manning and Whitehouse (2009) report the calibration of the mini-flume hydrodynamics. Each run was initiated at the fastest rotational velocity and decreased towards the slowest speed as the run progressed. Further details of the experimental protocols are outlined by Manning et al. (2007).

The mixed sediment slurries were sheared in the flume for 30 minutes at each stress level. This duration of shearing, which was pre-determined in accordance with theoretical flocculation time (T_F), allowed each sediment suspension to attain floc equilibrium. Van Leussen (1994) defines T_F as the time required to decrease the number of individual unflocculated particles in a suspension, to just 10% of the initial number as a result of flocculation.

Floc population sampling comprised careful extraction of a suspension sample from the same height in the water column as the ADV calibration using a bespoke glass pipette. To obtain a floc sample, the rotation was stopped for approximately 6-8 seconds, although flow in the flume still continued through inertia, maintaining particles in suspension throughout this period. Manning and Whitehouse (2009) showed that the flow does not significantly slow until at least 15-20 seconds after stopping the drive motor. The floc sample was then transferred to the LabSFLOC Perspex settling column, whereby each individual floc was observed by the video camera as it was settling. Parameters of individual floc size (D) and settling velocity (Ws) were recorded during settling and the values obtained by video image post-processing. The experimental flow speeds generated in the flume were sufficient to keep the fine sand in suspension. The aperture of the pipette was brought into contact with the settling column water surface and held in place (vertically) allowing the captured flocs to undergo gravitational settling through the still water column. Extensive testing of this sampling protocol during the EC COSINUS project (e.g. Gratiot and Manning, 2004) revealed that this technique created minimal floc disruption during acquisition. Once floc samples were extracted, the flume lid rotation continued at the next selected velocity.

5.4. LabSFLOC data processing

Parameters D and Ws, for all settling flocs viewed by the LabSFLOC video camera (for each sample), were measured simultaneously from the video recordings. Digitisation of the calibrated images resulted in a pixel resolution of 6.3 μm to determine floc size and position, from which settling velocity is determined by analysis of sequential images at a sampling rate of 25 Hz. The effective density (ϱ_e) of each floc was calculated by applying Stokes' Law relationship; ϱ_e is the difference between the floc bulk density (ϱ_f) and the water density (ϱ_w). To apply Stokes' Law, it is assumed that each sampled floc that fell through the still water enclosed within the settling column was within the viscous Reynolds region; i.e. when the individual floc Reynolds number (R_e) was less than 0.5. For instances where R_e exceeded 0.5, the Oseen modification, as advocated by ten Brinke (1994), was applied in order to correct for the increased inertia during settling. It is assumed that the measured particle is spherical; that is, it is as 'deep' as the measured D size.

The observed flocs were measured within a reference volume of water. By implementing a sequence of algorithms, originally derived by Fennessy et al. (1997) and modified by Manning (2004b), the dry mass of a floc population could be compared with the measured SPM concentration. This provides an estimate of the efficiency of the sampling procedure, and yielded corresponding rates of MSF. By definition, the data obtained from LabSFLOC are both of qualitative and quantitative value.

The floc data is presented as individual scatterplots and also as spectral size-banded (SB) distributions of floc mass and MSF; SB1 represents microflocs less than 40 μm in size and SB12 are macroflocs greater than 640 μm in diameter. Sample mean values are quoted. To provide a quantitative framework for population comparisons, the macrofloc and microfloc range of properties were assessed (Eisma, 1986; Manning, 2001), as these parameters are often used in flocculation modelling. The demarcation point for the macrofloc:microfloc fractions was a floc size of 160 μm (Manning, 2001) and was chosen for two main reasons: i) this was found to be the most statistically significant separation point for the majority of the mixed sediment floc populations in terms of mass settling properties; ii) it also provides computational continuity with previously derived flocculation algorithms for pure mud suspensions, such as the Manning Floc Settling Velocity (MFSV) algorithms which describe floc settling at different concentrations within turbulent flow (Manning and Dyer, 2007). Strictly it should be noted that microflocs are cohesive sediment flocs resistant to break-up by shear, however, in this study, many pure sand particles fall within the microfloc size range. Therefore, in this chapter microflocs refer to the 'fine particle population' < 160 μm in diameter. The sand used in the tests also contains a fraction with grains greater than 160 μm (around 10% by mass). Therefore, the macrofloc fraction may also contain a number of pure sand grains.

5.5. Floc microstructure

In order to examine the floc internal microstructure (matrix) at a sub-micron level (1-2 nm; Buffle and Leppard, 1995), use of transmission electron microscopy (TEM) was employed in a separate series of experiments (see Spencer et al., 2010). In addition, energy dispersive spectroscopy (EDS) was used to provide the elemental composition of the floc components. Samples were prepared for TEM analysis by first stabilising the samples in glutaraldehyde and embedding the samples in Spurr resin. The samples were polymerised at 60 °C overnight. Ultrathin sections of the polymerised resins (50 nm) were obtained by sectioning with a diamond knife mounted in an ultramicrotome (RMC Ultramicrotome MT-7) and were then mounted on formvar copper grids for analysis. The ultra-thin sections were then observed in transmission mode at an accelerating voltage of 80 kV using a JEOL 1200EXIITEMSCAN scanning transmission electron microscope (STEM). The scanning mode of the STEM was used to generate a microprobe beam for EDS of individual floc components in sections allowing observation of minerals across the aggregates. A Princeton Gamma Tech (PGT) Si[Li] X-ray detector and Imix multichannel analyser provided spectra of all elements, with an atomic number greater than 10, on a "per colloid" basis.

6. Experimental results

Sections 6.1-6.5 report findings from the laboratory studies with pre-determined (PD) mud:sand mixtures conducted by Manning et al. (2007). Sections 6.5-6.6 report a selection of tests on naturally occurring mud and sand mixtures (NM), and analysis of a mixed sediment microfloc internal structure, respectively.

6.1. Sediments (PD)

The sand used in the pre-determined mixtures was named Redhill 110, which is a well-rounded and closely graded silica sand used by HR Wallingford for model testing with mobile sediment beds. Redhill 110 has a d_{50} of about 110 μm, with a d_{10} of 70 μm and a d_{90} of approximately 170 μm (Redhill 110 size values quoted are from independent analysis conducted at HR Wallingford). The experimental mud sample was obtained from the surface down to a depth of about 50 mm from the Calstock region of the upper Tamar Estuary (UK) and had an average organic content of approximately 10%. Fitzpatrick (1991) found Tamar Estuary mud to be generally high in kaolinite clay minerals and Fennessy et al. (1994) also report microscopic fragments of Tourmaline and Hornblende minerals present in Calstock mud. This particular mud was used as its floc properties are widely reported from earlier studies (e.g. Manning and Dyer, 2002b ; Mory et al., 2002 ; Bass et al., 2006). The mud was collected only a few days before the flume experiments were conducted, and cold stored (frozen) in a wet form to maximise organic matter preservation.

6.2. Overview of experimental runs (PD)

These experiments comprised a series of three main flume runs, A to C, based on pre-deter-mined mud:sand (M:S) ratios (i.e. Run A = 75M:25S, Run B = 50M:50S and Run C = 25M:75S; units expressed as percentages). These main runs were each divided into 12 minor runs (based on concentration). This produced a total of 36 mixed sediment floc spectral samples. Three nominal total SPM concentrations were used: 200 mg.l^{-1}, 1000 mg.l^{-1} and 5000 mg.l^{-1}. Four shear stresses were used per run and these were determined by the ADV records as nominal clearwater τ values of: 0.06, 0.35, 0.6 and 0.9 Pa. The experimental conditions are summarised in Table 1.

Run	Sample	Mud (%)	Sand (%)	τ (Pa)	SPM (mg/l)
A	1	75	25	0.9	200
A	2	75	25	0.6	200
A	3	75	25	0.35	200
A	4	75	25	0.06	200
A	5	75	25	0.9	1000
A	6	75	25	0.6	1000
A	7	75	25	0.35	1000
A	8	75	25	0.06	1000
A	9	75	25	0.9	5000
A	10	75	25	0.6	5000
A	11	75	25	0.35	5000
A	12	75	25	0.06	5000
B	1	50	50	0.9	200
B	2	50	50	0.6	200
B	3	50	50	0.35	200
B	4	50	50	0.06	200
B	5	50	50	0.9	1000
B	6	50	50	0.6	1000
B	7	50	50	0.35	1000
B	8	50	50	0.06	1000
B	9	50	50	0.9	5000
B	10	50	50	0.6	5000
B	11	50	50	0.35	5000
B	12	50	50	0.06	5000
C	1	25	75	0.9	200
C	2	25	75	0.6	200
C	3	25	75	0.35	200
C	4	25	75	0.06	200
C	5	25	75	0.9	1000
C	6	25	75	0.6	1000
C	7	25	75	0.35	1000
C	8	25	75	0.06	1000
C	9	25	75	0.9	5000
C	10	25	75	0.6	5000
C	11	25	75	0.35	5000
C	12	25	75	0.06	5000

Table 1. Overview of experimental runs & samples.

During a pilot study to design and refine experimental protocols on the floc population evolution of a few pre-selected slurries, observations indicated that at a τ of 0.06 Pa the sand in the upper part of the water column settled to the channel base. However, this preliminary inspection indicated that there was still sufficient fine sand in suspension in the lower half of the flume to maintain the nominal mud to sand ratio in the floc sampling region. Furthermore, during the pilot study, checks were made on mixture homogeneity during suspension and revealed a nominal 8% mixture deviation (in terms of the sand) for a 75% sand slurry, reducing to less than 5% for a 75M:25S mixture. These nominal deviations are deemed acceptable for these mixed sediment flocculation experiments, but are taken into consideration when interpreting the study results.

During the main flume run, the total suspended concentrations were monitored by gravimetric analysis of samples withdrawn at the floc sampling point. This analysis indicated that the 200 mg.l^{-1} total SPM varied the least at ±3%; the higher 5000 mg.l^{-1} varied by ±4.7%; and the 1000 mg.l^{-1} slurry nominally varying by ±4.3% by the time of floc sampling. Therefore, these relatively small deviations demonstrate that the majority of the mixed sediment mass was remaining in suspension for the shearing duration. Therefore the floc population characteristics were related closely to the initial total concentrations and mud:sand ratios. Further details on the homogeneity of mud:sand mixing within the mini-annular flume is reported by Manning et al. (2009).

6.3. Floc size and settling velocity spectra with mixtures of mud and sand (PD)

To demonstrate the floc properties for suspensions comprising 75M:25S, 50M:50S and 25M:75S, a number of examples of the individual detailed spherical-equivalent dry mass weighted floc sizes vs. settling velocity spectra are presented (Figs 4Ai-4Av). The plots represent the mass-balance corrected floc distributions, thus an individual point on each graph may represent several flocs with very similar floc characteristics. The diagonal lines on each scatterplot represent contours of constant floc effective density, ϱ_e, (units = kg.m^{-3}), i.e. the bulk density minus the water density.

For completeness the full set of D vs. Ws floc distributions for all experiments can be found in Figs. 5, 6 and 7. By following the plots in each column, starting at the lower plot, one can track the evolution of the floc populations formed in a constant SPM concentration as the shear stress rises through the various increments. Similarly, by following the plots from left to right, the effect of rising concentration on the floc dynamics can be observed. Sections 6.3 and 6.4 summarise some of the key observations from a selection of the populations.

6.3.1. Run A (75M:25S) (Fig. 5)

The flocs from the lower SPM concentration (200 mg.l^{-1}), A1-A4, appear to produce three separate clusters: a sub-70 μm group, a fraction greater than 160 μm; with a third group sandwiched in between. For example, the 204 individual flocs that comprised sample A3 (Fig. 5 box A3) ranged from 42 μm to 182 μm in diameter (also Fig. 4Ai). Corresponding settling velocities spanned 0.3 mm.s^{-1} to 3.4 mm.s^{-1} for sample A3.

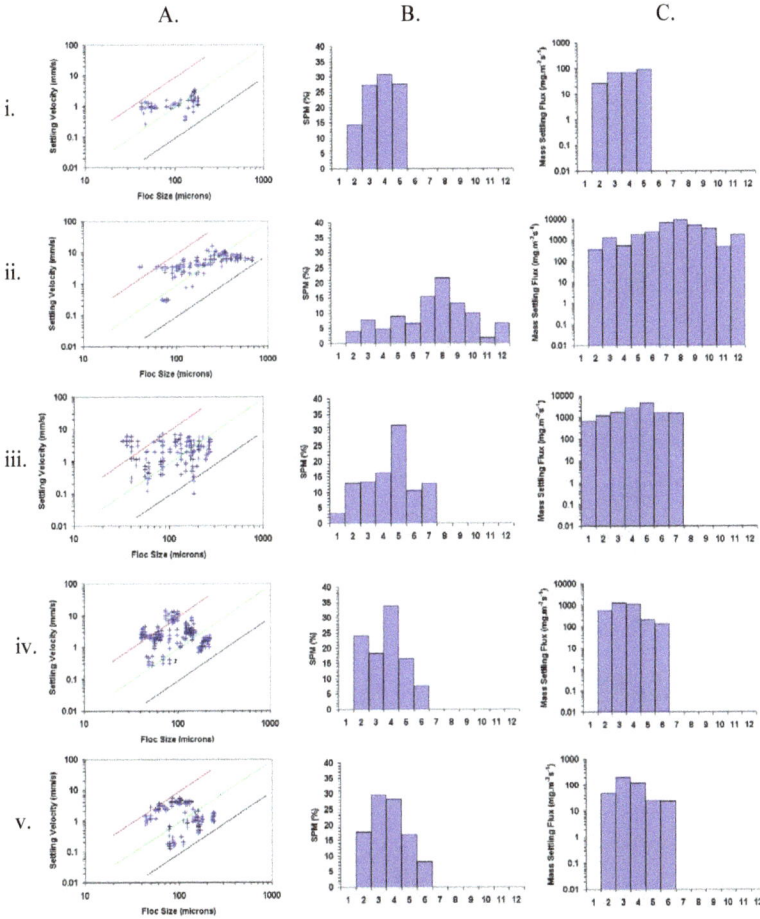

Figure 4. The floc size vs. settling velocity scatter plots (A, left-hand column) for five selected samples: i) A3, ii) A11, iii) B9, iv) C6 and v) C3. Diagonal lines on figures in column A represent contours of constant Stokes equivalent effective density: red = 1600 kgm⁻³, green = 160 kgm⁻³, and black = 16 kgm⁻³. The centre (B) and right-hand (C) columns represent the corresponding size-banded SPM% and mass settling flux distributions (units = mg.m⁻²s⁻¹). The size bands are illustrated in the table below the plots.

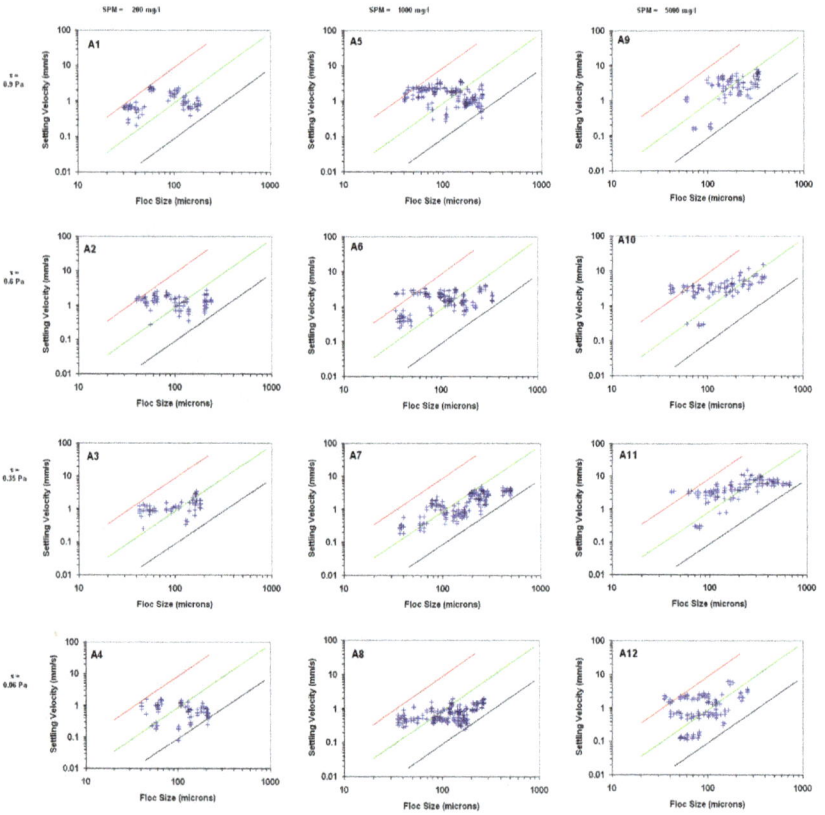

Figure 5. Distribution floc/aggregate size and settling velocity characteristics for the Run A (75M:25S) samples. Diagonal lines represent contours of constant Stokes equivalent effective density: red = 1600 kgm⁻³, green = 160 kgm⁻³, and black = 16 kgm⁻³.

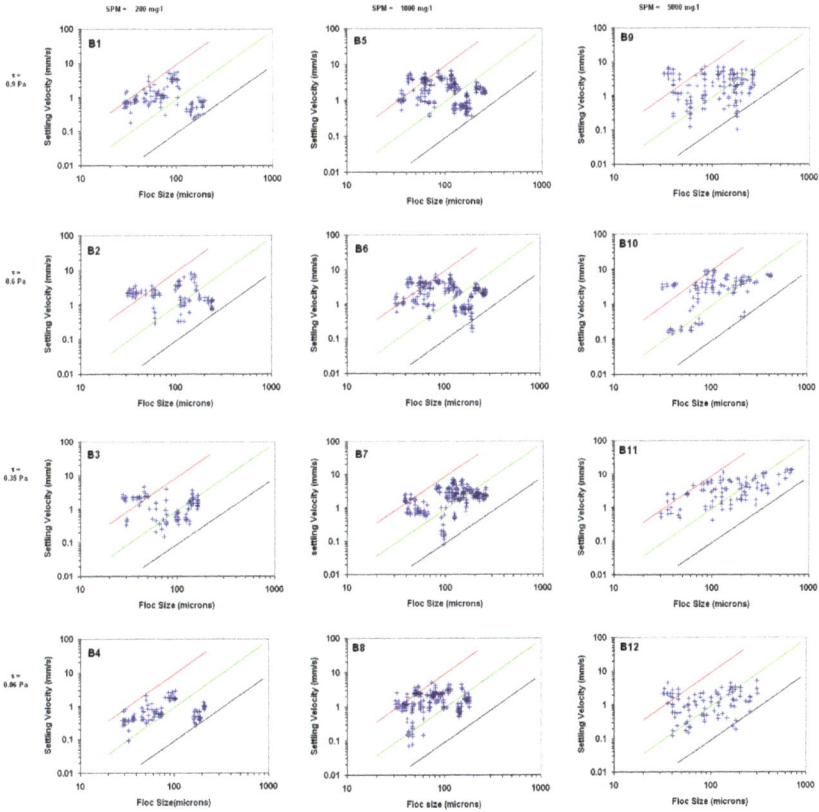

Figure 6. Distribution floc/aggregate size and settling velocity characteristics for the Run B (50M:50S) samples. Diagonal lines represent contours of constant Stokes equivalent effective density: red = 1600 kgm⁻³, green = 160 kgm⁻³, and black = 16 kgm⁻³.

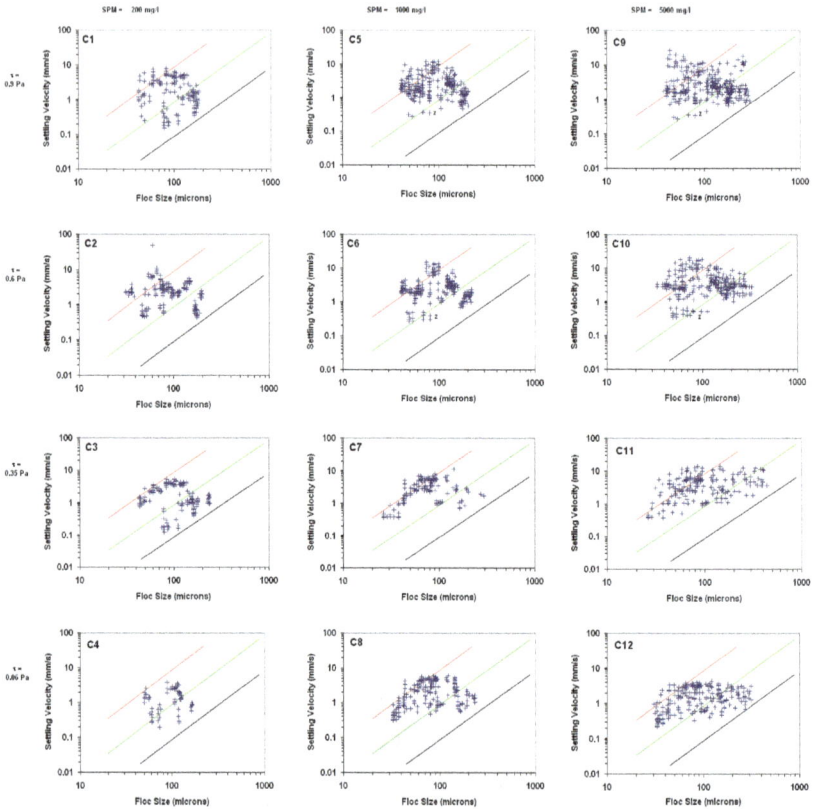

Figure 7. Distribution floc/aggregate size and settling velocity characteristics for the Run C (25M:75S) samples. Diagonal lines represent contours of constant Stokes equivalent effective density: red = 1600 kgm^{-3}, green = 160 kgm^{-3}, and black = 16 kgm^{-3}.

The Run A floc growth was potentially stimulated by a greater abundance of sediment, with D_{Max} (maximum floc diameter) nearly reaching 700 μm at peak turbidity (5000 mg.l^{-1}). The floc growth signified a corresponding quickening in Ws with rising SPM, producing Ws_{Max} (maximum settling velocities) of 7-8 mm.s^{-1} at 5000 mg.l^{-1}; approximately double the speed exhibited by the dilute sandy mud suspensions. This is demonstrated by A11 (Fig. 5 box A11 and Fig. 4Aii) where the shear stress was the same as A3 (0.35 Pa), but the particle mass in suspension were raised by a factor of twenty five. Flocs greater than 160 μm comprised 61% of the total population. In terms of the effects of shear stress, 0.35 Pa seems to produce the largest, fastest settling macroflocs at 75M:25S. These inter-relationships will be further examined in the Discussion (Section 7).

6.3.2. Run B (50M:50S) (Fig. 6)

Increasing the sand content to equal the mud fraction (50M:50S), brought about a general decrease in the macrofloc settling velocity across the entire shear stress range at each base concentration increment.

In contrast to the macroflocs, the smaller 50M:50S microfloc fractions all displayed quicker fall rates when compared to the 75% mud in settling rate for each mixed suspension run, with the 5000 mg.l^{-1} 50M:50S mixed suspension (sample B9, Fig. 6) microfloc fraction settling velocity peaking at a highly turbulent τ of 0.9 Pa.

The B9 size vs. settling velocity floc scatter plot (Fig.6 box B9 and Fig. 4Aiii) shows a "W" or "double-V" pattern to the aggregates distribution. By this we mean there are small, fast settling microflocs (nominal 20-40 μm), whose settling velocity range expands at the mid-size microfloc fraction (nominal 40-80 μm). Then, for the microflocs nominally greater than 80 μm in size, the spread in the microfloc Ws again reduces, thus producing a "V" shaped distribution. This "V" pattern is repeated for the macroflocs, with their largest Ws scatter occurring between 185-230 μm for Sample B9.

The microflocs forming the first "V" spanned from 32 μm and up to 114 μm where they form the apex with the adjacent "V" to form the "W". At each end of the size range there are aggregates settling at 5-7 mm.s^{-1}, whilst the middle part of the "V" sections shows flocs falling as slowly as 0.1 mm.s^{-1}. In the upper left part of the D vs. Ws scatterplot, there are a number of aggregates which appear to be between 35-50 μm in diameter, settling at 3-6 mm.s^{-1} and exhibiting effective densities of 2000-5000 kg.m^{-3}, which is up to three times the effective density of a sand grain. It is most probable that these are individual fragments of either Tourmaline or Hornblende; minerals native to the Tamar Estuary and its catchment. The majority of the aggregate population between 45-90 μm appears to be dominated by sand grains, with a minimum amount of cohesive matter (i.e. mud content) attached to the sand grains. These would form very basic, dense, lower order floc structures, which would trap very little interstitial water. This is indicated by high effective densities (ρ_e ~1200-1400 kg.m^{-3}), large fractal dimensions (nf of 2.8-2.9) and low porosities (~10-20%), but they are still not characteristic of pure (i.e. unflocculated) sand grains.

6.3.3. Run C (25M:75S) (Fig. 7)

Reducing the mud content to 25%, meant the microfloc size fraction tended to dominate the size and settling dynamics as the total concentration rose throughout Run C. At dilute conditions, the microflocs represented less than one quarter of the individual flocs for the A1-4 samples; for example C3 (Fig 7. Box C3 and Fig. 4Av). However, with many of the sub-160 μm C1-4 flocs settling at 4-7 mm.s^{-1}, they were falling significantly quicker than their muddier Runs A and B counterparts.

A five-fold rise in the total SPM concentration increased the production of smaller flocs, with the macrofloc size fractions only accounting for 10-20% of the individual aggregates. For example, nearly 90% of the C6 flocs (τ = 0.6 Pa, SPM = 1000 mg.l^{-1}) were within the microfloc range (Fig. 7 box C6 and Fig. 4Aiv). This was approximately 15-20% more microflocs when compared to the more cohesive B6 and A6 samples (see relevant boxes in Fig. 6 and Fig 5. respectively).

The accuracy of effective density values is crucial to the determination of when mixed sediment particles are flocculating, or if the fine sand particles remain as individual inert entities. The reliability of the LabSFLOC effective density estimates are demonstrated by their observation of pure sand grains (Fig. 8). The D and Ws fine sand observations produce a distribution which closely follows the 1600 kg.m^{-3} density contour, generally not deviating by no more than ±100 kg.m^{-3} for over three hundred sand grain observations.

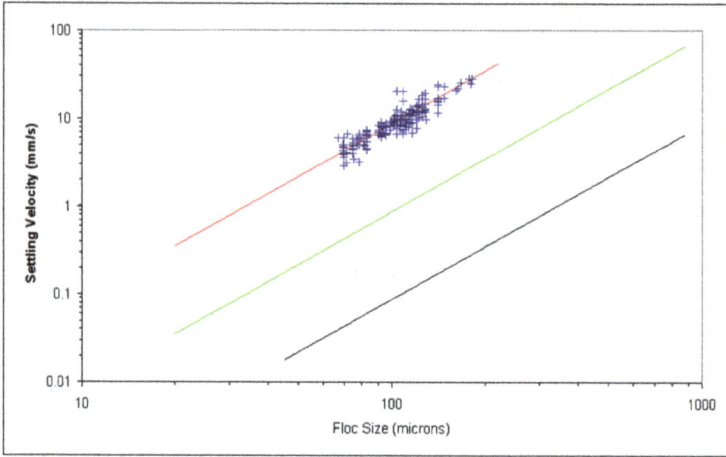

Figure 8. Settling settling vs. floc size for a 100% sand sample. Diagonal lines represent contours of constant Stokes equivalent effective density: red = 1600 kgm^{-3}, green = 160 kgm^{-3}, and black = 16 kgm^{-3}.

6.4. Floc composition with mixtures of mud and sand (PD)

To illustrate how the floc structure varies at different mud:sand ratios, a few examples will be presented with the compositional properties (effective density and SPM) as size band distributions. We start with sample A3 which represents a muddier dilute concentration and the D vs. Ws scatterplot (see Fig. 4Ai) shows that the macrofloc and microfloc fractions formed three distinctively separate groups. From Fig. 4Ai we can determine that the microfloc effective densities (ϱ_{e_micro} ranging from 200-1580 kg.m^{-3}) were generally an order of magnitude greater than the macroflocs (ϱ_{e_macro} from 30-100 kg.m^{-3}). This suggests that together with some individual sand grains, some of the sand grains may have also been included into the microfloc structure during the flocculation process.

In terms of the mass distribution across the dilute concentration floc population, the small microflocs for A3 represented three quarters of the mass (Fig. 4Bi). This is similar to fully cohesive suspensions within a moderately-high shear zone (τ of 0.6-1 Pa) which suggests the mixture is still behaving as a cohesive suspension, even with 25% sand present in the initial mixture. For this sample, the denser, more compact microflocs represented two thirds of the total 254 mg.m^{-2}s^{-1} mass settling flux (Fig. 4Ci).

At a concentration of 5000 mg.l^{-1}, the 75M:25S macroflocs of sample A11 (Fig. 4Aii) the macroflocs were observed to be delicate, low in density (ρ_e ranging from 20-200 kg.m^{-3}) entities. The A11 macroflocs now represented 84% of the mass (Fig. 4Bii), which was more than double the A3 macrofloc mass. Higher turbidity stimulated floc growth in A11, resulting in the largest flocs (D$_{max}$) growing to 670 μm.

The A11 microfloc fraction consisted of higher density flocs, with the smallest flocs (40-80 μm) demonstrating effective densities of over 1100 kg.m^{-3}, which are indicative of sand-laden microflocs or sand grains (where the effective density is greater than 1600 kg.m^{-3}). With sand accounting for one quarter of the total suspension and the microflocs representing 16% of the A11 mass, continuity of mass dictates that a reasonable portion of the sand must have been incorporated in many of the macrofloc structures during the flocculation process. This is very different from some segregational theories (e.g. van Ledden, 2003) which regard suspensions of sand and mud as completely independent entities.

Collectively, the fast settling A11 macroflocs contributed 94% of the total mass settling flux (33 g.m^{-2}s^{-1}; Fig. 4Cii); a result of a macrofloc settling velocity of 7.2 mm.s^{-1}, which was nearly three times quicker than the corresponding Ws$_{micro}$. To put this all into context, the A11 total MSF was 13 times greater than the value computed by the use of an estimated mean settling velocity of 0.5 mm.s^{-1}; a typical parameterised cohesive sediment Ws value derived from the gravimetric analysis of Owen tube (Owen, 1976) samples. Dearnaley, (1996) summarised the primary drawback associated with the Owen tube and other field settling tube devices, including the disruptive nature on flocs of the instrument sampling. Even the A11 microflocs were settling five times quicker than a 0.5 mm.s^{-1} parameter value (A11 Ws$_{micro}$ = 2.5 mm.s^{-1}).

Examination of the 50M:50S sample B9 D vs. Ws scatterplot (Fig. 4Aiii), reveals the presence of a high density sub-group of flocs (upper left-hand section). These flocs, which are only 35-50 μm in diameter, are settling at 3-6 mms^{-1} and exhibiting effective densities of up to 2000-3500 kg.m^{-3}. This is up to three times the typical effective density of a sand grain. It is proposed that these are individual fragments of either Tourmaline or Hornblende; minerals native to the Tamar Estuary and its catchment (Fennessy et al., 1994). However, given the Tamar's history for shipping copper out of Calstock, and the rich mining history for everything from tin to silver, these heavier particles could be from a number of sources. The majority of the floc population between 45-90 μm appears to be dominated by sand grains as their effective densities are typically greater than 1600 kg.m^{-3}, with a minimum amount of cohesive matter (i.e. mud content) attached to the sand grains. These would form very basic, dense, lower order floc structures, which would trap very little interstitial water. We could ask the question; if these high density particles were included in the mud used for all mixtures, why are they observed only in this case? It is possibly due to uncertainty made when estimating size and settling velocity of flocs rises as the particles become smaller (i.e. they are harder to detect as their images are formed from less pixels). Furthermore, these very dense mineral fragments only constitute a few percent of the total mass.

To reiterate, the microflocs tended to dominate the less cohesive Run C samples (25M:75S). The C6 (τ = 0.6 Pa and SPM = 1000 mg.l^{-1}) macroflocs did not grow larger than 215 μm (Fig. 4Aiv). This was a 40% reduction in size when compared to the corresponding 75M:25S sample (A6). The low density (effective densities of less than 70 kg.m^{-3}) C6 small macroflocs fell at a combined average Ws$_{macro}$ of 1.35 mm.s^{-1}, whilst the Ws$_{micro}$ was 3.6 mm.s^{-1}. The C6 microflocs also represented three quarters of the SPM and 90% of the C6 MSF of 3.2 g.m^{-2}s^{-1} (see Fig. 4Biv and Fig. 4Civ, respectively). To place this MSF observation into perspective: it was approximately double the flux produced either by pure mud or a 75% mixed mud suspension; 31% greater than a 50:50 mixture could produce, and six times greater than the flux obtained by using a constant 0.5 mm.s^{-1} Ws (a typical settling parameter used in cohesive sediment transport modelling).

The 'clustered' appearance depicted by the lower concentration (SPM = 200 mg.l^{-1}) 25M:75S C3 sample (Fig. 4Av) is similar to Sample C6 (Fig. 4Av). The shear stress was less turbulent (τ = 0.35 Pa) than C6, so one would assume the floc settling dynamics would improve. However, the removal of three quarters of the cohesive matter meant that the Ws$_{macro}$ was only 0.9 mms^{-1}; half the Ws$_{macro}$ for the 75M:25S run A3. As with the 1000 mg.l^{-1} C6 suspension, the C3 macroflocs only represented a quarter of the SPM (Fig. 4Bv). The main difference between the lower and the higher Run C suspension was fewer individual unflocculated sand grains in the suspension at the lower turbidity.

6.5. Analysis of macrofloc: Microfloc trends (PD)

This section will look at the macrofloc and microfloc (Eisma, 1986) settling velocity trends (i.e. Ws$_{macro}$ and Ws$_{micro}$ respectively) calculated from the pre-determined mud:sand mixture data presented earlier in Section 6.3. A dual-modal approach is advised when assessing parameterised floc settling and floc mass population data, as it tends to be more realistically representative than a single sample average (Dyer et al., 1996; Mietta, 2010), especially when considering the effects of mass settling fluxes to the bed (Baugh and Manning, 2007). This approach also permits quantitative inter-comparisons with previous pure mud flocculation studies.

The density contours superimposed on the Ws vs. D scatterplots presented in Section 6.3 indicate that only a minimum number of sand grains remained in an unflocculated state. This was confirmed from an assessment of both the effective density and SPM distributions. Therefore these few grains were included in the microfloc analysis presented in this section, as they form part of the total suspension and this provides the continuity of mass when comparing the different samples. However, to make these assessments fully rigorous, the mud fraction of the samples will be isolated and examined independently in the 'modelling implications' section (see Section 9).

6.5.1. Run a using 75% mud: 25% sand

Fig. 9 shows the macrofloc and microfloc averaged settling velocity plots which cover both the pre-determined mixtures experimental concentration and shear stress ranges. The solid lines on Figs 9.A and 9.B correspond to the 25% sand mixed suspensions; the dotted curve lines are the contrasting 100% mud suspension outputs from the MFSV (this prediction was calibrated principally for Tamar mud extracted from the same study location). The straight dotted lines represent the d_{50} and d_{10} settling rates of pure sand grains determined by the *SandCalc* sediment transport computational software package (HR Wallingford, 1998).

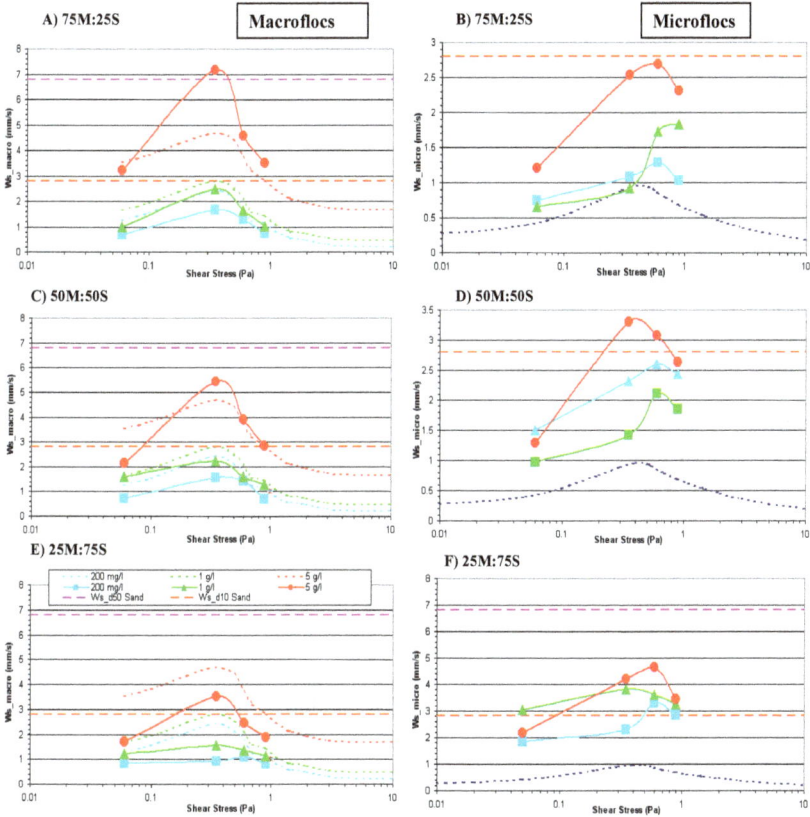

Figure 9. Ws_{macro} (left column, y-axis, units = mm.s^{-1}) & Ws_{micro} (right column, y-axis, units = mm.s^{-1}) values for runs A (75M:25S), B (50M:50S) and C (25M:75S), plotted against shear stress (x-axis, units = Pa). Solid lines + symbols indicate mixed sediment floc data points. Dashed lines indicate predicted behaviour of 100% mud macroflocs at three concentrations, and 100% mud microflocs at a single concentration. Lines indicating SandCalc estimated settling velocities of unhindered d_{10} and d_{50} pure sand grains are also plotted.

Substituting 25% of the pure mud suspension for sand produced a distinct change to the macrofloc settling velocity (Fig. 9.A). Starting at the lowest concentration (200 mg.l^{-1}), the quiescent conditions of 0.06 Pa only produced a Ws_{macro} of 0.65 mm.s^{-1}: nearly half the settling rate of pure mud. As the shear stress increased, the floc dynamics respond and the settling velocity increased to a maximum of 1.7 mm.s^{-1} at 0.35 Pa, which was 0.8 mm.s^{-1} slower than pure mud at the same concentration. The intermediate concentration (1000 mg.l^{-1}) Ws_{macro} closely mimicked the settling profile of pure mud macroflocs at the less turbid 200 mg.l^{-1}. This is primarily a result of the 75M:25S suspension lacking sufficient cohesion because it only comprises 75% mud and the potential level of flocculation is more restricted than pure mud. The mixed sediment macroflocs also demonstrated lower effective densities (~30-50 kg.m^{-3}) than their pure mud counterparts.

The smaller mixed sediment microfloc fractions all settled faster than the pure mud equivalents, at each stress increment (Fig. 9.B). Where the macrofloc mixed fraction showed settling peaks at 0.35 Pa, similar to natural muds (Manning, 2004b), the mixed Ws_{micro} tended to produce a maximum at the higher turbulent shear stress of 0.6 Pa.

At high turbidity (5000 mg.l^{-1}), the macroflocs were nearly three time more dense than at lower turbidity. This saw the Ws_{macro} peaking at 7.2 mm.s^{-1}, which was 2.5 mm.s^{-1} faster than the 100% mud equivalent, and 0.4 mm.s^{-1} quicker than a d_{50} pure sand. The corresponding Ws_{micro} was 2.7 mm.s^{-1}, which was similar to a d_{10} sand grain and 1.7 mm.s^{-1} quicker than pure mud microflocs.

6.5.2. Run B using 50% mud: 50% sand

Increasing the sand content to equal the mud fraction (50M:50S), brought about a general decrease in the macrofloc settling velocity across the entire shear stress range at each base concentration increment (Fig. 9.C). For the 200 mg.l^{-1} slurries sheared at 0.35 Pa, the equally mixed sediment produced a Ws_{macro} of 1.6 mm.s^{-1}, a reduction of 0.1 mm.s^{-1} from the 75% mud, and was 0.8 mm.s^{-1} slower at settling than the pure mud benchmark.

At the highest suspended concentration (5000 mg.l^{-1}), and again at a turbulent stress of 0.35 Pa, the 50M:50S slurry produced a Ws_{macro} of 5.4 mm.s^{-1}. This was 0.8 mm.s^{-1} faster than pure mud, but 1.8 mm.s^{-1} slower than the 75M:25S macroflocs. This large Ws_{macro} difference exhibited between the 75M:25S and 50M:50S mixtures, decreased as the TKE dissipated to a lesser level. However, both mixed suspension macroflocs at the low shear stress were still slower than pure mud, which settled considerably faster.

In contrast to the macroflocs, the smaller 50M:50S microfloc fractions (Fig. 9.D) all displayed quicker settling velocities when compared to 75M:25S. The one main exception was the 5000 mg.l^{-1} concentration, where Ws_{micro} achieved a maximum speed of 3.3 mm.s^{-1}; which was 2.3 mm.s^{-1} faster than pure mud and 0.75 mm.s^{-1} quicker than the corresponding 75M:25S microflocs.

6.5.3. Run C using 25%mud: 75% sand

The addition of a greater amount of sand particles in suspension significantly enhanced the settling dynamics at their respective shearing stresses which stimulate maximum flocculation. All 25M:75S values of Ws_{micro} exceeded the purely cohesive suspensions by more than a factor of two (Fig. 9.F), and the majority of the microfloc samples also exceeded the settling rate of a d_{10} sand grain. At an SPM concentration of 200 mgl^{-1}, the Ws_{micro} at 0.06 Pa was 1.8 mm.s^{-1} and increased to a peak of 3.3 mm.s^{-1} at 0.6 Pa. By increasing the SPM concentration to 5000 mg.l^{-1}, the Ws_{micro} maximum peaked at 4.7 mm.s^{-1}. This was approximately five times faster than the value for 100% mud, and nearly double the equivalent 75M:25S Ws_{micro} (Fig. 9.F).

Conversely, all macrofloc fractions settled significantly slower within the less cohesive suspensions. At peak turbidity, the macrofloc fraction fell at 3.5 mm.s^{-1}; this was the sole macrofloc fraction to exceed the settling velocity of d_{10} sand. In fact, this 25M:75S macrofloc fraction was 1.2 mm.s^{-1} slower than the corresponding Ws_{micro} from the same run.

In terms of the particle mass distribution: as the percentage content of non-cohesive sediment rose (i.e. mud content decreased), the relative contribution of the microfloc fraction to the total SPM concentration in each population increased.

6.6. Comparative data for sediment from Portsmouth Harbour – Natural Mixture (NM)

To support the data derived from the pre-determined mud:sand slurries, a selection of naturally occurring mixed sediment samples collected from within Portsmouth Harbour (a tidal inlet on the southern coast of the UK) were also assessed using the same type of laboratory flume runs (Pidduck and Manning, in prep.). The same protocols used for the pre-determined mixture experiments, were adopted for these runs. Sediment transport in Portsmouth Harbour has been studied by Hydraulics Research (1959), Lonsdale (1969) and Harlow (1980). Regular dredging activities for military vessel access to the Royal Naval Base, combined with an ebb-dominant macrotidal regime, mean that the fine mud and coarser sands that reside in the Harbour can become mixed.

Two Portsmouth Harbour samples at a constant SPM concentration of 2000 mg.l^{-1} and sheared at 0.35 Pa are described. The first suspension, 4_A (Fig. 10a), was a low cohesive sediment composed of 38M:62S (including coarse silts). Loss-on-ignition tests indicated that sediment 4_A was approximately 6% organic. The 4_A flocs ranged in size from 29-313 µm, although there is an absence of particles in the 33 to 69 µm range. The smallest microflocs (2% of the population) all demonstrate effective densities of quartz and beyond, which suggests the presence of some very dense minerals; possibly some metallic particles. The larger microflocs were less dense (~ 700 kg.m^{-3}).

The 4_A microflocs comprised just over half of the SPM, with their settling velocities spanning three orders of magnitude from 0.36-34 mm.s^{-1}. This resulted in a Ws_{micro} of 5.4 mm.s^{-1}, which was 1.3 mm.s^{-1} quicker than the larger macroflocs. This was due to the macroflocs demonstrating effective densities predominantly below 200 kg.m^{-3}, which are more indicative of cohesive flocs.

The second sample, 6_B (Fig. 10b), was more cohesive as it contained only 30% sand (70M:30S) and the sediment mixture had 8.4% organic matter present within its matrix. Where the sample 4_A D vs. Ws distribution favoured the smaller size fractions, 6_B depicts a population more characteristic of a pure mud. The microflocs were distinctly slower in settling, ranging from

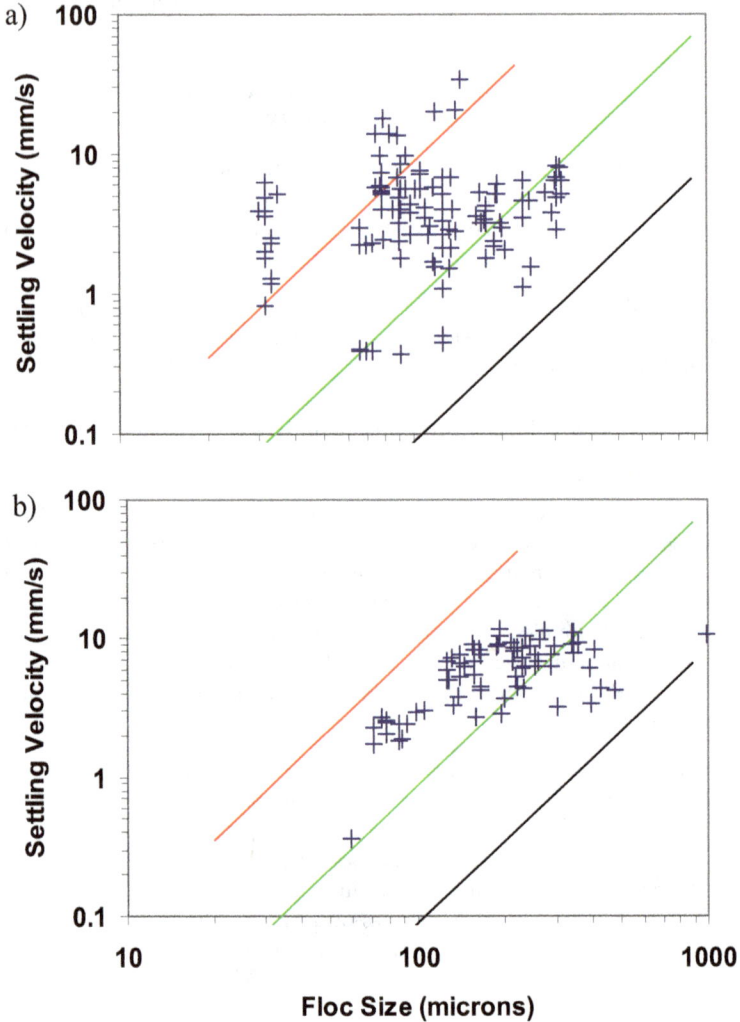

Figure 10. Settling settling vs. floc size for Portsmouth Harbour samples: a) 4_A (38M:62S); b) 6_B (38M:62S). Both samples had nominal 2 g.l⁻¹ total SPM concentrations and were sheared at a stress of 0.35 Pa. Diagonal lines represent contours of constant Stokes equivalent effective density: red = 1600 kgm⁻³, green = 160 kgm⁻³, and black = 16 kgm⁻³.

2-8 $mm.s^{-1}$. All flocs were also less dense than 4_A; effective densities under 740 $kg.m^{-3}$, with the largest flocs having a ϱ_e of just 20 $kg.m^{-3}$.

The macroflocs comprised nearly two thirds of 6_B population and over three quarters of the mass. The macrofloc and microfloc settling dynamics of the Portsmouth Harbour samples, at the three induced shear stresses (0.06, 0.35 and 0.6 Pa; 0.9 Pa was not available for the Portsmouth Harbour tests), are illustrated in Fig. 11. The data reveals some interesting settling velocity trends and these will be discussed in Section 7.

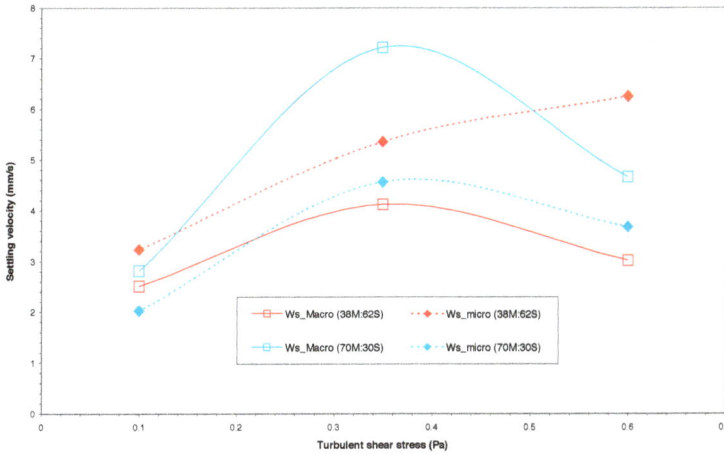

Figure 11. Ws$_{macro}$ and Ws$_{micro}$ values plotted against shear stress for Portsmouth Harbour samples 4_A (38M:62S) and 6_B (70M:30S). Both samples had nominal 2 g.l^{-1} total SPM concentrations and were sheared at a stress of 0.35 Pa.

6.7. Floc microstructure

To illustrate how both non-cohesive and cohesive sediments components can combine in natural microflocs, electron micrographs of cross-sections through natural microflocs from the Tamar Estuary (UK) are shown in Fig. 12. The low resolution TEM image which encompasses the entire microfloc (Fig. 12a) shows the complex matrix of structurally interdependent components of a typical floc section. Both organic and inorganic particles are present creating a highly porous, high water content, three-dimensional sedimentary matrix.

7. Discussion of experimental findings

7.1. Settling velocity

This section addresses issues relating to research questions i-iii listed in Section 5. A number of generalised trends, in terms of the settling velocity, can be deduced from the macrofloc and microfloc data. The macrofloc settling velocities generally slowed as the sand content rose. These macroflocs fell slightly quicker than the microflocs at low turbidity, but almost three-times as quick at the higher suspended concentration. However, as the mud content decreased,

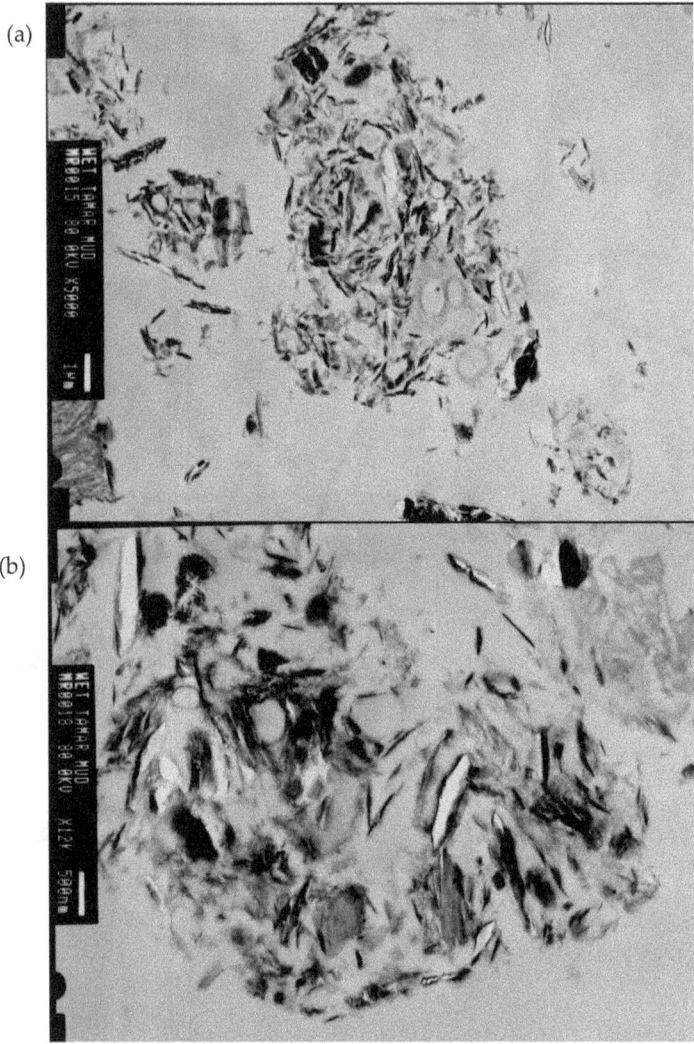

Figure 12. Low resolution (a) and high resolution (b) TEM images of a natural microfloc composed of a mud:sand mixture.

the particle cohesion efficiency would also reduce and could potentially limit the floc growth potential, curbing the equilibrium floc size of the macrofloc fraction.

The microfloc settling responded to a greater abundance of sand, whereby the greater the sand content in a mixed fraction - the faster the Ws_{micro}. For example, for a 25M:75S suspension, the microfloc settling velocities demonstrated a three-fold increase at low turbidity and nearly doubled in settling speed at high turbidity to produce Ws_{micro} of 3.5 mm.s^{-1} and 4.7 mm.s^{-1}, respectively. The effective density data of many of the microfloc fractions from the pre-determined mixtures tests ranged from 800-1200 kg.m^{-3}. This would suggest that the finer sand grains tended to interact and bond better with the smaller floc structures, accounting for the quicker microfloc settling velocities observed.

The flocs produced from the natural Portsmouth Harbour sediments showed similar general settling velocity patterns to those of the pre-determined Tamar mixed suspensions. For the less cohesive 38M:62S slurry (sample 4_A), the microfloc fraction settled quicker than the macroflocs. By taking into account differences in SPM and M:S ratio, one can deduce that the 4_A microflocs were settling approximately 1.5 mm.s^{-1} quicker than their manufactured slurry equivalents, whilst the Portsmouth macroflocs fell nearly twice as quick as their pre-determined slurry equivalent. This could be a result of slightly larger sand grains present in the Portsmouth 4_A sediment and also stronger bio-film coatings present in the 4_A mixture providing extra adhesion for the sand grains permitting greater uptake within the macrofloc fraction.

It is interesting to observe that the microflocs in 4_A produced their fastest settling velocities at a τ of 0.6 Pa, whilst the Ws_{macro} peaked at a less turbulent 0.35 Pa. This can be explained by the denser microflocs being stronger than the weaker macroflocs, hence they can survive larger stresses. The ratio of a floc's diameter to the corresponding dissipating eddy size, such as the Kolmogorov microscale (1941a, b), in turbulent flow is a fundamental governing condition for estuarine flocculation dynamics (Tomi and Bagster, 1978; Tambo and Hozumi, 1979; McCave, 1984). Furthermore, if settling velocities are large, more turbulent energy is required to keep those flocs in suspension.

7.2. Composition and SPM distribution

Aspects relating to research question iv are now discussed. The LabSFLOC data has provided evidence of how sand grains can be potentially included within a floc matrix. The Ws vs. D spectra show that only a minimal amount of potentially unflocculated pure sand particles are present in a few of the samples; this is in terms of both individual numbers and the percentage of the total SPM (typically less than 1-2% of the total mud:sand concentration). An accurate mass balance between the predetermined mixed suspension introduced into the flume at the commencement of each run and the filtered SPM obtained from each sample promotes confidence in the mixed sediment LabSFLOC floc observations.

The LabSFLOC sampling protocol of measuring D and Ws simultaneously means that data on individual floc effective density is available. The latter provides important information about the composition of each floc (Dyer, 1989). The data identifies that there is a wide range in effective densities exhibited across each spectrum, particularly in the microfloc range, but most are less than pure quartz (~1600 kg.m^{-3}). Transmission electron microscopy (TEM) images have also visually identified the presence of both clay minerals and quartz mineral fragments within natural microfloc structures (Spencer et al., 2010). This leads to the suggestion that when mixed sediments flocculate, the sand particles favour the microfloc fraction, which is logical reasoning: microflocs tend to have the stronger bonding potential due to the closeness of the bonds.

Uptake of individual sand particles will probably be much less in the macroflocs. This is consistent with the order of aggregation theory (Krone, 1962; Eisma, 1986) which states that microflocs will flocculate into macroflocs when the ambient conditions are favourable. This provides a more efficient mechanism / pathway for the fine sand grains to move into the macrofloc fractions.

The EDS floc structural analysis of the TEM floc images presented in Section 6.7, identified that the microfloc inorganic constituents primarily comprised planar clay minerals (identified by the thin dark grey objects in Fig. 12) and fine quartz fragments (all much smaller than the mean sand grain size), evident from concoidal fracturing (the black marks in Fig. 12a and 12b). Other minerals present included Fe and Mn oxides and opaque sub-cubic minerals (probably pyrite), which are all typical of estuarine sediments. The organic constituents are predominantly observed to be bacteria and their EPS (extracellular polymeric substance; see Underwood and Paterson, 2003; Tolhurst et al., 2002) fibrils, which are produced by the bacteria for attachment, assimilation of food (dissolved organic carbon) and for protection from predation and contaminants. In the high resolution TEM image of the microfloc (Fig 12b.), the EPS can be seen linking the biological and inorganic particles and represents a micro-structural framework of the floc matrix (Fig. 12b). The EPS matrix is considered to be the component of the floc that enhances floc building and provides it with its strength.

For the Tamar mixtures, with a sand d_{50} of 0.11 mm, it is geometrically possible that only one sand grain may form a microfloc. The data shows that many of the microflocs exhibited effective densities significantly less than pure quartz, but higher than most pure mud microfloc. This suggests that the mixed sediment microflocs could be either combined mixtures of very fine quartz fragments and mud, as illustrated by the TEM images, or they could be individual larger quartz particles which are coated in organic mud. For example, Whitehouse et al. (2000) offer a scenario where mud can create a 'cage-work' structure which can fully encompass the sand grains, thus trapping the sand within a clay floc envelope. Mehta et al. (2009) observed flocs of various sizes in Lake Apopka (Florida, USA) where the inorganic particles are held together by embayment within a spacious exopolymeric biofilm (e.g. organic mucus) (Fig. 13). Such flocs do not conform to the mathematical fractal description typically attributed to predominantly inorganic flocs (e.g. Winterwerp and van Kesteren, 2004, Winterwerp et al, 2006), because there is no floc formation that can be described as the primary structure. All these cases would produce microflocs which are both less dense than their constituent minerals, but would have the potential to bond with a macrofloc due to their part-biological matrix.

7.3. MSF distributions

By combining the settling velocity and mass distribution findings, it is possible to assess the mass settling flux (i.e. the product of the concentration and the Ws); this enables aspects of research question v to be discussed.

The combined effects of particle concentration and turbulent shearing have long been attributed to the growth of mud flocs (e.g. Tsai et al., 1987; Burban, 1987; Puls et al., 1988; Kranck and Milligan, 1992). Under optimum flocculation conditions, Mehta and Lott (1987) suggested

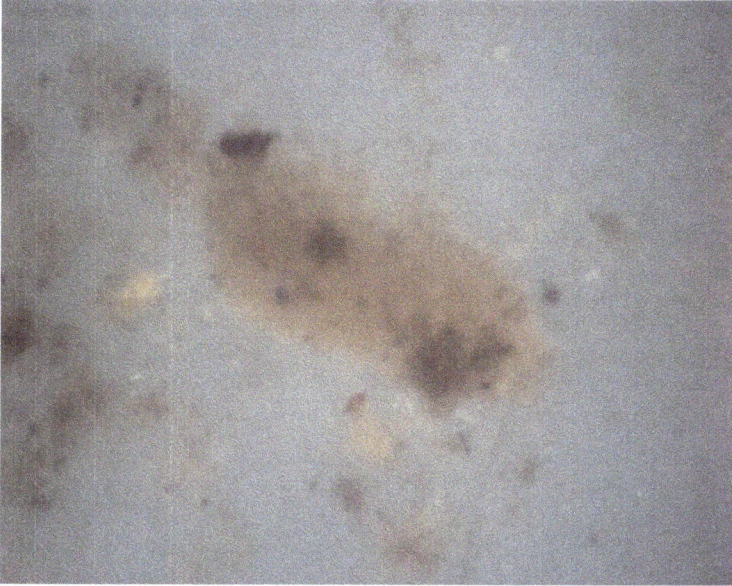

Figure 13. A very porous (low density) floc, composed from a translucent organic coating eveloping a solid (opaque) core (from Mehta et al., 2009).

that pure mud macroflocs tend to contribute most to the MSF, on account of high instability (van Leussen, 1994) due to floc growth potential producing a greater number of larger macroflocs with fast settling velocities. Observations in estuaries reveal these pure mud macroflocs can typically grow to mean a diameter > 400 µm, exhibiting effective densities of less than 40-50 kg.m^{-3} and becoming more than 95% porous. These macroflocs are highly delicate entities and are easily progressively broken apart as they pass through regions of higher turbulent shear stress (Glasgow and Lucke, 1980). However, the data presented in this chapter indicates a trend whereby an increase in sand content, and a subsequent decrease in mud, favours the microflocs as the dominant flux contributor.

For example, if we consider a flocculating mixture comprising 25% mud and 75% sand, at a nominal concentration of 1000 mg.l^{-1} and sheared at a τ of 0.6 Pa (i.e. Sample C6 ; see Fig. 4Civ), this results in the microflocs representing three quarters of the SPM. Therefore, the microfloc fraction would be contributing 88% of the total MSF (3.08 g.m^{-2}s^{-1}). To place this MSF value into perspective: it is approximately double the flux estimated for either a pure mud or a 75% mixed mud suspension; nearly 30% greater than the flux for a 50M:50S mixture; and six times greater than the MSF obtained by using a constant 0.5 mm.s^{-1} settling velocity.

In contrast, by maintaining the ambient SPM concentration at 1000 mg.l^{-1}, but making the suspension 75% cohesive (i.e. 75M:25S), when it is sheared at 0.35 Pa (Sample A7) the total MSF (2.2 g.m^{-2}s^{-1}) would be weighted 73%:27% in favour of the macroflocs. This settling flux distribution is more characteristic of a fully cohesive suspension (Manning and Bass, 2006). This suggests that with just an 8% lower MSF than pure mud, the 75M:25S mixture is behaving, to some degree, predominantly as a cohesive suspension, even with 25% fine sand present in the mixture.

The data shows that the greater the sand content of a mixed suspension, the higher the total MSF. Although it is not possible to state how much, or even when, cohesive material attaches to individual sand grains, the effective density distributions (see Figs 4Ai-v) indicate that many of the microflocs are less dense than quartz (for a nominally mass-balanced mud:sand mixture). The Ws_{micro} generally rose with rising sand content. One can see that this smaller size fraction is extremely important in terms of the total MSF for less cohesive suspensions. By averaging the MSF over the entire concentration and shear stress ranges for a nominally constant ratio of mud and sand, the data reveals that for a predominantly sandy suspension (Run C - 25M: 75S), the microflocs represented the majority (~80%) of the total MSF. In contrast, the microflocs contributed less than half (~42%) of the settling flux for the muddier 75M:25S slurry (Run A).

With the sandier 4_A Portsmouth microflocs (see Fig. 10a) representing over half of the total 2000 mg.l^{-1} suspension and the macroflocs comprising three quarters of the more cohesive sample 6_B flocs (see Fig. 10b), the Portsmouth samples displayed a similar mass distribution to those of the Tamar pre-determined slurries. In terms of the MSF, Sample 4_A produced a resultant 9.9 g.m^{-2}s^{-1}, which was approximately 50% greater than the Tamar manufactured suspension. Whilst the Sample 6_B depositional flux, 13.6 g.m^{-2}s^{-1}, was more than three times the settling flux of the Tamar equivalent mixtures. The higher mass settling fluxes were a function of the quicker settling velocities demonstrated by the Portsmouth Harbour suspensions.

A direct comparison of the mass settling fluxes and their associated dynamics, can also provide a practical way to illustrate the enhanced / increased flocculation with respect to turbulent intensity. If we consider the 5000 mg.l^{-1} B9 floc sample from the 50M:50S suspension, the very turbulent (τ = 0.9 Pa) environment produced a net MSF of 13.8 g.m^{-2}s^{-1} (see Fig. 4Ciii), with just half the flux attributed to the macroflocs. In comparison the more advanced flocculation of the less turbulent (τ = 0.35 Pa) Sample B11, resulted in a MSF of 26.2 g.m^{-2}s^{-1}. This was nearly double the Sample B9 flux and was primarily due to the B11 macroflocs contributing 80% of the total flux. The fast settling (Ws of 6-14 mm.s^{-1}) macroflocs ranging from 482 to 650 µm produced nearly one quarter of the B11 MSF.

8. Parameterisation of mixed sediment flocculation

Since the mid-1990s, much research has been conducted in Europe on the parameterisation of the natural flocculation process, through projects such as COSINUS - *Prediction of COhesive Sediment transport and bed dynamics in estuaries and coastal zones with Integrated NUmerical Simulation models* (see Berlamont, 2002). A significant degree of progress has been achieved on the practical modelling of flocculation (e.g. Winterwerp et al., 2006; Baugh and Manning, 2007, Soulsby and Manning, 2012). In terms of general modelling applicability, these flocculation advancements are still limited to the modelling of solely pure cohesive sediment estuaries. Due to the complexity of the mixed sediment flocculation process (as demonstrated in this chapter), statistical relationships between floc properties acquired from direct empirical observations can be used to quantify the response of flocculation to different environmental conditions (Manning and Dyer, 2007). Fig. 14 shows a conceptual representation of the 25M: 75S data compared to pure mud suspensions, at a SPM concentration of 5000 mg.l^{-1}. The mixed sediment macrofloc settling curve, within a turbulent shear stress (τ) region of 0.06-0.6 Pa, can be quantified by the following algorithm:

$$Ws_{macro} = 0.259 + 5.76^*\tau - 7.61^*\tau^2 + 0.000317^*SPM \tag{1}$$

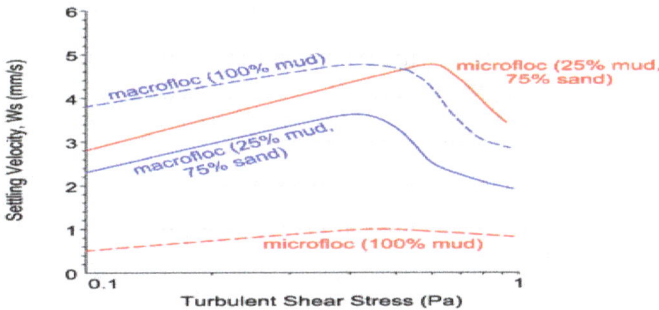

Figure 14. Conceptual illustration of Ws_{macro} (blue lines) & Ws_{micro} (red lines) trends for a mixed sediment suspension of ratio 25M:75S (solid lines) and a pure mud (dotted lines) suspension, all for a total concentration of 5 gl^{-1}, plotted against shear stress.

A parametric multiple regression was used to generate Eqn 1. For this particular type of multi-regression derivation we are using non-homogeneous dimensions, therefore the units used are as follows: Ws_{macro} = mm.s^{-1}, τ = Pa, and SPM = mg.l^{-1}. Demonstrating an R^2 = 0.84, the algorithm is a close approximation of the parameterised observations covering the 200-5000 mg.l^{-1} laboratory experimental SPM concentration range. Eqn 1 is just one form of algorithm and others can be generated from the data depending upon the modelling input variables.

The general structure of Eqn. 1 is similar to the pure mud macrofloc settling velocity relationship derived by Manning (2004a) as part of the Estuary Processes Research Project – EstProc (Estuary Process Consortium, 2005). The general shape of the Eqn. 1 curve is similar to the flocculation schematic proposed by Dyer (1989), with an increase in settling velocity at low stress due to flocculation enhanced by shear, and floc disruption at higher stresses for the same concentration. Also, the combined influence of concentration and turbulent shear on the control of the macrofloc properties, as listed in Eqn. 1, agrees with the hypotheses offered by both Puls et al. (1988) and Kranck and Milligan (1992).

However, the relative magnitudes and peaks in the mixed sediment conceptual curves (illustrated in Fig. 14) differ from the pure mud representations in a number of ways. The microflocs in the 25M:75S mixed suspension microflocs settle at a maximum velocity of 4.7 mm.s^{-1}; this is 380% quicker than the equivalent pure mud and nearly double the Ws_{micro} for a 75% mud suspension. Interestingly, the mixed suspension Ws_{micro} is virtually the same as the macrofloc settling velocity for pure mud. The peak 25M:75S Ws_{micro} occurred at a shear stress of 0.6 Pa, which falls within the "moderately-high" shear stress zone (Manning, 2004a); 0.2-0.3 Pa above the shear stress region typically recognised as producing optimum stimulation for pure mud flocculation (Manning, 2004a).

Manning and Dyer (2007) demonstrated that, for varying levels of suspended concentration, mud microfloc settling in turbulent flows could be represented by a single algorithm curve. In contrast, mixed sediment microfloc settling velocities appear to be dependent upon both concentration and shear stress variations, as well as the proportion of mud and sand. This is indicated by different curves representing Ws_{micro} throughout a shear stress range at varying concentration levels, even when the mud:sand ratio is constant. From this we can deduce that the mixed sediment Ws_{micro} parameter is far more sensitive to changes in SPM concentration, compared to pure mud microflocs whose dynamics only seem to vary with turbulent shear stress.

If we now examine the macrofloc settling for the conceptual curve for a 5000 mg.l^{-1} 25M:75S mixed suspension (Fig. 14), one can observe that the maximum Ws_{macro} of 3.5 mm.s^{-1} occurs at a shear stress of about 0.35-0.4 Pa; the same stress range as pure mud macroflocs. However, the 25M:75S macroflocs are settling 1.2 mm.s^{-1} (or 25%) slower than both the Ws_{micro} peak for the mixed sediments and the Ws_{macro} for pure mud.

If we consider the mixed sediment settling velocity variations in terms of Krone's (1963) classic hierarchical order of aggregation theory, the smaller microflocs (D < 160 μm) are generally considered to be the building blocks from which the macroflocs are composed. The microflocs tend to display a much wider range in effective densities and settling velocities than the macrofloc fraction. It is highly plausible that for mixed sediments, the microfloc fraction samples may comprise both flocculated mud and some unflocculated sand grains depending on mud:sand ratio, concentration and shear stress. This could account for the faster microfloc settling velocities with rising sand content and concentration.

The macroflocs are deemed to be composed of microflocs, so this fraction will also contain both cohesive and non-cohesive particulates. The intra-bonding of microfloc to microfloc is usually far weaker than the closer internal particle bonds of individual microflocs. This means that macrofloc bonding relies heavily on the sediment cohesional properties (primarily those from extra-cellular polymeric substances), and these will exponentially decrease with muddy sediments being replaced by non-cohesive sands.

The parameterisation of biological process for inclusion in numerical sediment transport models is notoriously difficult, and algorithms such as Eqn. 1 do not include a specific "biological" term. However, where the algorithms are based on data derived from natural sediments which would include some of the biological effect. A limitation of many mixed sediment laboratory studies, is that the mud:sand matrix is over-simplified through the use of a pure clay mineral (e.g. kaolinite) devoid of any biology. As clay minerals only flocculate through electrostatic (i.e. salt) flocculation, at best a segregated environment may be simulated if the water is brackish, but resultant mixed sediment flocculation effects will never be observed.

9. Modelling implications of mixed sediment flocculation

The prediction and modelling of mud:sand segregation effects on processes such as deposition are very useful from an estuarine management perspective. Numerical models are typically the chosen tools with which estuarine management groups attempt to predict sediment transport rates. In order for these models to provide sufficiently accurate results, a good scientific understanding of the flocculation process and interactions between mud and sand is required (e.g. Chesher and Ockenden, 1997; van Ledden, 2002; Waeles et al., 2008), and these processes need to be adequately described mathematically.

The complexity of mud:sand suspensions and a general lack of suitable experimental data which can describe the resultant dynamics of different mixtures of mud and sand, means that most numerical sediment transport models treat mud and sand as entirely separate entities. These conditions may exist for a segregational environment. However, if the mud:sand particles interact as a combined matrix, it has the potential to flocculate (as demonstrated in this chapter). This research has indicated that when mud and sand are mixed in different ratios and interact, the level of inter-particle cohesion can also vary and this is reflected in the macrofloc:microfloc mass settling flux distributions. Therefore it may be important for modellers to consider potential flocculation effects when parameterising mixed sediment deposition in turbulent flows that are conducive to flocculation.

When faced with a potential mixed sediment regime, an estuarine sediment transport modeller has two initial basic choices. The first and most simple option, is to assume that the mud:sand mixtures act solely as one sediment type when suspended, thus entirely demonstrating either cohesive or non-cohesive settling characteristics. If all sediment is assumed to be non-cohesive, e.g. pure sand grains devoid of any cohesive matter, the SPM would behave as inert particles as their dynamic settling spectrum would not alter greatly with increasing concentration as they do not flocculate. Similarly pure sand grain dynamics are not affected by shear stresses in the same way muddy sediments are. Thus, the settling properties of pure sand suspensions are similar over the SPM concentration range (200-5000 mg.l^{-1}) encompassed by the flume experimental data reported in this chapter; this is also because the influence of hindered settling is not important in this range of concentration. In contrast, if all SPM present is deemed to be pure mud, flocculation will completely dominate the settling process.

The second option acknowledges the presence of a mud:sand mixed environment; the issue is then how this is handled. For example, Van Ledden's (2002) mixed sediment model employed the segregational criteria for low concentration depositional simulations in which flocculation effects are ignored. However, if it is assumed that the mixed suspensions are acting in a segregated manner, when in fact they are demonstrating a degree of flocculation, a wide range in predicted settling flux errors may arise from the modelling output.

To illustrate the potential pitfalls of solely using either a sand or mud settling parameterisation, when there is actually a flocculating mud:sand mixture present, we compare fraction-maximum settling velocities for: pure mud, pure sand and a 50M:50S ratio suspension, all at an SPM concentration of 200 mg.l^{-1}. For the 100% mud condition, the respective macrofloc and microfloc settling velocities are 2.4 mm.s^{-1} and 0.9 mm.s^{-1}. The contrasting pure sand settling velocity values are Ws$_{macro\_sand}$ = 20.1 mm.s^{-1} and Ws$_{micro\_sand}$ = 7.4 mm.s^{-1}; this was a comparative 7 to 8 -fold settling velocity rise for the two respective pure sand fractions, over the pure mud. An equal division of mud and sand resulted in an observed mixed sediment macrofloc settling velocity of 1.6 mm.s^{-1}, which was more than twelve times slower than the pure sand macrofloc-equivalent sized fraction and two thirds the velocity of the pure mud macroflocs. However, the observed 50M:50S microflocs fell at 2.2 mm.s^{-1}, which was three-times slower than pure sand, and twice as fast as pure mud suspensions. This example demonstrates the importance of obtaining high quality temporal and spatial settling velocity data of mixed sediments in suspension. It is anticipated that the effects of mixed sediment flocculation on numerical sediment transport modelling, will be the topic of future research and publication.

10. Conclusion

The aim of this chapter was to provide an overview of mixed sediment flocculation dynamics and how they can influence sediment transport. It has drawn on key literature and new data to address this aim. The theoretical aspects relating to the flocculation of mud:sand mixtures include flocculation processes, segregation versus flocculating suspensions, and biological influences on mixed sediment flocculation.

In order to demonstrate the flocculation potential and characteristics of mud:sand mixtures, the second part of the chapter has drawn on the findings from recently completed laboratory studies that examined the flocculation dynamics for mud:sand (M:S) mixtures primarily using Tamar estuary mud and silica sand at different concentrations and shear rates in a mini-annular flume. Turbulent shear stresses during the experimental runs ranged from 0.06-0.9 Pa (±5%), with maximum flow speeds in the annular flume of about 0.7 m.s^{-1}, for three total suspended sediment concentrations representative of estuarine concentrations, namely 200, 1000 and 5000 mg.l^{-1}. The video-based LabSFLOC instrument was used to determine floc properties including size, settling velocity, density, and mass.

The experiments showed that as mud content decreased, the particle cohesion efficiency reduces which can limit the growth potential of the macrofloc fraction (sizes > 160 μm). For a 75M:25S suspension, the settling velocity Ws$_{macro}$ was slightly quicker than the microflocs at 200 mg.l^{-1}, but almost three-times as fast at the higher suspended concentration (5000 mg.l^{-1}). Parameterised data indicated that by adding more sand to a mud:sand mixture, the settling velocity of the macrofloc fraction slows and the settling velocity of microflocs (sizes < 160 μm) increases.

In terms of floc composition, effective density data of many of the microfloc fractions ranged between 800-1200 kg.m^{-3}. This would suggest that the finer sand grains tended to interact and bond better with the smaller floc structures, accounting for the quicker microfloc settling velocities observed.

The general trends revealed by the pre-determined (Tamar mud and silica) mixtures were also observed with independent tests on naturally mixed Portsmouth Harbour sediments. However, compositionally, the Portsmouth sediment matrix produced differences in the absolute settling velocities of the macrofloc and microfloc fractions from those of the Tamar mixtures. Both fractions of the Portsmouth sediment tended to fall quicker than their Tamar mixed sediment equivalents. It is proposed that this could be a result of a different sand grain size distribution combined with stronger bio-film coatings producing added cohesion in the Portsmouth sediment mixtures. This would permit a greater uptake of the sand grains within the macrofloc fraction, whilst also potentially forming the faster settling microflocs observed.

The data showed that the greater the sand content of a mixed suspension, the higher the total mass settling flux (MSF). As the microflocs have been seen to be more conducive at flocculating with the finer sand grains, and the Ws_{micro} rose with rising sand content, one can see that this smaller size fraction is extremely important in terms of the total MSF for less muddy suspensions. By averaging the MSF over the entire concentration and shear stress ranges for a constant ratio of mud (M) and sand (S), the data revealed that for a predominantly sandy suspension (25M:75S), the microflocs represented the majority of the total MSF. In contrast, the microflocs contributed less than half of the settling flux for a much muddier mixture (75M:25S).

Biology is considered to be extremely important in the mixed sediment flocculation process. For example, the presence of sticky extracellular polymeric substances (EPSs) produced by epipelic and epipsammic diatoms could significantly enhance particle bonding. Energy dispersive spectroscopy analysis confirmed the presence of both clay minerals and quartz mineral fragments within a natural microfloc. A high resolution transmission electron microscopy (TEM) image revealed EPS fibrils linking the biological and inorganic particles within a micro-structural framework of a microfloc matrix.

Since estuaries may have mixed or segregational mud:sand environments and numerical models are used to inform management decisions, some issues relating to the parameterisation of mud:sand flocculation and their implementation in sediment transport models have been discussed. It is anticipated that these two topics will be the subject of future research and publication on mixed sediment flocculation.

Acknowledgements

The mini-annular flume experiments were primarily funded by the HR Wallingford Company Research Programme as part of the 'Mud:Sand Transport' projects (DDD0301 and DDD0345), and completed during the 'Sediment in Transitional Environments' – SiTE project (DDY0427).

Author details

Andrew J. Manning[1,2], Jeremy R. Spearman[1], Richard J.S. Whitehouse[1], Emma L. Pidduck[2], John V. Baugh[1] and Kate L. Spencer[3]

1 HR Wallingford, Howbery Park, Wallingford, Oxfordshire, UK

2 School of Marine Science & Engineering, University of Plymouth, Plymouth, Devon, UK

3 Department of Geography, Queen Mary – University of London, Mile End Road, London, UK

References

[1] Alvarez-Hernandez, E. (1990). The influence of cohesive sediment on sediment movement in channels of circular cross-section. PhD thesis, University of Newcastle-upon-Tyne.

[2] Bass, S.J., Manning, A.J. and Dyer, K.R. (2006). Preliminary findings from a study of the upper reaches of the Tamar Estuary, UK, throughout a complete tidal cycle: Part I. Linking sediment and hydrodynamic cycles. In: J.P.-Y. Maa, L.P. Sanford and D.H. Schoellhamer (eds), Coastal and Estuarine Fine Sediment Processes - Proc. in Marine Science 8, Amsterdam: Elsevier, pp. 1-14, ISBN: 0-444-52238-7.

[3] Baugh, J.V. and Manning, A.J. (2007). An assessment of a new settling velocity parameterisation for cohesive sediment transport modelling. Continental Shelf Research, doi:10.1016/j.csr.2007.03.003.

[4] Berlamont, J.E. (2002). Prediction of cohesive sediment transport and bed dynamics in estuaries and coastal zones with integrated numerical simulation models. In: Winterwerp, J.C., Kranenburg, C. (eds), Fine Sediment Dynamics in the Marine Environment - Proc. in Mar. Sci. 5., Elsevier, Amsterdam, pp. 1-4.

[5] de Brouwer, J.F.C., Wolfstein, K., Ruddy, G.K. (2005). Biogenic stabilization of intertidal sediments: The importance of extracellular polymeric substances produced by benthic diatoms. Microbial Ecology, 49, 501-512.

[6] Buffle, J. and Leppard, G.G. (1995). Characterisation of aquatic colloids and macromolecules. 2. Key role of physical structures on analytical results. Environmental Science and Technology, 29, 2176-2184.

[7] Burban, P.Y. (1987). The flocculation of fine-grained sediments in estuarine waters. MSc. thesis, Dep. of Mech. Eng. Univ. of Calif., Santa Barbara, USA.

[8] Cahoon, L.B. (1999). The role of benthic microalgae in neritic ecosystems. Oceanography and marine biology: an annual review, 37, 47-86.

[9] Chesher, T.J. and Ockenden, M.C. (1997). Numerical modelling of mud and sand mixtures. In: N. Burt, R. Parker and J. Watts (Eds), Cohesive Sediments – Proc. of IN-TERCOH Conf. (Wallingford, England), Chichester: John Wiley & Son, pp. 395-406.

[10] Dankers, P.J.T., Sills, G.C. and Winterwerp, J.C. (2007). On the hindered settling of highly concentrated mud-sand mixtures. In: T. Kudusa, H. Yamanishi, J. Spearman and J.Z. Gailani (Eds), Sediment and Ecohydraulics - Proc. in Marine Science, INTER-COH 2005, Amsterdam: Elsevier, pp. 255-274.

[11] Dearnaley, M.P. (1996). Direct measurements of settling velocities in the Owen Tube: a comparison with gravimetric analysis. Journal of Sea Research, Vol. 36, Nos. 1-2, 36, 41-47.

[12] Dyer, K.R. (1986). Coastal and Estuarine Sediment Dynamics. Wiley & Sons, Chichester, 342p.

[13] Dyer, K.R. 1989. Sediment processes in estuaries: future research requirements. J. Geophys. Res., 94 (C10): 14,327-14,339.

[14] Dyer, K.R., Cornelisse, J.M., Dearnaley, M., Jago, C., Kappenburg, J., McCave, I.N., Pejrup, M., Puls, W., van Leussen, W. and Wolfstein, K. (1996). A comparison of in-situ techniques for estuarine floc settling velocity measurements. Journal of Sea Research 36, 15-29.

[15] Dyer, K.R. and Manning, A.J. (1998). Observation of the size, settling velocity and effective density of flocs, and their fractal dimensions. Journal of Sea Research 41, 87-95.

[16] Edzwald, J.K. and O'Melia, C.R. (1975). Clay distributions in recent estuarine sediments. Clays and Clay Minerals, 23:39-44.

[17] Eisma, D. (1986). Flocculation and de-flocculation of suspended matter in estuaries. Neth. J. Sea Res., 20 (2/3), 183-199.

[18] Estuary Process Consortium (2005). Final Report of the Estuary Process Research Project (EstProc) – Algorithms and Scientific Information. Integrated Research Results on Hydrobiosedimentary Processes in Estuaries, R & D Technical Report prepared by the Estuary Process Consortium for the Fluvial, Estuarine and Coastal Processes Theme, co-funded by Defra & Environment Agency, Report FD1905/TR3, 140p.

[19] Feates, N.G. and Mitchener, H.J. (1998). Properties of dredged material: measurement of sediment properties of dredged material from Harwich Harbour. HR Wallingford Report TR 46.

[20] Fennessy, M.J., Dyer, K.R. and Huntley, D.A. (1994). Size and settling velocity distributions of flocs in the Tamar Estuary during a tidal cycle. Netherlands Journal of Aquatic Ecology, 28: 275-282.

[21] Fennessy, M.J., Dyer, K.R., Huntley, D.A. and Bale, A.J. (1997). Estimation of settling flux spectra in estuaries using INSSEV. In: N. Burt, R. Parker and J. Watts (Eds), Cohesive Sediments – Proc. of INTERCOH Conf. (Wallingford, England), Chichester: John Wiley & Son, pp. 87-104.

[22] Fitzpatrick, F. (1991). Studies of sediments in a tidal environment. Ph.D. Thesis, Department of Geological Sciences, University of Plymouth, 221p.

[23] Förstner, U. and Wittmann G. T. W. (1983). Metal Pollution in the Aquatic Environment. Springer Verlag, Berlin, Heidelberg et New York, 2nd revised edition, 486p.

[24] Gerbersdorf, S.U., Bittner, R., Lubarsky, H., Manz, W. and Paterson, D.M. (2009). Microbial assemblages as ecosystem engineers of sediment stability. J Soils Sediments, 9, 640–652.

[25] Grabowski, R.C., Droppo, I.G. and Wharton, G. (2011). Erodibility of cohesive sediment: The importance of sediment properties. Earth-Science Reviews, 105, 101-120.

[26] Glasgow, L.A. and Lucke, R.H. (1980). Mechanisms of deaggregation for clay-polymer flocs in turbulent systems. Ind. Eng. Chem. Fundam., 19: 148-156.

[27] Gratiot, N. and Manning, A.J. (2004). An experimental investigation of floc characteristics in a diffusive turbulent flow. In: P. Ciavola and M. B. Collins (Eds), Sediment Transport in European Estuaries, Journal of Coastal Research, SI 41, 105-113.

[28] Harlow, D. (1980). Sediment Processes, Selsey Bill to Portsmouth, PhD thesis, Department of Civil Engineering, University of Southampton.

[29] Harper, M.A. and Harper, J.F. (1967). Measurements of diatom adhesion and their relationship with movement. British Phycological Bulletin, 3, 195-207.

[30] Hickman, M. and Round, F.E. (1970). Primary production and standing crops of epipsammic and epipelic algae. British Phycological Journal, Vol 5 (2), pp. 247-255.

[31] HR Wallingford (1998). SandCalc: Marine Sands Calculator Interface. Version 2.0 for Windows. Software by Tessela & HR Wallingford.

[32] Hydraulics Research (1959). Portsmouth harbour investigation, parts i and ii. Reports 213b and 214, Technical report, Hydraulics Research.

[33] Kamphuis, W and Hall, K.R. (1983). Cohesive material erosion by unidirectional current. J. Hyd. Eng., ASCE, 109, 49-61.

[34] Kennish, M.J. (1986). "Ecology of estuaries Volume 1: Physical and chemical aspects." Boca Raton Florida, CRC Press, 1.

[35] Koglin B. (1977). Assessment of the degree of aggregation in suspension. Powder Technology 17, 219-227.

[36] Klimpel R.C. and Hogg R. (1986). Effects of flocculation conditions on agglomerate structure. Journal of Colloid Interface Science 113, 121-131.

[37] Kolmogorov, A.N. (1941a). The local structure of turbulence in incompressible viscous fluid for very large Reynolds numbers. C. R. Acad. Sci. URSS, 30: 301.

[38] Kolmogorov, A.N. (1941b). Dissipation of energy in locally isotropic turbulence. C. R. Acad. Sci. URSS, 32: 16.

[39] Kranck, K. and Milligan, T.G. (1992). Characteristics of suspended particles at an 11-hour anchor station in San Francisco Bay, California. Journal of Geophysical Research, 97, 11373-11382.

[40] Krone, R.B. (1962). Flume studies of the transport of sediment in estuarial shoaling processes. Final report. Hyd. Eng. Lab. and Sanitary Eng. Lab., University of California, Berkeley.

[41] Krone, R. B. (1963). A study of rheological properties of estuarial sediments. Report No. 63-68, Hyd. Eng. Lab. and Sanitary Eng. Lab., University of California, Berkeley, 63-68.

[42] Little, C. (2000). The biology of soft shores and estuaries. Oxford University Press (UK), 252p., ISBN: 978-0-19850-426-9.

[43] Lonsdale, B. J. (1969). A sedimentary study of the eastern Solent, Master's thesis, Department of Oceanography, University of Southampton.

[44] Manning, A.J. (2001). A study of the effects of turbulence on the properties of flocculated mud. Ph.D. Thesis. Institute of Marine Studies, University of Plymouth, 282p.

[45] Manning, A.J. (2004a). The observed effects of turbulence on estuarine flocculation. In: P. Ciavola and M. B. Collins (eds), Sediment Transport in European Estuaries, Journal of Coastal Research, SI 41, 90-104.

[46] Manning, A.J. (2004b). Observations of the properties of flocculated cohesive sediment in three western European estuaries. In: P. Ciavola and M. B. Collins (Eds), Sediment Transport in European Estuaries, Journal of Coastal Research, SI 41, 70-81.

[47] Manning, A.J. (2006). LabSFLOC – A laboratory system to determine the spectral characteristics of flocculating cohesive sediments. HR Wallingford Technical Report, TR 156.

[48] Manning, A.J., Bass, S.J. and Dyer, K.R. (2006). Floc Properties in the Turbidity Maximum of a Mesotidal Estuary During Neap and Spring Tidal Conditions. Marine Geology, 235, 193-211.

[49] Manning, A.J., Baugh, J.V., Spearman, J. and Whitehouse, R.J.S. (2009). Flocculation Settling Characteristics of Mud:Sand Mixtures. Ocean Dynamics, PECS2008 SI, DOI: 10.1007/s10236-009-0251-0.

[50] Manning, A.J. and Dyer, K.R. (1999). A laboratory examination of floc characteristics with regard to turbulent shearing. Marine Geology 160, 147-170.

[51] Manning, A.J. and Dyer, K.R. (2002a). The use of optics for the in-situ determination of flocculated mud characteristics. J. Optics A: Pure and Applied Optics, Institute of Physics Publishing, 4, S71-S81.

[52] Manning, A.J. and Dyer, K.R. (2002b). A comparison of floc properties observed during neap and spring tidal conditions. In: J.C. Winterwerp and C. Kranenburg (Eds), Fine Sediment Dynamics in the Marine Environment - Proc. in Marine Science 5, Amsterdam: Elsevier, pp. 233-250, ISBN: 0-444-51136-9.

[53] Manning, A.J. and Dyer, K.R. (2007). Mass settling flux of fine sediments in Northern European estuaries: measurements and predictions. Marine Geology, 245, 107-122, doi:10.1016/j.margeo.2007.07.005.

[54] Manning, A.J., Spearman, J. and Whitehouse, R.J.S. (2007). Mud:Sand Transport – Flocculation & Settling Dynamics within Turbulent Flows, Part 1: Analysis of laboratory data. HR Wallingford Internal Report, IT 534, 32p.

[55] Manning, A.J. and Whitehouse, R.J.S. (2009). UoP Mini-annular flume – operation and hydrodynamic calibration. HR Wallingford Technical Report, TR 169.

[56] McCave, I.N. (1984). Size spectra and aggregation of suspended particles in the deep ocean. Deep-Sea Res., 31: 329-352.

[57] Mehta, A.J., Jaeger, J.M., Valle-Levinson, A., Hayter, E.J., Wolanski, E. and Manning, A.J. (2009). Resuspension Dynamics in Lake Apopka, Florida. Final Synopsis Report, submitted to St. Johns River Water Management District, Palatka, Florida, June 2009, Report No. UFL/COEL-2009/00, 158p.

[58] Mehta, A.J. and Lott, J.W. (1987). Sorting of fine sediment during deposition. Proc. Speciality Conf. Advances in Understanding Coastal Sediment Processes. Am. Soc. Civ. Eng., New York, pp. 348-362.

[59] Mietta, F. (2010). Evolution of floc size distribution of cohesive sediments. Ph.D. Thesis, Delft University of Technology, Faculty of Civil Engineering and Geosciences, The Netherlands, 169p.

[60] Migniot, C. (1968). Study of the physical properties of various very fine sediments and their behaviour under hydrodynamic action. La Houille Blanche, 23 (7). (Translation of French text).

[61] Mitchener, H.J., Torfs, H. and Whitehouse, R.J.S. (1996). Erosion of mud/sand mixtures. Coastal Engineering, 29, 1-25 [Errata, 1997, 30, 319].

[62] Mory, M., Gratiot, N., Manning, A.J. and Michallet, H. (2002). CBS layers in a diffusive turbulence grid oscillation experiment. In: J.C. Winterwerp and C. Kranenburg (Eds.), Fine Sediment Dynamics in the Marine Environment - Proc. in Mar. Science 5, Amsterdam: Elsevier, pp.139-154, ISBN: 0-444-51136-9.

[63] Nowell, A.R.M., Jumars, P.A. and Eckman, J.E. (1981). Effects of biological activities on the entrainment of marine sediments. Mar. Geol., 42, 133-153.

[64] Ockenden, M.C. and Delo, E.A. (1988). Consolidation and erosion of estuarine mud and sand mixtures – an experimental study. HR Wallingford Report, SR 149.

[65] Owen, M.W. (1976). Determination of the settling velocities of cohesive muds. Hydraulics Research, Wallingford, Report No. IT 161, 8p.

[66] Panagiotopoulus, I., Voulgaris, G. and Collins, M.B. (1997). The influence of clay on the threshold of movement of fine sandy beds. Coastal Eng., 32, 19-43.

[67] Parker, D.S., Kaufman, W.J. and Jenkins, D. (1972). Floc break-up in turbulent flocculation processes. J. Sanitary Eng. Div., Proc. Am. Soc. Civil Eng., 98 (SA1): 79-97.

[68] Paterson, D.M. (1989). Short-term changes in the erodibility of intertidal cohesive sediments related to the migratory behaviour of epipelic diatoms. Limnol. Oceanogr. 34: 223-234.

[69] Paterson, D.M., Crawford, R.M. and Little, C. (1990). Subaerial exposure and changes in the stability of intertidal estuarine sediments. Estuarine Coastal and Shelf Science, 30, 541-556.

[70] Paterson, D.M. and Hagerthey, S.E. (2001). Microphytobenthos in contrasting coastal ecosystems: Biology and dynamics. In: Ecological comparisons of sedimentary shores (K. Reise, Ed.), Ecological studies, pp. 105-125.

[71] Pidduck, E.L. and Manning, A.J., in prep. A Laboratory Examination of Flocculation Properties Exhibited by Natural Sediment Mixtures from Portsmouth Harbour, UK. HR Wallingford Technical Report, TR 182.

[72] Puls, W., Kuehl, H. and Heymann, K. (1988). Settling velocity of mud flocs: results of field measurements in the Elbe and the Weser Estuary. In: J. Dronkers, and W. van Leussen, (eds), Physical Processes in Estuaries. Berlin: Springer-Verlag, pp. 404-424.

[73] Raudkivi, A.J. (1998). Loose boundary hydraulics. 3rd Edition. Balkema, Rotterdam.

[74] Reid, G.K. and Wood, R.D. (1976). "Ecology of inland water and estuaries." New York: D. Van Nostrand Company.

[75] Soulsby, R.L. and Manning, A.J. (2012). Cohesive sediment settling flux: settling velocity of flocculated mud. Technical Note DDY0409-01, HR Wallingford, Wallingford, UK.

[76] Spearman, J.R., Manning, A.J. and Whitehouse, R.J.S. (2011). The settling dynamics of flocculating mud:sand mixtures: Part 2 – Numerical modelling. Ocean Dynamics, IN-TERCOH 2009 special issue, DOI: 10.1007/s10236-011-0385-8.

[77] Spencer, K.L., Manning, A.J., Droppo, I.G., Leppard, G.G. and Benson, T. (2010). Dynamic interactions between cohesive sediment tracers and natural mud. Journal of Soils and Sediments, Volume 10 (7), doi:10.1007/s11368-010-0291-6.

[78] Tambo, N. and Hozumi, H. (1979). Physical characteristics of flocs – II. Strength of flocs. Water Research, 13, 409-419.

[79] Tambo, N. and Watanabe, Y., (1979). Physical characteristics of flocs-I. The floc density function and aluminium floc. Water Research 13, 409-419.

[80] ten Brinke, W.B.M. (1993). The impact of biological factors on the deposition of fine grained sediment in the Oosterschelde (The Netherlands). Ph.D. Thesis, University of Utrecht, The Netherlands, 252p.

[81] Tolhurst, T.J., Gust. G. and Paterson, D.M. (2002). The influence on an extra-cellular polymeric substance (EPS) on cohesive sediment stability. In: J.C. Winterwerp and C. Kranenburg (Eds), Fine Sediment Dynamics in the Marine Environment - Proc. in Marine Science 5, Amsterdam: Elsevier, pp. 409-425, ISBN: 0-444-51136-9.

[82] Tomi, D.T., Bagster, D.F. (1978). The behaviour of aggregates in stirred vessels: Part I - Theoretical considerations on the effects of agitation. Trans. Inst. Chem. Eng., 56, 1-8.

[83] Torfs, H. (1994). Erosion of layered sand-mud beds in uniform flow. Proc. 24th Int. Conf. Coastal Eng., Kobe, Japan, 23-28 October 1994.

[84] Torfs, H., Mitchener, H.J., Huysentruyt, H. and Toorman, E. (1996). Settling and consolidation of mud/sand mixtures. Coastal Engineering, 29, 27-45.

[85] Tsai, C.H., Iacobellis, S. and Lick, W. (1987). Flocculation of fine-grained sediments due to a uniform shear stress. J. Great Lakes Res., 13: 135-146.

[86] Uncles, R.J., Stephens, J.A. and Harris, C. (1998). Seasonal variability of subtidal and intertidal sediment distributions in a muddy, macrotidal estuary: the Humber-Ouse, UK. In: Sedimentary Processes in the Intertidal Zone, Black, K.S., Paterson, D.M. and Cramp, A. (Eds), Geological Society, London, Special Publications, 139, 211-219.

[87] Underwood, G.J.C. and Kromkamp, J. (1999). Primary production by phytoplankton and microphytobenthos in estuaries. Advances in ecological research, 29, 93-153.

[88] Underwood, G.J.C. and Paterson, D.M. (2003). The importance of extracellular carbohydrate production by marine epipelic diatoms. Advances in Botanical Research (incorporating Advances in Plant Pathology), Vol. 40, Elsevier, Amsterdam, pp.183-240, ISBN: 0-12-005940-1.

[89] Van de Ven, T.G. and Hunter, R.J. (1977). The energy dissipation in sheared coaggulated soils. Rheologica Acta, 16, 534-543.

[90] van Ledden, M. (2002). A process-based sand-mud model. In: J.C. Winterwerp and C. Kranenburg (Eds.), Fine Sediment Dynamics in the Marine Environment - Proc. in Mar. Science 5, Amsterdam: Elsevier, pp.577-594, ISBN: 0-444-51136-9.

[91] van Ledden, M. (2003). Sand-mud segregation in estuaries and tidal basins. Ph.D. Thesis, Delft University of Technology, The Netherlands, Report No. 03-2, ISSN 0169-6548, 217p.

[92] van Leussen, W. (1988). Aggregation of particles, settling velocity of mud flocs: a review. In: Dronkers, J., van Leussen, W. (Eds), Physical Processes of Estuaries, Berlin: Springer, pp. 347-403.

[93] van Leussen, W. (1994). Estuarine macroflocs and their role in fine-grained sediment transport. Ph.D. Thesis, University of Utrecht, The Netherlands, 488p.

[94] van Wijngaarden, M., Venema, L.B., De Meijer, R.J., Zwolsman, J.J.G., Van Os, B. and Gieske, J.M.J. (2002a). Radiometric sand-mud characterisation in the Rhine-Meuse estuary, Part A: Fingerprinting. Geomorphology, 43, 87-101.

[95] van Wijngaarden, M., Venema, L.B., and De Meijer, R.J. (2002b). Radiometric sand-mud characterisation in the Rhine-Meuse estuary, Part B: In situ mapping. Geomorphology, 43, 103-116.

[96] Waeles, B., Le Hir, P. and Lesueur, P. (2008). A 3D morphodynamic process-based modelling of a mixed sand/mud coastal environment : the Seine Estuary, France. In: T. Kudusa, H. Yamanishi, J. Spearman and J.Z. Galiani, (eds.), Sediment and Ecohydraulics - Proc. in Marine Science 9, Amsterdam: Elsevier, pp. 477-498, ISBN: 978-0-444-53184-1.

[97] Whitehouse, R.J.S., Soulsby, R., Roberts, W. and Mitchener, H.J. (2000). Dynamics of Estuarine Muds. Thomas Telford Publications, London, 232p.

[98] Widdows, J., Blauw, A., Heip, C.H.R., Herman., P.M.J., Lucas, C.H., Middelburg, J.J., Schmidt, S., Brinsley, M.D., Twisk, F. and Verbeek, H. (2004). Role of physical and biological processes in sediment dynamics of a tidal flat in Westerschelde Estuary, SW Netherlands. Mar. Ecol. Prog. Series, 274, 41-56.

[99] Williamson, H.J. (1991). Tidal transport of mud / sand mixtures: Sediment distributions – a literature review. HR Wallingford, Report SR 286.

[100] Williamson, H.J. and Ockenden, M.C. (1993). Laboratory and field investigations of mud and sand mixtures. In: Sam S.Y Wang (Ed.), Advances in Hydro-science and Engineering, Proceedings of the First International Conference on Hydro-science and Engineering, Washington D.C. (7-11 June 1993), volume 1, pp. 622-629.

[101] Winterwerp, J. C. (1998). A simple model for turbulence induced flocculation of co-
        hesive sediment, J. Hyd. Eng., 36 (3), 309-326.

[102] Winterwerp, J.C. and van Kesteren, W.G.M. (2004). Introduction to the physics of co-
        hesive sediment in the marine environment. Developments in Sedimentology, 56,
        van Loon, T. (Ed.), Amsterdam: Elsevier, 466p.

[103] Winterwerp, J.C., Manning, A.J., Martens, C., de Mulder, T., and Vanlede, J. (2006). A
        heuristic formula for turbulence-induced flocculation of cohesive sediment. Estuar-
        ine, Coastal and Shelf Science, 68, 195-207.

[104] Wolanski, E. (2007). Estuarine Ecohydrology. Elsevier (Amsterdam, The Nether-
        lands), 157p, ISBN: 978-0-444-53066-0.

Chapter 6

Composition and Transport Dynamics of Suspended Particulate Matter in the Bay of Cadiz and the Adjacent Continental Shelf (SW - Spain)

Mohammed Achab

Additional information is available at the end of the chapter

1. Introduction and objective

The quantification of suspended particulate matter (SPM) and the investigation of its dynamics are of major importance to understand sediment transport dynamics and many land-shelf-ocean interaction processes [1, 2, 3, 4, 5, 6]. The analysis of SPM transport processes in the marine environment requires a simultaneous study of water masses dynamic and movement, the direction and intensity of the currents as well as the characteristics of the suspended matter and the characteristics of the bottom sediments [7, 8, 9]. The application of remote sensing techniques to the study of suspended matter dynamics allows model for marine and coastal water circulation, based on the use of "turbidity patterns" as natural tracers; relating parameters of water quality to satellite images [10, 11, 12]. The utility of satellite images lies in the high frequency with which data can be taken on a point of the earth. This allows a large volume of information for various meteorological situations, that would be very difficult to obtain using conventional sampling methods. However, images must be calibrated and evaluated using "in situ" data [13].

Studies on the behaviour of suspended particulate matter have been made by many oceanographers and sedimentologists [14, 15, 16]. Several authors have studied the mineralogy of suspended matter, in order to investigate the influence of tides in estuarine systems and the relationship between the minerals in suspension and those deposited [17, 18]. Others studies have been focused on the possibility of using clay minerals in suspension as tracers, allowing to follow the progression of the river flows path in marine environment [19, 20, 21]. In the Gulf of Cadiz, Several works have been realized with the objective of determining the suspended matter content and their influence on the recent marine sedimentation [8, 22, 23].

Studies about the dynamics of fine sediments and clay minerals in the bay of Cadiz and the adjacent marine deeper zones have been approached [24, 25, 26]. The paths by which fine sediment are transported from different source areas to the marine environment have been deduced, using clay minerals as dynamic tracers [27, 28]. Others studies have been focused on the sedimentary exchange dynamics between inner area of Cadiz bay and the continental shelf [29, 30, 31]. The main objective was the determination of the nature and origin of the suspended particulate matter in the bay of Cadiz, as well as the hydrodynamic behaviour of turbidity plumes and the dispersal of SPM and its effects on the inner continental shelf. The proposed chapter is based on data of suspended particulate matter concentration, minera-logical compositions, degrees of turbidity, Landsat images analysis and complemented by grain sizes and hydrodynamic data. The combined study of dispersal of suspended sediment and degrees of turbidity by analyzing Landsat images, allows to recognize the transport paths followed by the fine sediments in the surface marine waters from the inner areas of Cadiz bay to the external zones and the adjacent continental shelf.

2. Area description

The study area is located at the Southwest of the Iberian Peninsula, between the mouth of the Guadalquivir River and the Trafalgar Cape (Fig.1).

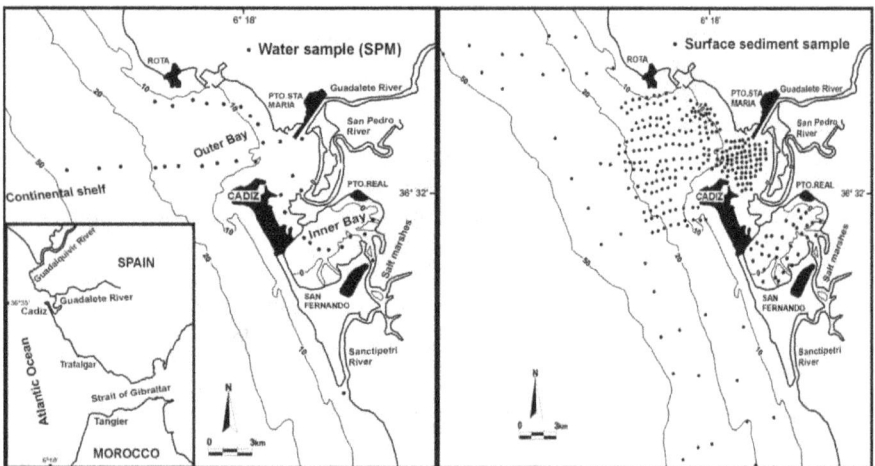

Figure 1. Geographical setting and location of surface sediment and water samples in the study area.

The Bay of Cadiz is about 28.5 km long and 13.5 km wide. Three sedimentary environments are distinguished: The outer bay (surface of about 118km^2), located to the north, is divided into two zones, a western and an eastern one, with presence of rocky outcrops in the North (Rota) and south (Cádiz), resulting slopes of 2°. The outer bay, is well connected to the continental shelf, and is very affected by the waves, currents and storms. The inner bay (Surface ≈ 40 km2) or Lagoon system located to the South and protected from waves and storms of the West and Southwest. Characterized by shallow water (<5 m depth and slope of 0.15 °), except in the navigation channel connecting the inner and outer bay (12 m depth and slope of 2.2 °). The salt marshes and tidal flat (Surface ≈ 227km²), occupy the most internal and sheltered areas, drained by a complex system of tidal creeks and channels of great importance in the hydrodynamic. This wide marshes zone occupied by halophyte vegetation is isolated from the open sea by sandy beach ridges and littoral spits.

2.1. Oceanographic setting

The zone is affected by a mesotidal regime, where the mean tidal range is 2.39 m. In spring tides the highest range reaches 3.71 m and in neap tides the lowest range is 0.65 m [32]. In the Bay of Cadiz, the bottom morphology influences the behaviour of the tidal wave, causing a time delay of 12 minutes between the outer and the inner bay. The speed of the tidal current is highly variable, with the highest values reached inside the Bay of Cadiz, where the highest speeds have been determined along the Strait of Puntales [33], reaching values over 1.5 m/s during spring tide ebbs. The input and output flows, are controlled by the tidal regime. The tidal currents oriented SW-NE are directed into the bay, while the ebbs (NW-SE) outwards [34, 35]. They are responsible of fine sediment transport from inner zones toward the external bay and continental shelf [36]. Wind and waves action are also essential factor in the sedimentary dynamics. Waves present seasonal character and the storm average frequency is of 20 days/year. The strongest storms occur in the fall-winter season and the calm periods during the summer. The prevailing swell is from the west. The data from the point WANA 1054046 in the WANA network, show that the waves from the W and SW represent 70% of the time for the period 2006- 2012 (Fig. 2). Storm waves are related to southwest Atlantic storms [37]. Mean significant wave height (H1/3) is 0.85 m and represent 45%. The most common periods range from 3 to 4 seconds. The maximum values of H1/3 reached during storms are 4 m. with periods of 6.24 s [38].

The main littoral-drifts spreads towards the southeast, generating transport in the same direction because of the coastal orientation, facing westerly and SW winter storms. Easterly winds also generate littoral-drifts towards the North and NW. In the continental shelf, the hydrodynamic regime is controlled by the North Atlantic Surface Water (NASW) moving towards the southeast, and is responsible for the dispersal of fine sediments from the Guadalquivir and Guadiana Rivers (annual flux of 9200km³ and 5500 km³ respectively) [39,40, 27, 28]. The Mediterranean Outflow Water (MOW) moving west to deeper water [41, 42] and do not have an influence on the present day sedimentation in our study area.

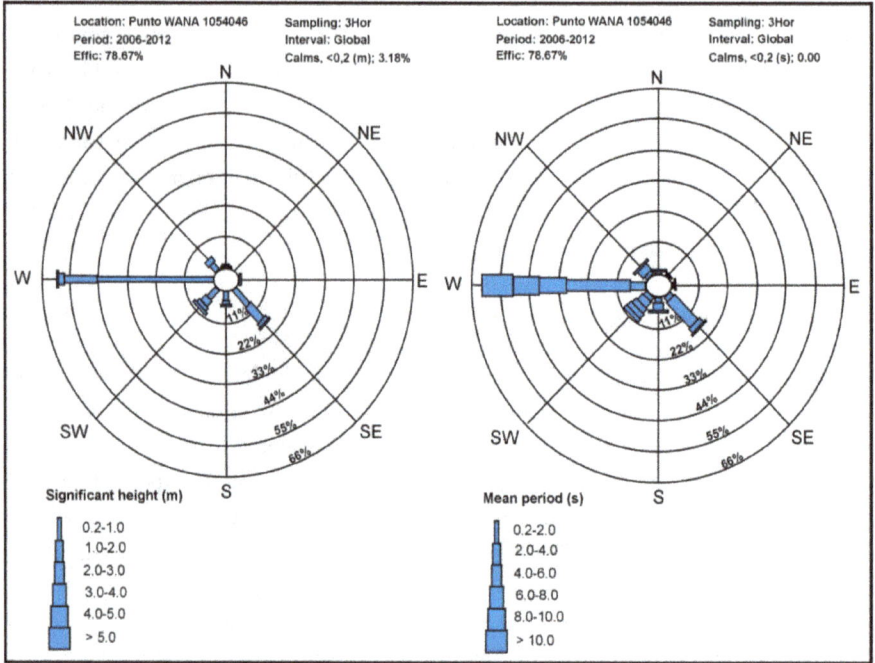

Figure 2. Significant height and peak period rose in the study area (Wana point 1054046).

2.2. Physiographic and Geologic framework

The coastline physiography is oriented NNW-SSE, with some sectors facing E-W that give the coast a stepped outline, strongly controlled by tectonic fractures [43]. The influence of tectonic structures and coastal morphology constitute an essential factor in the hydrodynamic control of the sedimentation and distribution of facies in the bay of Cadiz and the continental shelf [44, 45]. An Early Quaternary compressive tectonic episode has been deduced from reverse faults observed in marine sediments of the continental margin [46]. The continental shelf has a gentle slope and a slight inclination toward the west, with an average width of 40 km and is oriented from NNW-SSE with NNE-SSW sectors. The physiography of the sea-bottom shows a close concordance with the shoreline, the isobaths generally running parallel to the coast. The slope break occurs at 150-200 m water depth and shows significant variation north to south in cross section [26, 47]. The most important geological formations present in the surrounding areas of Cadiz bay are mainly pre-orogenic units from the Betic Mountain Range. Other units outcropping are Post-orogenic formation from the Neogene Guadalquivir basin. Upon all those materials, appear Quaternary deposits constituted by muddy marshes, beach sands and continental deposits [48, 49, 50, 51, 52, 53, 54].

3. Materials and methods

The study of concentration of suspended particulate matter (SPM) was based on the analysis of 36 water samples obtained in different zones of the study area (tidal creeks and channels, river mouths, lagoon, etc). The study of surface sediments was carried out on 250 samples collected from different sectors of the bay of Cadiz and the adjacent continental shelf (Fig.1). The sample position was determined by Differential Global Position System (DGPS) and errors made in the horizontal DGPS positioning were verified as being less than 2m. The extraction of surface sediments was performed by means of *Van Veen* dredge and gravity cores. The water samples were taken 3hours after high tide, wind from the north and northeast, average speed of 55km/hr and mean wave height of 0.6 and maximum of 1.5m. The extraction of samples was executed to specific depths with oceanographic bottles, simultaneously with the passage of the Landsat satellite over the study zone. The purpose of this operation is to obtain a synoptic picture of the turbid flumes, by comparing the data obtained from direct methods (water samples) and those of indirect methods (satellite images). The separation of SPM has been achieved following the method of [55], which consists to the filtration of a volume of five litters of water through pre-weighed filters by MILLIPORE (0.45 microns). Filters were washed with distilled water, dried at about 60Â° and weighed. The <2μm fraction was separated by a standard sedimentation method [56, 57]. The use of satellite images in this study is based on the utilization of inorganic SPM as a natural tracer. Satellite images of the Bay of Cadiz have been recorded by the satellite TM Landsat, using bands 2 and 5, and a spatial resolution of 30x30 m. Landsat images has been analysed to obtain extent and direction of turbidity plumes in several hydrodynamic situation in Cadiz bay and inner shelf waters. The process of the images has been carried out according to the methodology described by [58].

Systematic granulometric and mineralogical analyses were carried out to establish facies distribution and mineralogical composition. The collected samples (approximately 250 g of surface sediments) were placed in plastic bags sealed and identified. The Grain size of coarse fraction was determined by dry sieving sediments, using sieve column ranging from 4 Phi (0.063 mm) to -2 Phi (4mm). The fine fraction analysis was made by use of laser diffraction analyser (AMD). Grain size data were processed by GRADIST software (version 4.0). The characteristic statistic indexes and parameters were calculated using standard method [59, 60]. The mineralogical analysis of suspended particulate matter (SPM) and surface sediment was made through X-ray Diffraction techniques (XRD) using a Philips PW-1710 diffractometer, equipped by Cu-Kα radiation, automatic silt and graphite monochromator. Quantification of different mineralogical phases was calculated by the classic method of area measurement of peaks, considering the different reflection capacities of the minerals [61, 62]. Factor Analysis (Principal Components Analysis, PCA) was used to determine the mineral assemblages and to establish possible sediment transport paths from the bay toward the continental shelf.

4. Results

4.1. Surface Sediment Characteristics

Grain size analysis of modern sediments of Cadiz bay bottoms show that samples are mainly composed of sand and mud, and subordinate amounts of gravel. The grain size distribution shows the predominance of the coarser fractions in the outer bay, more exposed to wave action and currents. The finer fractions appear in more sheltered zones of the bay.

Taking into account the textural characteristics of the marine deposits and the sedimentary environments, we can differentiate various types of sediment (Fig. 3).

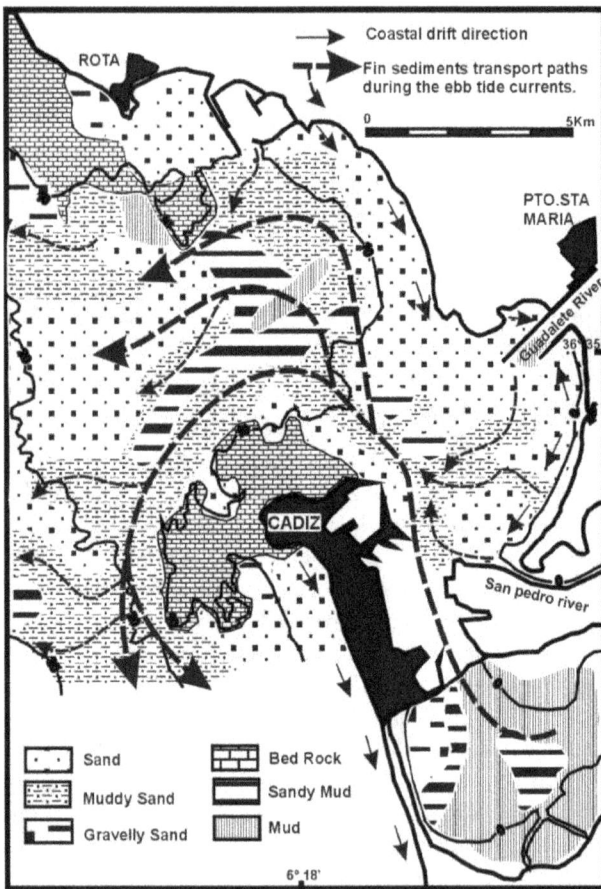

Figure 3. Surface sediments distribution on the Bay of Cadiz and transport paths of fine particules.

The sand is the dominant fraction of the outer bay sediments, with an average of 75% especially in the littoral zone. In some areas, this fraction changes laterally to muddy sand facies; due to the recent action of transport processes taking place in this area of the Cadiz bay. This facies extend into the 20-30 m deep inner shelf, being configured in two bands, one by the north margin of the bay and another one more to the South, bordering the city of Cadiz. The silt appears in low proportions (10%), giving the highest values in some sectors of the outer bay and the adjacent continental shelf. It distribution is of great importance to understand the modern sediment dynamics in the Bay of Cadiz, especially the sediment exchange between the inner and outer bay. The Clays are an important sedimentary fraction, especially in low-energy sedimentary environments where concentrations reach 90% such as inner bay and the tidal channels that drain the salt marshes. In the outer bay, the contents are very low (<5%). The gravels (less than 5%), are mainly composed of bioclasts and rock fragments, derived from erosion of rocky shoals and coastal cliffs.

Deposit environment	Bay of Cadiz and inner shelf	Salt marshes and tidal creeks
Clay minerals	Factor 1 (99% of the variance)	Factor 1 (100% of the variance)
Illite	2.15	2.16
Smectite	0.31	0.25
Interstratified I-S	0.28	0.19
Chlorite	0.29	0.38
Kaolinite	0.29	0.28
Mineral assemblage	I"/"/Sm"/Cl"/K"/"/I-Sm	I"/"/Cl"/K"/Sm"/I-Sm

Table 1. Factor scores of the clay minerals in the bay of Cadiz-inner shelf and the salt marshes- tidal creeks, Q-mode factor analysis.

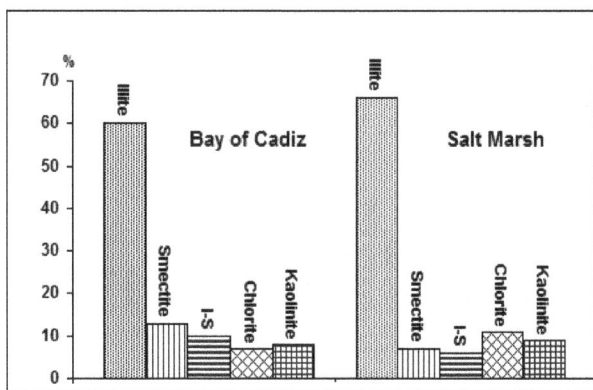

Figure 4. Clay mineralogy of surface sediments in the bay of Cadiz and the salt marsh.

In what concerns the mineralogical composition, the main minerals in terrigenous sediments of the bay of Cadiz are quartz (55%), calcite (20%) and plagioclases (5-10%). The mineralogical analysis of clay fraction indicate the predominance of Illite (60%), followed by Smectite (13%), interstratified Illite-smectite (10%), kaolinite (8%) and Chlorite (7%) (Fig. 4). In the salt marshes, the most abundant clay mineral is Illite (66%). Other clay minerals found in lesser quantities are Kaolinite (9%) and chlorite (10%) especially in the high tidal zone, and smectite (7%) in tidal creeks and tidal channels bottoms. Factor analysis was used to establish the relationships between different clay minerals and their associations. The Q-mode factor analysis results are summarized in Table 1.

4.2. Suspended sediment concentration

In the study area, significant concentrations of suspended matter are frequently observed, due to the existence of different sources of fine sediments (river mouths,e.g Guadalquivir and Guadalete rivers, tidal channels, etc.), and the action of the hydrodynamic regime (Winds, waves and tidal currents). The concentration of suspended particulate matter under the hydrodynamic conditions at the sampling time (winds of N and NE, average speed of 55km/hr) shows an average dry weight content of 6.5 mg/l (Fig.5). The highest values were found in the outer bay, especially in the oriental sector, near the Guadalete and San Pedro river mouths (16 and 25 mg/l respectively) and to the south in Sanctipetri tidal creek (13mg/l). The concentrations of SPM are also relatively high in the central part of the bay, particularly north of the city of Cadiz (14mg/l) and in the navigation channel (12.87 mg / l) connecting the inner and the outer bay of Cadiz. The higher concentrations of SPM are consistent with the pattern of tidal currents and the distribution of fine facies on the sea bottom. The lowest values (1.5-5 mg/l) appear in the adjacent continental shelf waters characterized by low sediment input, as well as in parts of the inner bay less affected by tidal currents.

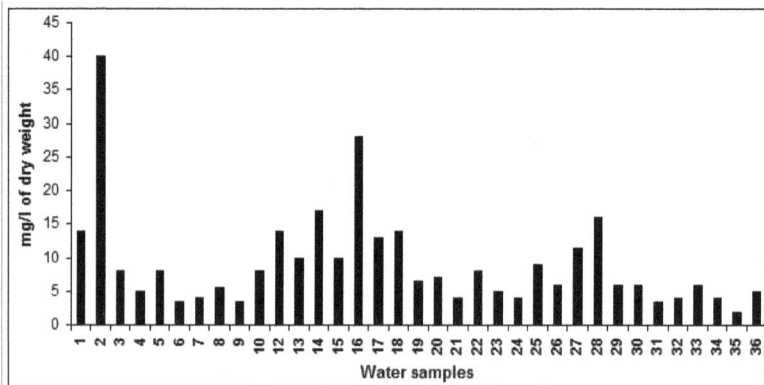

Figure 5. Distribution of suspended sediment concentration (in mg/l of dry weight) in the surface waters of the study area (see Fig. 6 for location of water samples).

4.3. Clay minerals composition of SPM

In the present study, clay minerals have been used as tracers, due to their small size making them the only particles susceptible of being transported away from the bay, out towards the continental shelf; and can be useful in the determination of the pathways of fine sediments and the suspended matter. Analysis of the mineralogical composition of suspended matter shows that the most abundant clay mineral consists mainly of illite (41%) followed by smectite (38%), chlorite (11%), interstratified llite-smectite (7%) and kaolinite (6%) (Fig.6 & table.2).

Clay minerals	Outer bay	Inner bay	Tidal creeks	Continental shelf
Illite	37	43	43	38
Smectite	36	40	35	40
Interstratified I-S	6	9	5	10
Chlorite	12	15	14	9
Kaolinite	7	5	6	7

Table 2. Means values in percentage of clay minerals in SPM of different deposit environments.

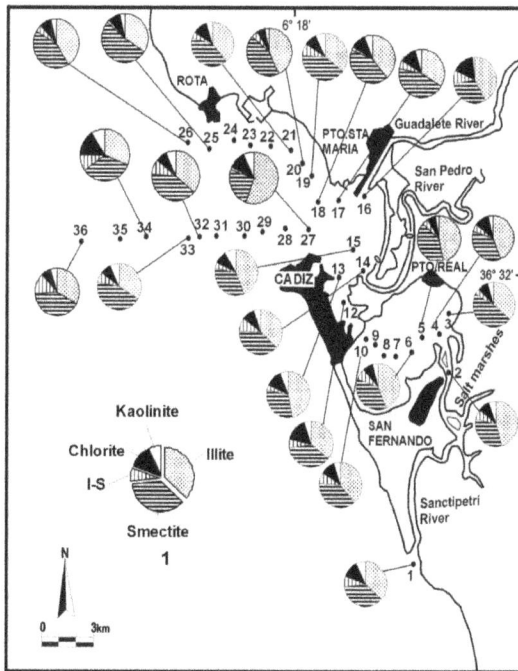

Figure 6. Distribution of principal clay minerals of suspended particulate matter in the study area.

The relations between the clay minerals were established by Q-mode factor analysis. The main clay minerals assemblage obtained in suspended matter is I>S>Cl>I-S>K (Fig. 7). The clay minerals distribution indicates high concentration of illite in the inner and protected part of Cadiz bay, while in the adjacent marine zones predominates the smectite.

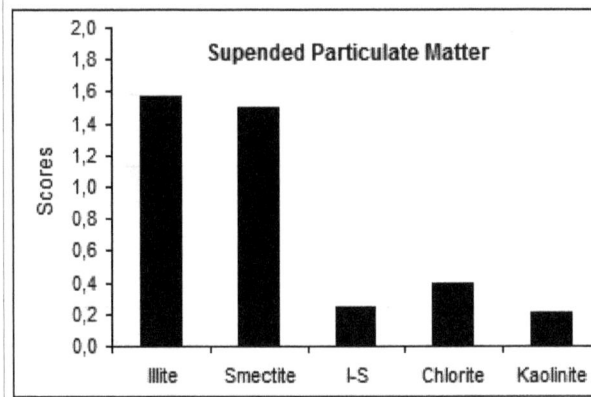

Figure 7. Factor scores (factor1: 99% of the variance) of the clay minerals in SPM obtained by Q- mode factor analysis.

The R-mode factor analysis results provide three factors explaining the 100% of the total variance. Factor 1 (52%) groups exclusively smectite (negative loading: -0. 95), is well represented on all bottoms of the bay of Cadiz. Factor 2 (28%) associates interstratified IS (positive loading:0.89) to kaolinite (0.4). This factor scores better in the western sector of the inner bay and central part of the external as well as in front of the river mouths of the Guadalete and San Pedro. Factor 3 (20%) with only the mineral chlorite (0.95), dominates next to the mouths of the Sancti Petri tidal creek and the San Pedro and Guadalete Rivers.

4.4. Landsat images analysis

The use of satellite images in studies of transport dynamics of sediment in suspension is of great interest. The usefulness of this technique is based on the monitoring of turbidity plumes, consisting mainly of suspended inorganic particles. Geologists, geographers and oceanographers, also applied this technique, for coastal processes study [10].

Landsat TM images used in this study are property of the Spain Ministry of Environment (Junta de Andalucía). The analyses of these images for different hydrodynamic and meteorological situations illustrate the existence of water masses of different degrees of turbidity, oriented from the inner zone towards the outer bay extending to the continental shelf (Fig. 8). The highest turbidity is observed in the coastal areas of the bay, as well as in front of the mouths of the Sancti Petri tidal creek and the San Pedro and Guadalete Rivers.

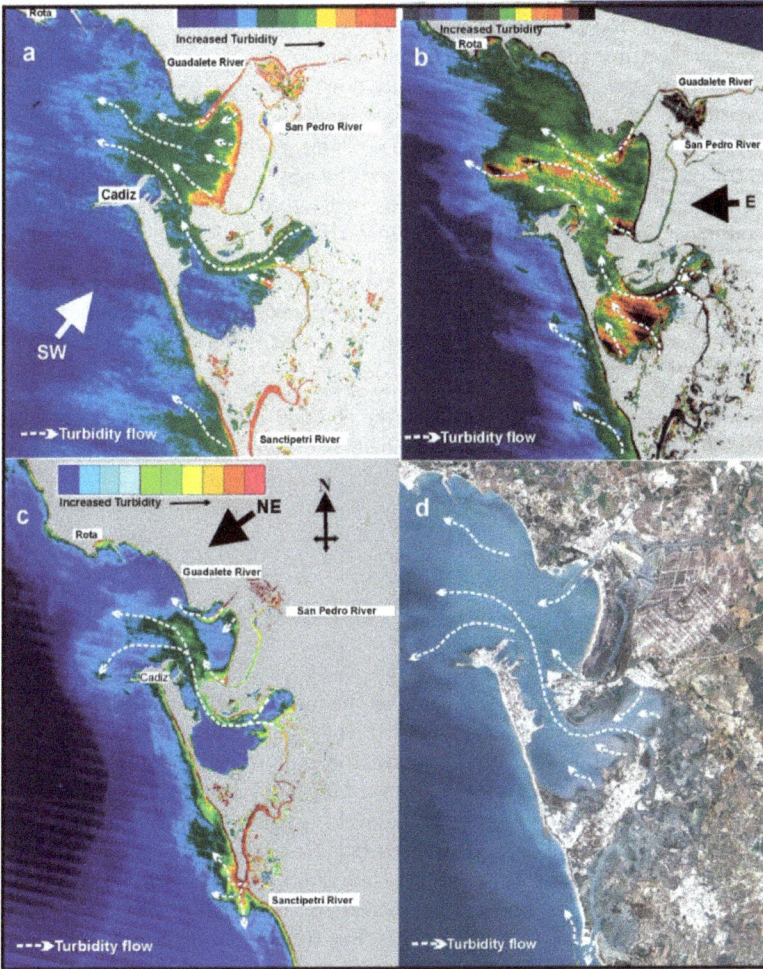

Figure 8. Turbidity plume trayectories in several hydrodynamic situations from the analysis of Landsat TM images of
Cadiz bay. 8a: date: 22-05-1996, Wind: Western-WWS, Speed: 39km/h, Hs: 0.6m, Tidal coef: 0.55 ; 8b: date:
13-08-1991, Wind: East, Speed: 37km/h, Hs: 0.54m, Tidal coef: 0.78; 8c: date: 07-06-1996, Wind: North-East, Speed:
55km/h, Hs: 0.6m, Tidal coef: 0.64; 8d: recent satellite image during ebb tide.

In situation of strong Easterly winds (Fig. 8a), suspended matter derived from resuspension
of the inner bay bottoms are subsequently transported by ebb tidal currents, with an average speed of 1.02 m /s, which is reduced to 0.77 m / s in the outer bay. These turbid flumes
(SPM flows) appear as branch oriented northwards along the eastern edge of the bay and
then turn west, influenced by coastline and bottom morphology. Part of the suspended matter tends to leave the bay being deposited in the inner continental shelf. In the continental

shelf, the North Atlantic Surface water current transport a large volume of fine sediments from the Guadalquivir rivers toward the SE. Part of this current can transport fine particles in suspension toward the bay of Cadiz. Satellite image obtained in situation of Northeast winds, shows the existence of several turbidity flows that appear to move from the inner-most zones of the bay to the outer one (Fig 8b). These flows are configured in three bands: one oriented towards the NNW following the north margin of the bay. Other band oriented towards the West and the third one goes towards the SSW bordering the Cadiz city. To the south of the bay, turbidity plumes are also observed and coming from the Sanctipetri tidal creek mouth that head towards the inner continental shelf.

On the other hand, during Western (SSW) Wind Conditions, turbidity in open coastal areas adheres to breaking waves zone, being highly concentrated within 20 meters near the beach and at depths less than 2 meters. outcrops rock located to the NW of the city cause, the existence of turbidity plumes in this area (Fig. 8c).

The observation of recent satellite image during ebb tide shows that turbid flumes extend from the inner bay out to the continental shelf, following the coastline and sea bottom morphology (Fig. 8d).

5. Discussion

5.1. Sedimentary processes

The distribution and transport of sediments on the marine environment is function of the wave-current and grain size interactions [63, 64]. They can also reflect the direction of water mass movement [65, 66]. In Cadiz bay bottoms, the grain size distribution, shows the predominance of fine fractions, (silt and clay), toward the more sheltered and internal zones of the bay. While the coarse fractions (sand and gravel), appear in external zones, more opened to the sea and exposed to waves and currents action. Grain size parameters show the prevalence of unimodale distribution (fine and very fine sand), however bimodal and polymodal distributions are present. The mode values increase in coastal areas and near rocky shoals, decreasing towards the central zone of the outer bay and in the inner bay. The general trend and the variability of different grain-size parameters, reflect the control that physiographic and hydrodynamic factors exert on different types of sediment [31].

The abundance and the progressive deposition of fine materials in the inner bay and the salt marshes is related to the existence of sedimentary environment of very low hydrodynamic energy. Their hydrodynamic regime is dominated by tidal currents, especially ebb tides and wind of the East sector [30, 36]. These fine sediments are also found in the outer bay, occupying the central and eastern sectors, covering sandy bottoms. They derive from resuspension of fine-grained materials in the marshlands of the bay and from fine materials supplied by the Guadalete River during periods of rainfall and floods. The presence of mud and sandy mud facies covering sandy bottoms, possibly will indicates actual processes of deposit and transport of fine sediments from the inner bay toward the external zone reaching the inner continental shelf (Fig.3). Muddy facies are also present in the continental shelf as a prodeltaic muddy zone situated to the north and deposited in low energy environment. These fine-grained sediments are related to supplies coming from the Guadalquivir River [26, 28, 67].

The grain size distribution allowed the differentiation of particles transport modes, according to its size and the hydrodynamic conditions. The suspended transport predominates in the inner part of the Bay (including tidal channels and marshland areas) and to the north of the continental shelf characterized by mud-clay bottoms deposited in low energy conditions. In general, this type of transport predominates on many marine environments including continental shelves [68]. In the outer bay, especially the western sector that coincide with sandy-mud and mud facies bottoms, the saltation-suspension transport mode predominates and may possibly indicate the transport paths of fine sediments and suspended matter in the study area. The distributions of these different grain size classes reflect the energy level of each depositional environment and the processes of sediment transport; as well as the action of hydrodynamic agents, which control recent sedimentary dynamics between different sedimentary environments of the bay of Cadiz and the continental shelf [33, 44,69].

5.2. Suspended matter dynamics

On coastal marine environments, rivers are the major sources of supplies of suspended particulate matter. Most of great rivers discharge important quantities of SPM, carried by longshore currents parallel to the coast, forming permanent turbidity plume. The spatial distribution of suspended sediments is a consequence of hydrodynamic forcing acting on the unconsolidated sediments of the shelf and the coastline [70, 71]. In general, waves are more important for resuspension in shallow water (< 10 m) and currents become more important in deeper water (> 10 m) [72, 73, 74].

The measures of concentration of suspended matter obtained in the Bay of Cadiz and the continental shelf waters, show values varying from one area to another. The highest values of concentration of suspended matter are given in the outer bay, due to the influence of Guadalete and San Pedro River mouths. These higher concentrations are consistent with the pattern of tidal currents and the distribution of fine facies on the sea bottom. In the inner bay, the concentrations of suspended matter have lower values, due to increased settling of particles in sheltered environments from waves. In the continental shelf, are given the lowest values found in the study area, corresponding to the lower turbidity of these waters. Numerous studies about the mineralogy of the clay fraction present in surface sediment and suspended matter, show the existence of a strict relationship between clay minerals transported by rivers and those in the sediments of the drainage basin [75, 76]. In our case, comparing the clay mineral composition of SPM and in surface sediment, a significanct change is observed in the particulate matter with an important increase of smectite content from 13% in surface sediment to 38% in SPM and decrease of illite content (from 60% to 41%). This effect can be explained by the tendency of illite to focus on the sea bottoms, while smectites tend to stay longer in suspension. The remaining minerals like chlorite, Kaolinite and I-

S has low contents, between 5 and 10%. The main clay mineral association established in SPM is I>S>>Cl>I-S>K, which basically coincides with that obtained for the bay and the continental shelf and differs significantly from that found only in the bay of Cadiz [28]. There is a greater importance of illite in samples of internal areas of the bay and tidal creeks. This mineral, decrease in open water and the continental shelf, favoring the increase of smectite contents. Those differences can be explained considering processes such as settling and selective transport of clay minerals in marine environments [77], in relation with hydrodynamics regime, and the location of the main sources of sediment supply that control the mineralogical composition and concentration of suspended matter.

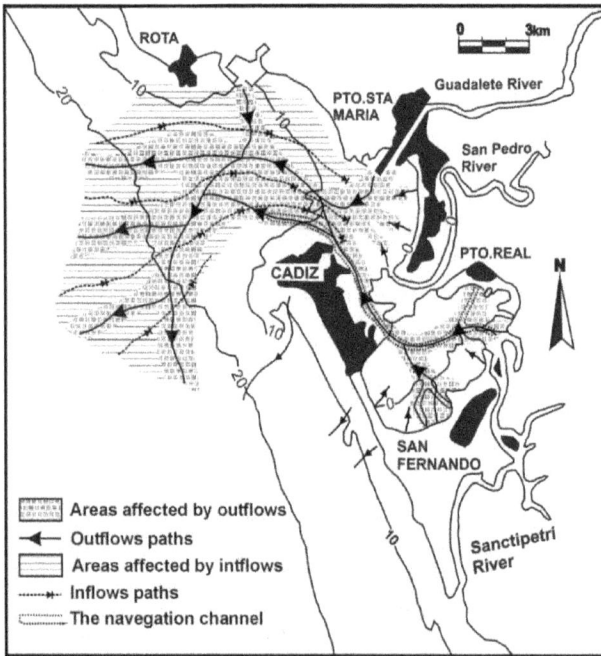

Figure 9. Transport dynamics model of fine sediments and suspended matter between the inner and the outer bay of Cadiz and the adjoining continental shelf.

Clay minerals and assemblages have been used as dynamic tracers to deduce transport path and the process of sedimentation of fine sediment and SPM [27, 28, 77, 78, 79, 80,]. Others studies indicate that clay minerals can be transported large distance by rivers, wind and currents, indicating the dominant trajectories of fine sediments and the suspended matter [22, 25, 42]. Data from the distribution of SPM concentration in the bay of Cadiz and clay minerals contents and assemblages have been used to establish the transport paths model of SPM in different area of the study zone, through the trace that different minerals have left in marine bottom surface. Two flows paths have been differentiated (Fig.8): i) The inflows coming from external marine areas located to the north, in particular the Guadalquivir river mouth and other sources. These flows can transport suspended matter and fine sediments to the Cadiz bay bottoms by the action of marine currents, specially the littoral and the Atlantic Surface water currents of SE direction. ii) The outflows coming from Cadiz bay and littoral zones; can reach the continental shelf by mean of ebb tide currents. Two possible transport paths can be deduced; the first one runs preferably by the northern margin of the bay of Cadiz and reaches the Rota city. Another flow oriented towards the west, bordering the city of Cadiz, eventually extending to the continental shelf. These bands might correspond to sea floor marks generated by flows between the bay and the continental shelf; agreement with the tidal flow pattern established by [33] in the Cadiz bay.

5.3. Turbidity flow patterns

The analysis of satellite images can provide information about size and direction of the turbidity plumes and the effect of winds and tidal currents in their distribution; and estimate the concentration of particles in water column, [81, 82, 83]. The turbidity caused by suspended particles is detectable by the reflective bands of Landsat satellite [84, 85]. Based on Landsat images, [34] and [86] show that In the Bay of Cadiz, depending to hydrodynamic conditions, Turbidity plumes follow different directions and can cross the bay area. They reach the inner shelf by the action of tidal ebb, depositing part of its SPM. Once in the open sea, SPM are moved by currents and interact with the general hydrodynamic system affecting coastal areas and the Gulf of Cadiz [27, 28]. According to information obtained from the analysis of Landsat-TM images and the concentration of suspended matter, four levels of turbidity has been differentiated in the sea area between the town of Rota (northwestern boundary of the Bay), and the mouth of Sanctipetri tidal creek (southeastern boundary of the bay) (Fig. 8):

i.   High and very high turbidity Waters appears in the eastern part of the outer bay, near the Guadalete and San Pedro river mouths, whose turbidity plumes are oriented toward the West and NW. Very high turbidity can be observed in the inner bay and the central part of the outer bay, when the spring tide and the southeast wind coincide. South of the bay, the Sanctipetri tidal creek shows also very high turbidity that is oriented toward the SSW.

ii.   Medium turbidity waters are found along the coast and occupy a variable band between 0.5 and 2 km. They are specially represented in the eastern part of the outer bay, configured in a large band following the morphology of the coast and the sea bottom. In situation of Easterly winds, these waters can exceed the environment of the bay of Cadiz and reach the inner shelf by the action of ebb tidal currents.

iii.   Low turbidity are observed in the western sector of the inner bay (except at moments of spring tide and east winds). The degree of turbidity was found to be related to a confined environment, less affected by the action of tidal currents leading to precipitation of suspended matter. The turbidity level in this area may be due to their muddy bottoms occupied by algae, whose roots and leaves act to retain the sediment and preventing their resuspension. Low turbidity can be observed in the outer bay, under certain conditions, in this case, suspended particulate matter coming from the inner bay, do not reach the outer bay, characterized by profound bottoms making difficult the process of resuspension of fine particles.

iv.   Very low turbidity waters, correspond to Atlantic waters, characterized by little influence of coastline and located at several kilometers from the coast. Under certain conditions, these water, can be mixed with other more turbid found near shore, because of the presence of suspended matter flows generated by the tides.

Taking into account the above data, three geographical sectors have been identified:

a.   Proximal sector. Is a shallow area of high and very high turbidity. Is affected by the physiography of the coast and the sea bottoms. The variation of degree of turbidity in this area are related to continental supply and sedimentary nature of the bottoms (muddy or sandy); as well as the direct action of waves and tidal currents.

b.   Middle Sector. Corresponds to the maritime area limited between the proximal and distal sectors. This is an area of high variability and is affected in its inner part by the action of tidal flows, which alter their limits and water quality up to 4 times daily. The outer boundary of this sector consists of less turbid water, flowing through this area of the Gulf of Cadiz.

c.   External sector. is the maritime zone located about 5 km away from the virtual line between the Rota and Cadiz cities, and characterized by low and very low turbidity. The internal limit is still somewhat affected by coastal dynamics, but less than in the middle and proximal sectors. The degree of turbidity present little change, except during periods of flood (espacially in Autumn and Winter) in which the rivers pouring into this part of the Gulf of Cadiz, transport lots of suspended matter, giving way to large turbidity plumes. water turbidity may be episodically affected by the action of tidal flows during spring tides, favored by Southeast winds.

6. Conclusion

i.   The high concentrations of suspended matter present in Cadiz bay waters are linked to major sources of contributions, like rivers and tidal creeks. Fine sediments coming from the remobilization of the inner bay bottoms and the marshlands by the effect of waves and winds, can supply large quantity of suspended material depending on the season. This material is then transported by the effect of tidal ebb currents to the external areas of the bay, and even continental shelf. The use of clay minerals as natural tracers has allowed the determination of suspended matter flows paths taking place in different sectors of the study area.

ii.   The general transport pattern of suspended particulate matter (SPM) is affected by local processes, which take place in littoral zones, in particular in Cadiz bay and the Guadalquivir estuary. Part of the Atlantic waters rich in SPM coming from the Guadalquivir Rivers reaches the bay of Cadiz and can be deposited in lagoons and salt marshes. The resuspension of fine-grained material in the inner zone of the bay during southeast wind and ebb tidal current generate suspended matter outflows towards the outer bay. Considerable quantity of this SPM is injected in the Atlantic waters.

iii.   Landsat images show that the turbidity pattern coincides generally with the area of the muddiest facies present on the outer bay bottoms and with the geographical locations of the sampling stations providing the highest contents of suspended solids. The combined study of suspended matter concentration and Lansat images analyses, allows recognize the transport paths followed by the sediments in the surface marine waters. these images show that turbidity plumes extend from the inner bay out to the continental shelf by action of the tidal ebb currents, following the coastline and sea bed physiography.

iv.   The transport dynamics of fine sediment and SPM from the inner bay toward the external zone and the inner continental shelf was found to be related to the location and diversity of supply sources, to the coastal and sea bottoms morphology; as well as to the influence of the hydrodynamic system, fundamentally ebb tide currents. Other transport agents are the currents generated by waves. In situation of easterly storm, currents are generated towards the West, which combined with the tidal ebb, give rise to large flumes of turbidity, which are directed offshore. In contrast, the wind and westerly waves are opposed to ebb tide currents favouring the inflow.

Acknowledgements

This paper was supported by projects 2002-01142/MAR of the Interministerial Commission of Science and Technology (CICYT) of the Spanish Government; as well as by the project SVT12-2010 of the Ministry of Higher Education of the Moroccan Government. The author thanks Dr. J.M Gutierrez-Mas of the University of Cadiz for his encouragement and availability.

Author details

Mohammed Achab*

Address all correspondence to: achab@israbat.ac.ma

Department of Earth Sciences, Scientific Institute, University Mohammed V-Agdal, Rabat, Morocco

References

[1] Allen, G. P. (1971). Etude des processus sédimentaires dans l'estuaire de la Gironde. *PhD thesis*, Univ. Bordeaux, I.(353), 314.

[2] Latouche, C. M. (1972). La sédimentation argileuse marine au voisinage de l'embo-chure de la Gironde. *Interprétation et conséquences. C. R. Acad. Sci. Paris*, t 274, 2929-2932.

[3] Castaing, P. (1981). Le transfer á l'ocean des suspension estuariennes. Cas de la Gironde. *PhD thesis*, Univ. Bordeaux, I.(701), 530.

[4] Jouanneau, J. M. (1982). Matiéres en suspension et oligo-élément métalique dans le systeme estuarien girodin: Comportement et flux. *PhD thesis*, Univ. Bordeaux, I.(32), 150.

[5] Webwro, O., Jouanneau, J. M., Ruch, Py., & Mirmand, M. (1991). Grain-size relation-ship between Suspended matter originating in the Gironde Estuary and shelf Mud-Patch deposits. *Mar. Geol*, 96, 195-165.

[6] Wegner, C. (2003). Sediment Transport on Arctic Shelves- Seasonal Variations in Sus-pended Particulate Matter Dynamics on the Laptev Sea Shelf (Siberian Arctic). *Ber. Polarforsch. Meeresforsch.*, 1618- 3193, 455.

[7] Maldonado, A., & Stanley, D. (1981). Clay mineral distribution patterns as influenced by deposicional processes in the southeastern Levantine Sea. *Sedimentology*, 28, 21-32.

[8] Palanques, A., Plana, F., & Maldonado, A. Estudio de la materia en suspensión en el Golfo de Cádiz. *Acta. Geol. Hisp*, t.21-22, 491-497.

[9] Chamley, H. (1993). Marine sedimentation of clay minerals. *in : Sedimentologie et geo-chimie de la surface*, Strasbourg (France), Academie des sciences, Paris (France), 27-28.

[10] Balopouls, E. T., Collins, M. B., & James, A. E. (1986). Satelite images and their use in the numerical modeling of coastal processes. *Inter. Jour. of Remote Sensing*, 7, 905-519.

[11] Fernández-Palcios, A., Moreira Madueño, J. M., Sánchez, Rodriguez E., & Ojeda Zu-jar, J. (1994). Evaluation of different methodological approaches for monitoring water quality parameters in the coastal waters of Andalusia (Spain) using Landsat.TM.Da-ta. *Earsel workshop on Remot Sensing and GIS for Coastal zone Managment*, Rijks Water-Staat, Survy Department. Delft, The Netherlands, 114-123.

[12] Shimwell, S. (1998). Remote sensing of suspended sediment off the Holderness Coast, UK. *ARGOSS Report A5.*, 23, unpublished.

[13] Baban, S. M. J. (1993). Detecting water quality parameters in the Norfolk Broads, U.K., using Landsat imagery. *Int. J. Remote Sensing*, 14(N17), 1247-1267.

[14] Manheim, F. T., Meade, R. H. y., & Bond, G. C. (1970). Suspended matter in surface waters of the Atlantic continental margin from Cape Cod to the Florida Keys. *Science*, 167, 371-376.

[15] Milliman, J. D., Summerhayes, C., P.y, , & Barretto, H. T. (1975). Oceanography and suspended matter of the Amazon River. *Jour. Sedim. Petrol*, 45, 189-206.

[16] Feely, R. A., Baker, E. T., Schumacher, J. D., Mascoth, G., J.y, , & Landing, W. M. (1979). Processes affecting the distribution and transport of suspended matter in the Notheast Gulf of Olaska. *Deep-Sea Res.*, 26(4A), 445-464.

[17] Rudert, M. y, & Müller, G. (1981). Mineralogy and provenance of suspended solids in estuarines and nearshores areas of the southern North Sea. *Senckenberg. Mar.*, 13, 57-64.

[18] Lixiangyun, Chen Hongxun., Lichuanrong, Wang Xiangzhen y Qiu Chuanzhu. (1989). Distribution characteristic of major minerals of suspended matters and sediments in Daya Bay. *Tropical Oceanology*, 8(1), 34-42.

[19] Monaco, A. (1970). Sur quelques phénoménes d'échange ionique dans les suspension argileuse an contact de léau de mer. *C. R. Acad. Sci.*, Paris, 270, 1743-1746.

[20] Arnoux, A. y, & Chamley, H. (1974). Mineraux des argiles et détergents des eaux interstitielles dans les sediments superficiels du Golfe de Lion. *C. R. Acad. Sci.*, Paris, 278, 999-1002.

[21] Aloisi, J. C., Monaco, A., Millot, C. y, & Pauc, H. (1979). Dynamique des suspensions et micanismes sédimentogénetique sur le plateau continental du Golfe du Lion. *C. R. Acad. Sci.*, Paris, ser. D, 189, 879-882.

[22] Gutiérrez Mas, J. M., Sánchez Bellón, A., Achab, M., Fernández Palacios, A., & Sánchez Rodríguez, E. (1998). Influence of the suspended matter exchange between the Cadiz Bay and the continental shelf on the recent marine sedimentation. Alicant, Spain. *15 th Inter. Sediment. Congr.*, Abstract, 404-405.

[23] Achab, M., Gutiérrez Mas, J. M., & Luna del Barco, A. (2000a). Concentration and mineralogic composition of suspended matter in the Bay of Cadiz and adjacent continental shelf. *Geotemas*, 1(14), 81-86.

[24] Gutiérrez Mas, J. M., Achab, M., Sánchez Bellón, A., Moral Cardona, J. P., & López Aguayo, F. (1996b). Clay minerals in recent sediments of the bay and their relationships with the adjacent emerged lands and the continental shelf. *Advances in Clay minerals*, 121-123.

[25] Achab, M., Gutiérrez Más, J. M., & Sánchez Bellón, A. (1998). Transport of fine sediments and clay minerals from the tidal flats and salt marshes in Cadiz bay towards outer marine zones. *European Land-Ocean Interaction Studies*, Huelva-Spain, 93-94.

[26] Lopez, Galindo A., Rodero, J., & Maldonado, A. (1999). Surface facies and sediment dispersal patterns: southeastern Gulf of Cadiz, Spanish continental margin. *Marine Geology*, 83-98.

[27] Gutiérrez Más, J. M., López Aguayo, F., & Achab, M. (2006). Clay minerals as dynamic tracers of suspended matter dispersal in the Gulf of Cadiz (SW Spain). *Clay minerals*, 727-738.

[28] Achab, M., Gutiérrez Más, J. M., & López Aguayo, F. (2008). Utility of clay minerals in the determination of sedimentary transport patterns in the bay of Cadiz and the adjoining continental shelf (SW-Spain). *Geodinamica Acta*, 5-6.

[29] Gutiérrez Mas, J. M., Luna del Barco, A., Achab, M., Muñoz Pérez, J. J., González Caballero, J. L., Jódar Tenor, J. M., & Parrado Román, J. M. (2000). Controlling factors

of sedimentary dynamics in the littoral domain of the Bay of Cadiz. *Geotemas*, 1(14), 153-158.

[30] Achab, M., El Moumni, B., El Arrim, A., & Gutierrez Mas, J. M. (2005b). Répartition des faciès sédimentaires récents en milieu marin côtier : exemple des baies de Tanger (NW-Maroc) et de Cadix (SW-Espagne). *Bull. Inst. Sci., sect. Sci. Terre* [27], 55-63.

[31] Achab, M. (2011). Dynamics of sediments exchange and transport in the Bay of Cadiz and the adjacent continental shelf (SW-Spain). *In : Manning, A. (Ed) Sediment Transport in aquatic environments*, Intech, Vienna, 19-44.

[32] Benavente, J. (2000). Morfodinámica litoral de la Bahía Externa de Cádiz. *Tesis Doctoral*, Univ de Cadiz, 533.

[33] Alvarez, O., Izquierdo, A., Tejedor, B., Mañanes, R., Tejedor, L., & Kagan, B. A. (1999). The influence os sediments load on tidal dynamics, a case study: Cadiz Bay. *Estuarine Coastal and shelf Science*, 48, 439-450.

[34] Guillemot, E. (1987). Teledetection des milieux litoraux de la Baie de Cadix. *PhD thesis*, Univ. Paris, 146.

[35] Muñoz Pérez, J. L. y, & Sánchez de Lamadrid, A. (1994). El medio físico y biológico de la bahía de Cádiz: Saco interior. *Informaciones técnicas*, 28/94. Consejeria de Agricultura y Pesca (Junta de Andalucía), 161.

[36] Achab, M., Gutiérrez Mas, J. M., Sanchez Bellon, A., & Lopez Aguayo, F. (2000b). Fine sediments and clay minerals embodiment and transport dynamics between the inner and outer sectors of Cadiz Bay. *Geogaceta*, 27, 3-6.

[37] Jodar, J. M., Voulgaris, G., Luna del Barco, A., & Gutiérrez-Mas, J. M. (2002). Wave and Current Conditions and Implications for the Distribution of Sediment in the Bay of Cadiz (Andalucia, SW Spain) En: Littoral. *The Changing Coast*, 3, 287-291.

[38] Rodríguez-Ramírez, A., Ruiz, F., Cáceres, L. M., Rodríguez Vidal, J., Pino, R., & Muñoz, J. M. (2003). Analysis of the recent storm record in the southwestern Spanish coast: Implications for littoral management. *The Science of the Total Environment*, 189-201.

[39] Meliéres, F. (1974). Recherche sur la dynamique sedimentaire du Golfe de Cadix. Espagne PhD thesis Univ. Paris.; , 235.

[40] Maldonado, A., & Nelson, C. H. (1988). Dos ejemplos de márgenes continentales de la península Ibérica: el margen del Ebro y el Golfo de Cádiz. *Rev. de la Soc. Geológica Esp*, 1(3-4), 318-325.

[41] Baringer, M. O., & Price, J. F. (1999). A review of the physical oceanography of the Mediterranean out-flow. *Marine Geology*, 55-63.

[42] Maldonado, A., & Stanley, D. (1981). Clay mineral distribution patterns as influenced by depositional processes in the southeastern Levantine Sea. *Sedimentology*, 21-32.

[43] Gracia, F. J., Rodríguez Vidal, J., Benavente, J., Cáceres, L., & López Aguayo, F. (1999). Tectónica cuaternaria en la bahía de Cádiz. *Avances en el estudio del cuaternario español*, Girona, 67-74.

[44] Achab, M., & Gutiérrez Mas, J. M. (1999a). Characteristics and controlling factor of the recent sedimentary infill on the bottom of the Bay of Cadiz. *Geogaceta*, 27, 3-6.

[45] Gutiérrez Más, J. M., Achab, M., & Gracia, F. J. (2004). Structural and physiographic control on the Holocene marine sedimentation in the bay of Cadiz (SW-Spain). *Geodinamica Acta*, 153-161.

[46] Maldonado, A. y, & Nelson, C. H. (1999). Interaction of tectonic and depositional processes that control the evolution of the Iberian Gulf of Cádiz margin. *Marine Geology*, 217-242.

[47] Lobo, F. J., Hernandez-Molina, F. J., Somoza, L., Rodero, J., Maldonado, A., & Barnolas, A. (2000). Patterns of bottom current flow deduced from dune asymetries over the Gulf of Cadiz Shelf (southwest Spain). *Marine Geology*, 164, 91-117.

[48] Mabesoone, J. M. (1966). Deposicional and provenance of the sediment in the Guadalete estuary (Spain). *Geologie en Minjbouw-Amesterdam*, 45, 25-32.

[49] Viguier, C. (1974). Le néogéne de l'andalousie Nord-occidentale (Espagne). Histoir geologique du bassin du bas Guadalquivir. *PhD thesis*, Bordeaux, 449.

[50] Zazo, C., Goy, J. L., & Dabrio, C. (1983). Medios marinos y marino-salobres en la bahía de Cádiz durante el pleistoceno. *Rev. Mediterránea. Ser. Geol*, 2, 29-52.

[51] Gutiérrez Mas, J. M., Martín Algarra, A., Domínguez Bella, S., & Moral Cardona, J. P. (1990). Introducción a la Geología de la provincia de Cádiz. Servicio de Publicaciones, Universidad de Cádiz , 315.

[52] Moral Cardona, J. P., Achab, M., Domínguez, S., Gutiérrez-Mas, J. M., Morata, D., & Parrado, J. M. (1996). Estudio comparativo de los minerales de la fracción pesada en los sedimentos de las terrazas del Río Guadalete y fondos de la Bahía de Cádiz. *Geogaceta*, 20(7), 14-17.

[53] Dabrio, C. J., Zazo, C., Goy, J. L., Sierro, F. J., Borja, F., Lario, J., Gonzalez, A., & Flores, J. A. (2000). Deposicional history of estuarine infill during the last postglacial transgression (Gulf of Cadiz, Southern Spain). *Marine Geology*, 162, 381-404.

[54] Achab, M., & Gutiérrez Más, J. M. (2005a). Nature and distribution of the sand fraction components in the Cadiz bay bottoms (SW Spain). *Revista de la Sociedad Geologica de España*, 18(3-4), 133-143.

[55] Greenberg, A. E., Clescer, L. S., & Eaton, A. D. (1992). Standard Methodes for the Examination of Water and Wastewater. 18th Edición, *American Public Heath Association*.

[56] Holtzapffel, T. (1985). Les minéraux argileux. Préparation. Analyse diffractométrique et détermination. *Bull. Soc. Géol. Nord*, 12, 136.

[57] Tucker, M. (1988). Techniques in Sedimentology. Blackwell, Oxford, UK, 394.

[58] Ojeda, J., Sanchez, E., Fernandez-Palacios, A., & Moreira, J. M. (1995). Study of the dynamics of estuarine and coastal waters using remote sensing: the Tinto-Odiel estuary, SW Spain. *J.Coastal Conserv*, 1, 109-118.

[59] Folk, R. L., & Ward, W. C. (1957). Brazos River bar: a study in the significance of grain-size parameters. *Journal of Sedimentary Petrology*, 27, 3-26.

[60] Martins, L. R. (2003). Recent sediments and grains-size analysis. *GRAVEL*, 1, 90-105.

[61] Pevear, D. R., & Mumpton, D. R. (1989). Quantitative Mineral Analysis of clays. Colorado. *CMS Workshop Lectures1. The Clay Mineral Society.*

[62] Ortega-Huertas, M., Palomo, I., Moresi, M., & Oddone, M. (1991). A mineralogical and geochemical approach to establishing a sedimentary model in a passive continental margin (Subbetic zone, Betic Cordilleras, SE Spain). *Clay Minerals*, 389-407.

[63] Van Rijn, L. C. (1984). Sediment transport part I: bed load transport. *Journal. Hydraul. Eng.*, 110(10), 1431-1456.

[64] Gao, S., & Collins, M. B. (1994). Análysis of grain size trends, for defining sediment transport pathways in marine environments. *Jour. Coast. Res*, 10(1), 70-78.

[65] Meade, R. H. (1972). Transport and deposition of sediments in estuaries. *Geological Society of America*, 133, 91-120.

[66] Poulos, S. E., Collins, M. B., & Shaw, H. F. (1996). Deltaic sedimentation, including clay mineral deposition patterns marine embayment of Greece (SE. Alpine Europe). *Journal of Coastal Research*, 12(4), 940-952.

[67] Gutiérrez-Mas, J. M., Hernández-Molina, J., & Lopez-Aguayo, F. (1996a). Holocene sedimentary dynamics on the Iberian continental shelf of the Gulf of Cadiz (SWSpain). *Continental Shelf Research*, 16(13), 1635-1653.

[68] Cacchione, D. A., & Drake, D. E. (1990). Shelf sediment transport: An overview with applications to the northem Califomia continental shelves. *In: Mehaute, B. L. & Hanes, D. M. (eds.) The Sea 7b. John Wiley and Sons*, New York, 729-773.

[69] Achab, M., Gutiérrez Más, J. M., Moral Cardona, J. P., Parrado Román, J. M., Gonzalez Caballero, J. L., & Lopez Aguayo, F. (1999b). Relict and modern facies differentiation in the Cadiz bay sea bottom recent sediments. *Geogaceta*, 27, 187-190.

[70] Mc Cave, I. N. (1975). Vertical flux of particles in the ocean. *Deep-Sea Research,,* 22, 491-502.

[71] Drake, D. E. (1976). Suspended sediment transport and mud deposition in continental shelves. *In: Marine sediment transport and environmental managment. Edited by D.J. Stantey and D.J.P. Swift*, Wiley, N.Y., 127-158.

[72] Cavaleri, L., & Malanotte-Rizzoli, P. (1981). Wind wave prediction in shallow water: theory and application. *J Geophys Res*, 86, 10961-10973.

[73] Holthuijsen, L. H. (2007). *Waves in Oceanic and Coastal Waters*, Cambridge Univesity Press, Cambridge.

[74] Karsten, A., Lettmann, Jörg, Olaf, Wolff, & Badewie, Thomas H. (2009). Modeling the impact of wind and waves on suspended particulate matter fluxes in the East Frisian Wadden Sea (southern North Sea). *Ocean Dynamics*, DOI 10.1007/s10236-009-0194-5.

[75] Griffin, G. M. (1962). Regional clay- mineral facies products of weathering intensity and current distribution in northeastern Gulf of Mexico. *Geol. Soc. Amer. Bull*, 73.

[76] Biscaye, P. E. (1965). Mineralogy and sedimentation of recent deep-sea in the Atlantic Ocean and adjacent seas and Oceans. *Geol. Soc. Amer Bull.*, 76, 803-831.

[77] Chamley, H. (1989). *Clay sedimentology,* Springer-Verlag, Berlin, 623.

[78] Sawheney, B. L., & Frink, C. R. (1978). Clay minerals as indicators of sediment source in tidal estuaries of long island sound. *Clays and Clay minerals,* 26(3), 227-230.

[79] Naidu, A. S., Han, M. W., Mowatt, T. C., & Wajda, D. (1995). Clay minerals as indicators of sources of terrigenous sediments their transportation and deposition: Bering Basin Russian-Alaskan Arctic. *Marine Geology,* 127(1-4), 87-104.

[80] Bhukhari, S. S., & Nayak, G. N. (1996). Clay minerals in identification of provenance of sediments of Mandovi Estuary, Goa, West Coast of India, Indian. *Journal of Marine. Science,* 25(4), 341-345.

[81] Lo, C. P. (1986). *Applied Remote Sensing,* Longman, London, 393.

[82] Kleman, Y., & Hardisky, M. A. (1987). Remote sensing of estuaries: An overview. Symp. *Remote sensing of envirinment,* Ann Arbour, Michigan, 183-204.

[83] Fiedler, P. C., & Laurs, R. M. (1990). Variability of the Columbia River plume observed in visible and infrared satellite imagery. *Inter. Jour. Remote Sensing,* 11(6), 999, 1010.

[84] Spitzer, D., & Driks, R. W. J. (1987). Bottom influence on the reflectance of the sea. *Int. Journal Remote Sensing,* 8, 779-290.

[85] Baban, S. M. J. (1995). The use of LandSat imagenery to map fluvial sediment discharge into coastal waters. *Marine Geology,* 1230, 263-270.

[86] Bernal, Ristori. E. (1986). Aplicaciones de la teledetección espacial al medio marino litoral. 1ª Jornadas de Ingeniería Geográfica. Madrid. *Bol. Inf. Serv. Geográf. Ejerc.,* 62, 37-56.

Chapter 7

Scour Caused by Wall Jets

Ram Balachandar and H. Prashanth Reddy

Additional information is available at the end of the chapter

1. Introduction

Scour that results directly from the impact of the hydraulic structure on the flow and which occurs in the immediate vicinity of the structure is commonly called local scour. It is very important to reduce local scour caused by impacting jets. A thorough understanding of the erosion of the bed due to local scour remains a challenge since it is associated with a highly turbulent flow field. The size, shape and density of sand particles, flow velocity, flow depth, turbulent intensity and shape of scour hole influence the complex processes of entrainment, suspension, transportation and deposition of sediment. This chapter briefly discusses transport and local scour mechanism of cohesionless sand caused by plane, two-dimensional and three-dimensional wall jets. The aim of the chapter is to facilitate the reader to understand scour characteristics, scour profile measurements, effects of tailwater depth, sand grain size, width of the channel, laboratory test startup conditions and scour that occurs under ice cover conditions.

2. Scour hole characteristics

The scour hole formation by various forms of jets (e.g., circular, two-dimensional (2D) and three-dimensional (3D)) can destabilize a hydraulic structure in the immediate vicinity of the jet. A typical visual example of the scour process caused by a 2D jet is presented herein to illustrate the complexity of the flow. As the flow is commenced, the jet exits the nozzle or the sluice gate, interacts with the bed and scouring action takes place. This is usually called the digging phase. During digging, the jet is directed towards the bed, a hole is formed and the excavated sand is deposited as a mound just downstream of the scour hole. The scouring process is very rapid in this phase. The location of the impingement point changes in the longitudinal direction during the digging process. Following the digging period, the jet flips towards the free surface and a refilling process is commenced. In this phase, the water surface tends to be wavy. The surface jet impinges on the mound region and some of the sediment deposited on the mound falls back and refills the scour hole. In time, the jet suddenly flips back towards the bed and once again causes rapid digging of the bed with lifting of the bed material into suspension. The free surface is not wavy during the rapid digging process. An intermediate hump is formed in the scour hole. In time, the jet is once again directed towards the free surface and refilling occurs. The alternate digging and refilling occurs only at low tailwater depths. The asymptotic condition (i.e., when changes to the bed profile is minimal) is attained after a very long time period. The above process is illustrated in Fig. 1 (Balachandar et al. 2000). Fig. 1(a) shows the end of a refilling cycle prior to the commencement of the digging process. Fig. 1(b) shows the beginning of digging phase, with a recirculating flow region rotating in a clockwise direction being formed upstream of impingement and a counter clockwise roller formed in the downstream section. Fig. 1(c) shows sand is primarily transported by advection of suspended particles. Fig. 1(d) shows a relatively calm water surface profile during the digging phase. The redirection of the jet towards the water surface (surface jet) after the end of digging phase causes the water surface to become wavy (Figs. 1e, 1f and 1g). Refilling takes place in Figs. 1(f) and 1(g). Fig. 1(h) shows the reoccurrence of digging phase. Studying the effects of tailwater depth, flow properties and sediment properties on the scour hole dimensions will help in better predicting the scour hole geometry.

The definition sketch of the scour hole geometry is shown in Fig. 2 (Sui et al. 2008). Several researchers including Dey and Sarkar (2006a, b), Pagliara et al. (2006) and Balachandar et al. (2000) studied the role of the following geometrical and hydraulic parameters governing the scour process:

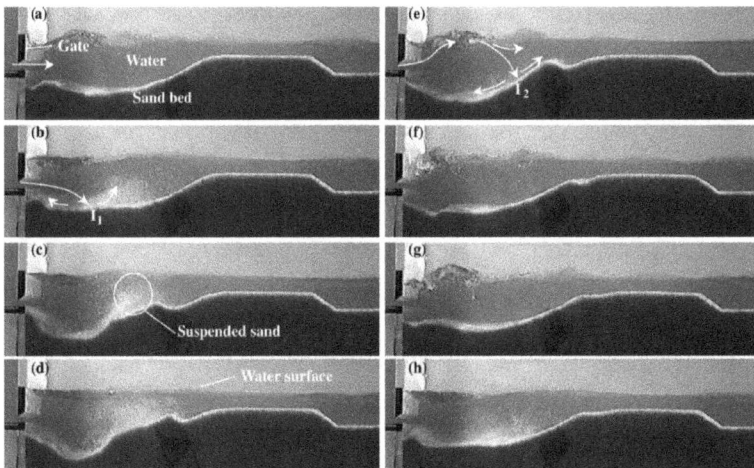

Figure 1. Demonstration of local scour cycle: (a) scour begins; (b) digging phase commences; (c) digging continues; (d) maximum digging; (e) filling phase begins; (g) maximum fill; (h) reoccurrence of digging phase. (Balachandar et al. 2000, copyright permission, NRC Research Press)

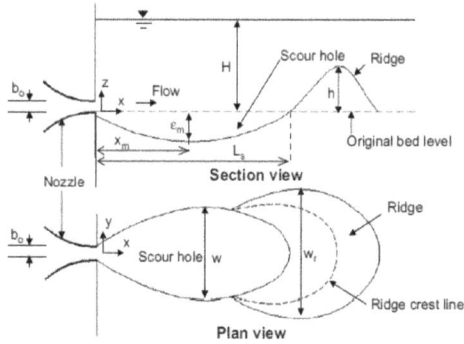

Figure 2. Definition of scour parameters (Sui et al. 2008, copyright permission of Science Direct)

i.   Nozzle hydraulic radius (defined as $b_0/4$)

ii.   Jet exit velocity (U_0) - the undiminished mean velocity in potential core

iii.   Grain size - generally, the median grain size (d_{50}) of the cohesionless bed is considered as the representative grain size

iv.   Densimetric Froude number (F_0) - $F_0 = U_0/\sqrt{g d_{50} \Delta \rho / \rho}$, $\Delta \rho$ is submerged density of bed material, and ρ is density of bed material.

v.   Tailwater depth (H or y_t) - defined as depth of water over the original bed

vi.   Channel width (W) - the width of downstream channel into which the jet is exiting

vii.   Expansion ratio (ER) - channel width to jet thickness,

viii.   Time evaluated from the start of the flow (t)

ix.   Submergence ratio - tailwater depth to the thickness of the jet at its origin.

The variables of interest include:

i.   Maximum scour depth (ε_m)

ii.   Location of maximum scour depth (x_m or δ) from the nozzle exit

iii.   Volume of scour (V) - measured by determining volume of the water needed to fill the scour hole.

iv.   Scour hole length (L_s) - maximum extension of the scour hole along the midsection.

v.   Scour hole width (w)- maximum extension of the scour hole measured perpendicular to the flow direction.

vi.   Ratio of the length of the scour hole at asymptotic conditions to jet exit velocity
        (L_s / U_o)

Rajaratnam and Berry (1977), Lim (1995), Aderibigbe and Rajaratnam (1996), Ade and Rajaratnam (1998) and Sui et al. (2008) studied wall jet scour and concluded that the key parameter that effects the scour hole dimensions is F_o. Sediment uniformity also influences the scour geometry. The formation of the armor layer in non-uniform beds leads to smaller scour geometry when compared to that formed in uniformly distributed sediment (Aderibigbe and Rajaratnam 1996, Mih and Kabir 1983). In general, it has been assumed that the effect of the sediment size on scour hole dimensions can be absorbed by the densimetric Froude number (Rajaratnam (1981), Mazurek and Rajaratnam (2002) and Rajaratnam and Mazurek (2003)). The influences of the various factors are described in the forthcoming sections. Sui et al. (2008) investigated the effect of the sediment size on scour hole dimensions in detail.

The effect of tailwater depth on the scour hole dimensions due to impinging and free falling jets have been investigated by Ghodsian et al. (2006) and Aderibigbe and Rajaratnam (1996). Critical values of the tailwater depth were introduced by Aderibigbe and Rajaratnam (1996) and Ghodsian et al. (2006) such that beyond the critical value (either by increasing or decreasing the tailwater depth), the scour hole dimensions decreased. However, Ali and Lim (1986), Faruque (2004), Sarathi et al. (2008) and Mehraein et al. (2010) have stated that scour hole dimensions due to 2D and 3D wall jets increase either by increasing or decreasing the tailwater depth from the critical value.

3. Scour due to wall jets

3.1. Characteristics of two–dimensional wall jets

Mohamed and McCorquodale (1992) investigated the local scour downstream of a rectangular opening with a swept-out hydraulic jump and identified two stages of local scour development.

i.   An initial stage of local scour which occurs rapidly (short-term scour).

ii.   A progressive stage which approaches equilibrium after a very long time (long-term scour).

Mohamed and McCorquodale (1992) observed that the equilibrium depth for short-term scour established rapidly in less than 1% of the time to reach the long-term scour depth. The short-term scour although not as deep as the long-term scour occurs much closer to the apron; the bed is more highly fluidized than in the regime that governs the long-term scour. They have also stated that short-term scour due to plane horizontal supercritical jets under low tailwater conditions is related to the energy dissipation regime that dominates the flow. They identified seven different jet forms:

1.   Attached jet: Bottom boundary of jet conforms to the bed and top boundary of jet forms the free surface from beginning of jet exit.

2. Moving jump: Hydraulic jump formed by the jet is propagating downstream.

3. Wave jump: Standing wave type of hydraulic jump is formed downstream of the jet exit.

4. Surface jet: Top boundary of jet forms the free surface and bottom boundary of jet is not confined.

5. Plunging jump: Jet is plunging down along the slope of the scour hole. Specifically, the bottom boundary is along the slope of scour hole and top boundary of jet is submerged.

6. Inverted jump: Top and bottom boundaries of jet try to attain the free surface and bed, respectively, beyond scour hole.

7. Classical jump (as in stilling basins)

The more rapid short-term scour was associated with regimes 1, 2, 3 and 6, while the deeper long-term scour was associated with regimes 4 and 5 indicated above.

Rajaratnam (1981) found that maximum depth and length of scour hole increased linearly with the logarithm of time. He noted that the scour hole reached an asymptotic stage. Rajaratnam (1981) concluded that maximum depth and length of scour are largely dependent on the densimetric Froude number. Johnston (1990) investigated three different scour hole regimes created in shallow tailwater conditions. Two scour hole regimes were formed when the jet permanently attaches itself to either the bed or free surface boundary whilst the third was formed when the jet periodically flips between the free surface and channel bed. Johnston (1990) found that the scour hole development in deep water conditions is orderly and invariably reaches a well-defined asymptotic state while in shallow conditions such a state is sometimes not reached. The depth of flow has a considerable influence on the near bed flow field and may promote the flipping of the jet from one boundary to another. Balachandar et al. (2000) suggested that tailwater depth was a key parameter when tailwater depth was less than 16 times the nozzle thickness or sluice gate opening (b_o). Chatterjee et al. (1994) found that maximum scour depth is a function of Froude number based on nozzle thickness ($F = U_o / \sqrt{g b_o}$). Kells et al. (2001) concluded that scour geometry is dependent on sediment bed grain size and initial jet exit velocity (U_o). Ali and Lim (1986) have developed the following power law relationship for time evolution of scour:

$$\frac{V}{R^3} = 187.72 \left(\frac{\varepsilon_m}{R} \right)^{2.28}$$

where, V is the volume of scour and R is hydraulic radius and ε_m is maximum scour depth.

3.2. Scour characteristics of three–dimensional wall jets

Scour caused by 3D and circular wall jets have been studied by Meulen and Vinje (1975), but to a lesser degree than that caused by plane and impinging jets. Rajaratnam and Berry (1977) studied the scour caused by circular wall jets on cohesionless soils. They concluded that the geometric characteristics of the scour hole mainly depend on the densimetric Froude number. Rajaratnam and Berry (1977) also found that the jet expands in an unconfined manner from the nozzle exit to the point of maximum scour depth. Ali and Lim (1986) concluded that for 3D jets, mean flow velocity for any section decreases continuously in the flow direction, whereas for two-dimensional flows, the mean velocity increases as the flow develops. In three-dimensional jet scour, no reverse flow was observed near the bed. It was observed in 3D flows that there were occasional turbulent bursts near the bed which moved sand particles from the upstream slope of the hole to the region of maximum scour. Ali and Lim (1986) have suggested the following power law relationship for time evolution of volume of scour:

$$\frac{V}{R^3} = 49.36 \left(\frac{\varepsilon_m}{R} \right)^{1.89}$$

Hoffmans and Pilarczyk (1995) developed a semi-empirical relation for the upstream scour slope and verified the equation using experimental results. Lim (1995) studied the effect of channel width (expansion ratio) on scour development of 3D jets and concluded that there was no effect of channel width on scour development for expansion ratios greater than 10. The width of scour hole was affected by the width of downstream channel only when normal diffusion of the 3D jet flow is restricted in transverse direction. Rajaratnam and Diebel (1981) found that relative tailwater depth and relative width of downstream channel affect the location of maximum scour depth. Chiew and Lim (1996) developed the following empirical equations by using the densimetric Froude number (F_o) as the main characteristic parameter to estimate scour dimensions caused by circular wall jet:

$$\frac{\varepsilon_m}{b_o} = 0.21 F_o \quad \frac{w}{b_o} = 1.90 F_o^{0.75} \quad \frac{L}{b_o} = 4.41 F_o^{0.75}$$

Ade and Rajaratnam (1998) further emphasized the use of F_o as the main parameter to analyze scour caused by circular wall jets. However, they noted that asymptotic dimensions of scour hole are dependent on tailwater depth for $F_o > 10$. Ade and Rajaratnam (1998) also found that the eroded bed profile near the nozzle attains asymptotic state earlier than locations away from nozzle.

4. Role of fluid structures on two–dimensional scour

Hogg et al. (1997) have pointed out that a comprehensive understanding of the scour remains elusive because of the complex nature of the flow field. Rajaratnam (1981), Rajaratnam and Macdougall (1983), Wu and Rajaratnam (1995) and Rajaratnam et al. (1995) have made significant contributions despite the complexity resulting from the hydrodynamic character-istics of the jet and the concave shape of eroded bed. Hopfinger et al. (2004) proposed new scaling laws relating time and the attainment of the quasi-steady scour depth. They suggested that turbulence created by Gortler vortices cause sediment transport and the associated scouring mechanism due to the destabilization of the turbulent wall layer by the concave curvature of the water sediment interface. Bey et al. (2007) studied the velocity field during asymptotic conditions in the scoured region to understand the role of turbulent flow structures that influence scour and evaluated higher-order velocity moments. Experimental findings of Bey et al. (2007) are presented here.

Bey et al. (2007) carried out two tests at exit velocities of 1.0 m/s and 1.27 m/s (Tests A and B) and at a tailwater depth (H) corresponding to 20 b_o continuously for 12 days duration. These tests can be classified under the high submergence flow regime. The high and low submergence is not clearly defined in literature, however high submergence can be defined as a state where no alternate flicking of the jet between the free surface and bed occurs as in low submergence cases (Balachandar et al. 2000). Generally, a value of $H/b_o > 10$ has been considered as high submergence. The time to reach asymptotic conditions was found to be as low as $t = 24$h depending on scour hole and ridge geometric parameter chosen. For example, the total length of the scour affected region (L_T) attains an asymptotic state at $t = 24$h, however the change in the maximum depth of the scour hole (ε_m) is less than 5% after $t = 48$h. However, it should be noted that beyond 72h, there is the presence of the turbulent bursts which cause local changes in the location of the maximum depth of scour profile (x_m) but no significant changes in the mean scour profile. Sectional and plan view of the scour region along with the definition of various variables are shown in Fig. 3(a).

Bey et al. (2007) divided the entire test duration from the start of flow to the attainment of asymptotic conditions into five time zones to study the influence of the different flow structures. A time scale $T = L_s/U_o$ is used to non-dimensionalize the five time zones, where L_s is length of scour hole at asymptotic conditions. Fig 3(b) shows variation of important scour parameters with time in five time zones. Each of the time zones had certain dominant flow features. Flow was characterized by the presence of longitudinal vortices and turbulent bursts at the start of the test and during early time periods, and movement of the jet impingement point during the later stages. Scour was very rigorous during first time zone ($t/T \leq 850$). A 2D hole was formed downstream of the nozzle and the bigger size particles were deposited further downstream as a 2D mound. In second time zone ($850 \leq t/T \leq 15 \times 10^3$), the large-scale suspension of the bed material seen in the earlier time zone was reduced significantly. No major scour happened in the negative slope region (Fig. 3a). Two or three longitudinal streaks and very prominent concave depressions were observed. Hopfinger et al. (2004) reported that vortices due to Gortler instability caused scour on the positive bed slope. During the third time zone ($15 \times 10^3 \leq t/T \leq 125 \times 10^3$), there appeared a "scoop-and-throw" like scouring action on either side of the flume axis, which caused longitudinal concave shaped depressions. A lifting spiraling motion of bed particles occurred near the end of concave shaped depression due to vortex activity. Dye injection confirmed the presence of spiral motions during the fourth time zone ($125 \times 10^3 \leq t/T \leq 375 \times 10^3$) and also scoop-and-throw like scouring action slowed down after 24 h. The scour hole attained an asymptotic state in the fifth time zone after 72 hours of scour process ($t/T \geq 375 \times 10^3$). In the asymptotic stage, turbulent bursts were noticed to occur in the near bed region across the section of the flume. In addition, two prominent scour mechanisms occurred on either side of the flume axis causing the bed particles to be spiraled toward the sidewalls from midsection and from sidewalls to the flume axis. However, no particle movement was observed along the nozzle axis.

At asymptotic conditions, scour profile in the scour hole region was nominally two-dimensional across the width of the flume. It should be remarked that the backward and forward movement of the jet contact point with bed, the frequent but random turbulent bursts, and the two prominent scour mechanisms (digging and refilling), all occurred at one time or the other. However, in the mound region, the lateral profile was not two-dimensional and had two distinct peaks. These peaks occurred closer to the sidewalls leaving a trough in center portion of the mound. Jet impingement point on the bed moved backward and forward, and the mean location of point of contact of jet with bed was close to the deepest point of scour hole. Wide range of variation in instantaneous velocity from positive to negative values was a clear indication of back and forth movement of jet. It was observed that most of the turbulence activity occurred in the region of near-zero slope of scour hole. Analysis of third-order moments and quadrant decomposition indicated sweep type events in the near bed region contributed to scour. Ejection events near the bed caused suspension of the bed particles to be carried away by the average flow velocity. Measured velocities at asymptotic conditions were extrapolated to other time periods to conclude that the sweep and ejections contributed significantly to the scour process.

Figs. 4(a) and (b) show the flow fields for the tests A and B in the scour hole region at asymptotic conditions. Velocity vectors show that the jet expands vertically to interact with the bed. A large-scale recirculating region was found in the region above the jet axis, which extended to about $35b_o$ downstream of the jet. The center of this region was located at about $16b_o$ along the x-axis and about $10b_o$ along the y-axis. It is important to recognize that this recirculation region caused significant negative vertical velocities immediate vicinity of nozzle. Impact of the jet on the bed also generated a flow separation near the region close to the nozzle. A similar flow field was also observed in test B, which is qualitatively similar to that obtained in test A, and is shown in Fig. 4(b). The mean impingement point in test B occurs farther from the nozzle and consequently has a larger near-bed recirculation zone.

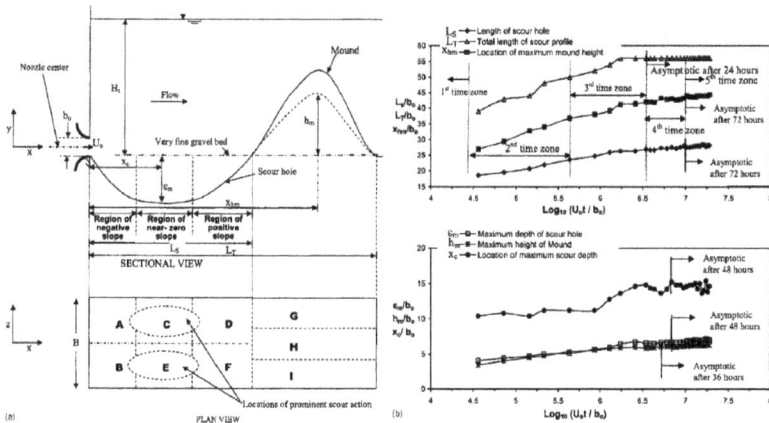

Figure 3. a) Definition sketch, (b) Variation of scour parameters with time (Bey et al. 2007, copyright permission of ASCE)

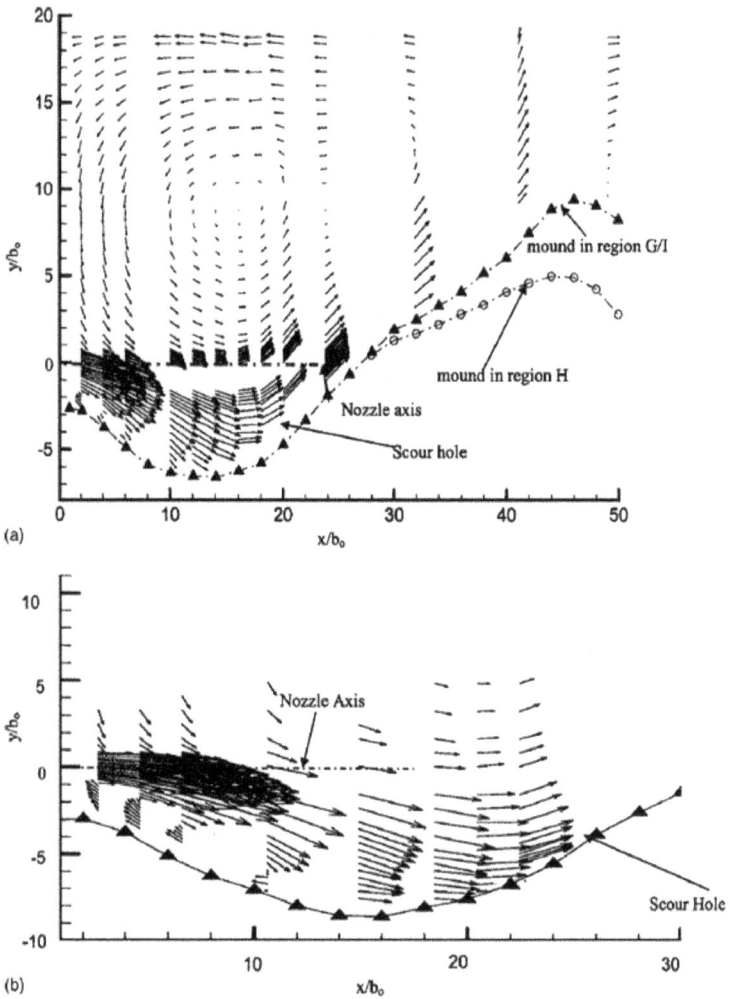

Figure 4. Velocity vector plot for (a) Test A; (b) Test B (Bey et al. 2007, copyright permission of ASCE)

5. Measurements using video image analysis

Figures 5(a) and 5(b) show a set of water surface profiles at about 840 seconds from the start of scour process. The profiles were obtained using a software called PROSCAN developed by Balachandar and Kells (1998). The time t_0 refers to the start of the cycle. Figure 5(a) shows the water surface profile during the digging phase and Fig 5(b) shows the water surface profiles during the refilling phase. Figure 5(c) shows bed scour profiles corresponding to the water surface profiles shown in Figs. 5(a) and 5(b). The bed profiles reflect the alternate digging and refilling phases as explained ealier. At $t = t_o$, the bed surface profile is similar to that provided by Rajaratnam (1981) where the water surface profiles are similar during digging and refilling phases. In the study of Balachandar and Kells (1998), the bed profile at $t = t_o+2$ s is quite different from initial bed profile at $t = t_o$. It is observed that at $t = t_o +2$ s, the dynamic flow field indicates the digging phase near the sluice gate and piling up of sand (hump) on downstream side. After $t = t_o + 24$ s, during the refilling phase, the sand particles in the hump part moved upstream and eventually refilled the trough. The refilling phase occurred over a longer period of time and the refilling process was nearing completion at $t = t_o+179$ s and profile eventually became similar to the profile shape recorded at $t = t_o$.

5.1. Computational stereoscopy

Ankamuthu et al. (1999) developed a stereoscopy scheme to measure the depth of scour in three-dimensional flow fields. The stereoscopy scheme makes use of an epipolar constraint and a relaxation technique to match corresponding points in two images. A correlation technique was developed to eliminate false matches. The depth of scour was calculated using the parallax between the matched points. However, the method is yet to be used to measure scour in a practical flow field.

6. The effect of tailwater depth on local channel scour

Ali and Lim (1986) studied scour caused by 3D wall jets at shallow tailwater conditions and noted that tailwater had an influence on the maximum depth of scour at asymptotic conditions. Tailwater depth which is under or above a critical condition causes an increase in maximum depth of scour (Ali and Lim, 1986). Further analysis of their data indicates that the critical value of tailwater depth increases with increasing densimetric Froude number. Ade and Rajaratnam (1988) stressed the use of F_o as the characteristic parameter to describe scour caused by circular wall jets and noted that the maximum depth of scour was found to be larger at high ratio of tailwater depth to nozzle width and higher values of F_o, which is consistent with the measurements of Ali and Lim (1986). However, they noted that the asymptotic dimensions of the scour hole were dependent on the tailwater conditions for only $F_o >10$. Rajaratnam and Diebel (1981) concluded that the relative tailwater depth and relative width of the downstream channel do not affect the maximum depth of scour, whereas the location of the maximum scour was affected.

Faruque et al. (2006) presented the results of clear water local scour generated by 3D wall jets in a non-cohesive sand bed at low tailwater depths. They indicated that extent of scour of 3D wall jets is collectively influenced by the densimetric Froude number, tailwater depth, and grain size-to-nozzle width ratio. Each parameter has a dominant influence compared to other parameters at different flow conditions. For $F_o < 5$, tailwater depth has no effect on the

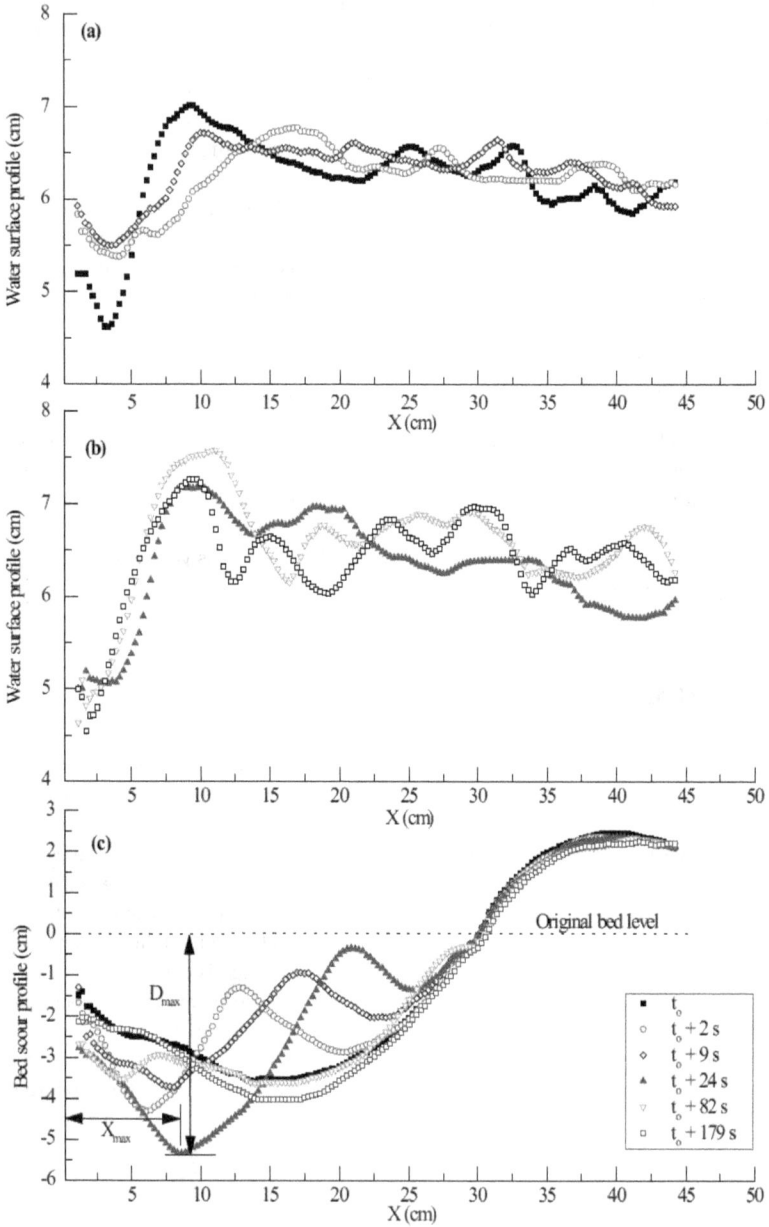

Figure 5. Variation of water surface profile and bed profile (after t = 823 s) (Balachandar and Kells, 1998, copyright permission of NRC Research Press)

maximum depth and width of scour. The effect of tailwater depth at higher densimetric Froude numbers appears to be important at larger values of non-dimensional sediment diameter (d_{50}/b_o). Previous observations have indicated that for $F_o > 10$, the effect of the tailwater depth was significant. However, the results of Faruque et al. (2006) clearly indicate that the effect of tailwater can be important at lower values of F_o depending on the value of d_{50}/b_o. At asymptotic conditions, comparing the volume of scour at $H/b_o = 6$ with that at $H/b_o = 4$, it was observed that the scour volume is greater at the higher tailwater depth. Faruque et al. (2006) also indicated that the maximum depth of scour is not necessarily deeper at higher values of tailwater depth in the lower range of submergences.

Experimental results of Faruque et al. (2006) are further discussed here. Three different tailwater depths (Tests A, B, and C) corresponding to $2b_o$, $4b_o$, and $6b_o$ where $b_o = 26.6$ mm and a jet exit velocity = 1.31 m/s (in the three tests) were considered in the experiments. The corresponding Reynolds number in the tests ($R_e = U_o b_o / v$) was 1.0×10^5 which shows preva-lence of fully turbulent conditions. The flow Froude number based on exit velocity and nozzle width was 1.5, grain size was 0.85 mm and corresponding densimetric Froude number was six. Velocity measurements were obtained using a single component fibre-optic laser Doppler anemometer and scour profiles at different time intervals were obtained using a point gauge with an electronic display unit. Fig. 6(a) illustrates the time development of scour profiles along the nozzle axis for tests A, B and C. The profiles gradually attain an asymptotic state around $t = 48$ h. Fig. 6(a) also illustrates the maximum depth of scour (ε_m) and the location of maximum depth of scour from the nozzle exit (x_m) increase with increasing time. It was also found that the difference between two consecutive profiles at a fixed distance from the nozzle increases with increasing x. It should be noted that Ade and Rajaratnam (1988) found that maximum depth of scour to be larger at deeper submergences. In Figs. 6(b-d), the extent of scour is consistently larger at $H/b_o = 6$ at all t, and therefore the volume of scour at asymptotic conditions was greater for the larger submergence. The corresponding top view of the perimeter of the scour hole is shown in Figs. 7(a-e) which indicates that the perimeter of the sour hole is consistently larger at $H/b_o = 6$ as compared to $H/b_o = 4$. However, the hole perimeter expands laterally as depicted in Fig. 7(e) at $H/b_o = 4$ as asymptotic conditions are reached.

7. Effect of grain size on local channel scour

Scour occurs when the high velocity jet produces bed shear stresses that exceed the critical shear stress to initiate motion of the bed material. Critical shear stress of the bed material is a function of grain size and therefore, scour is a function of grain size. This section summarizes the effect of grain size on local scour. Breusers and Raudkivi (1991) discussed achievement of equilibrium state of scour and the similarity of scour profiles for various sizes of bed material and jet velocity. Balachandar and Kells (1997) studied the scour profile variation with time using a video imaging process on uniformly graded sediments caused by flow past a sub-merged sluice gate. Kells et al. (2001) investigated the effect of varying the grain size and, to a lesser extent, the grain size distribution of the erodible bed material on the scour characteristics. Figs. 8 and 9 are drawn from their study.

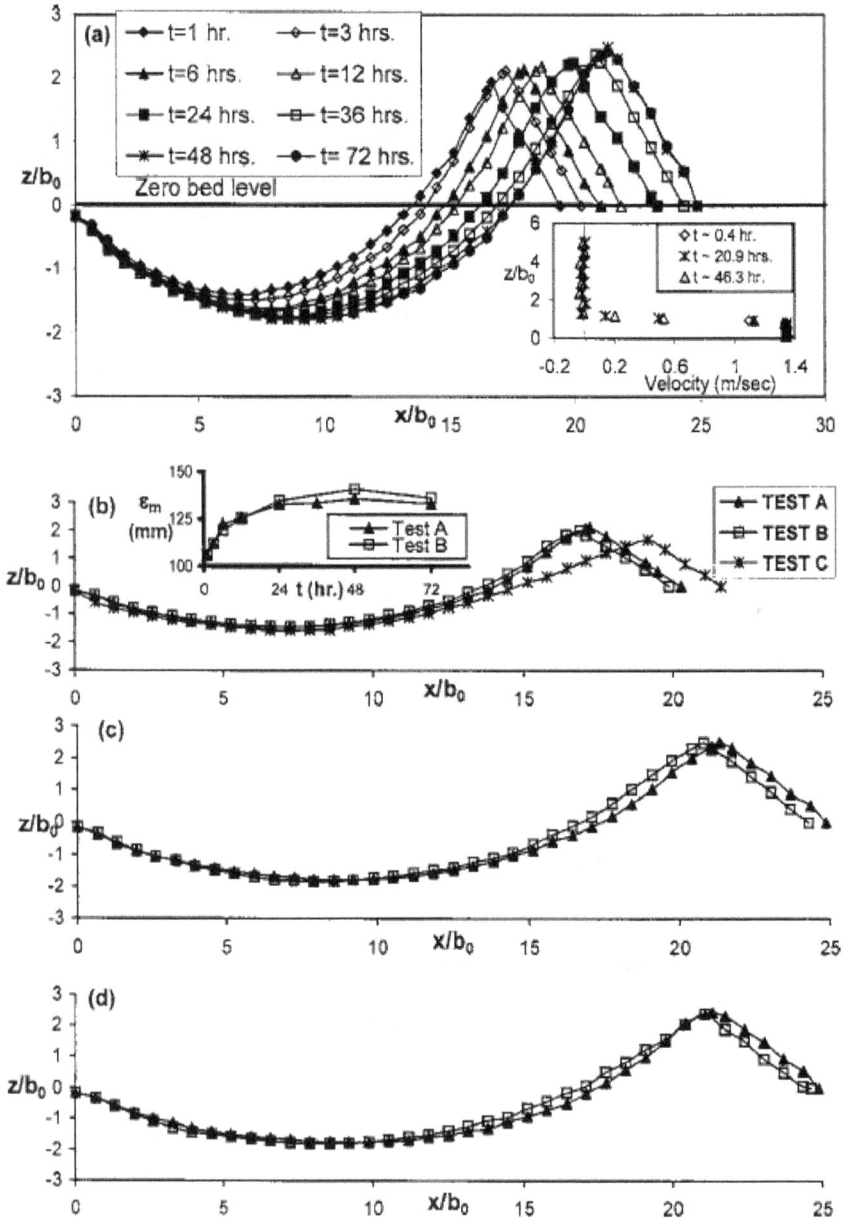

Figure 6. a) Development of scour profile along nozzle axis (b) comparison of scour profiles along nozzle centerline at $t = 3$ h; (c) $t = 48$ h; and (d) $t = 72$ h (Faruque et al. 2006, copyright permission of ASCE)

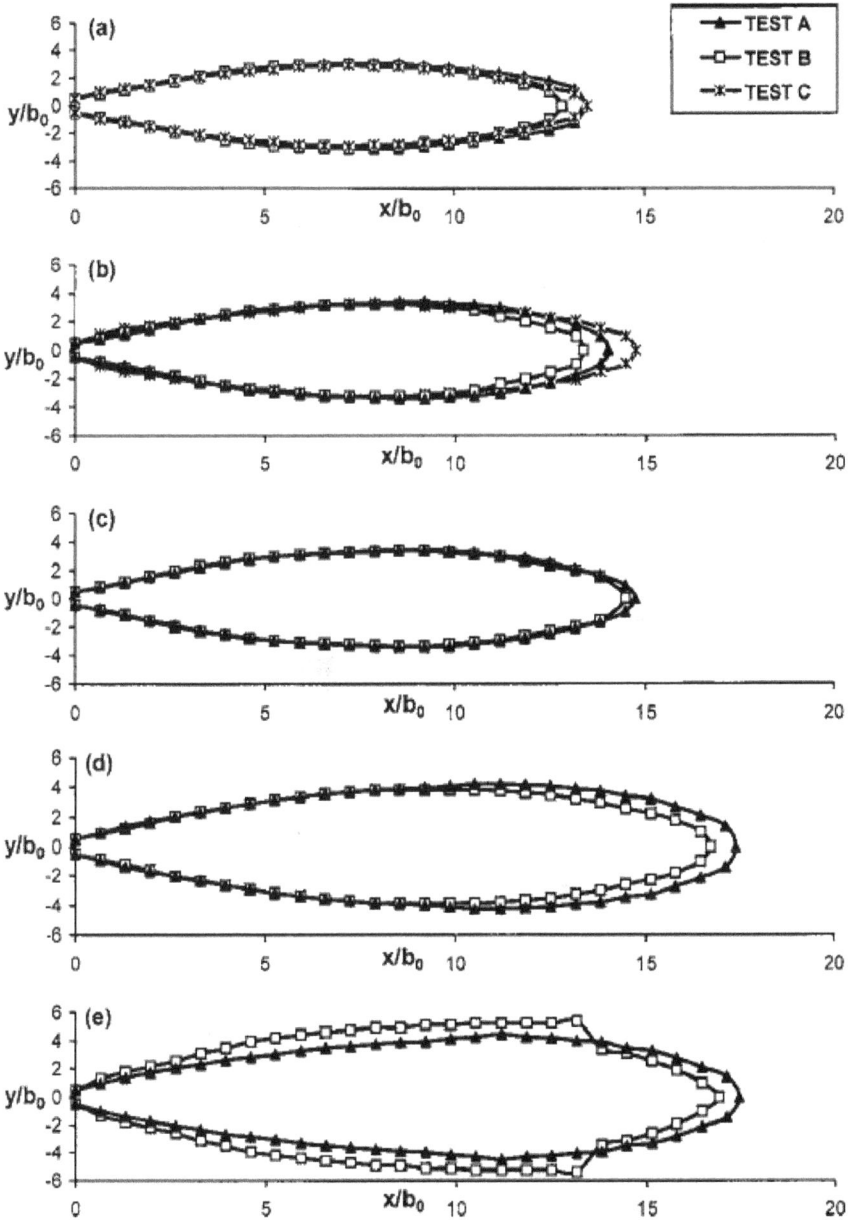

Figure 7. Comparison of plan view of perimeter of scour hole at various instances of time (a) $t = 1$ h; (b) $t = 3$ h; (c) $t = 6$ h; (d) $t = 48$ h; and (e) $t = 72$ h (Faruque et al. 2006, copyright permission of ASCE)

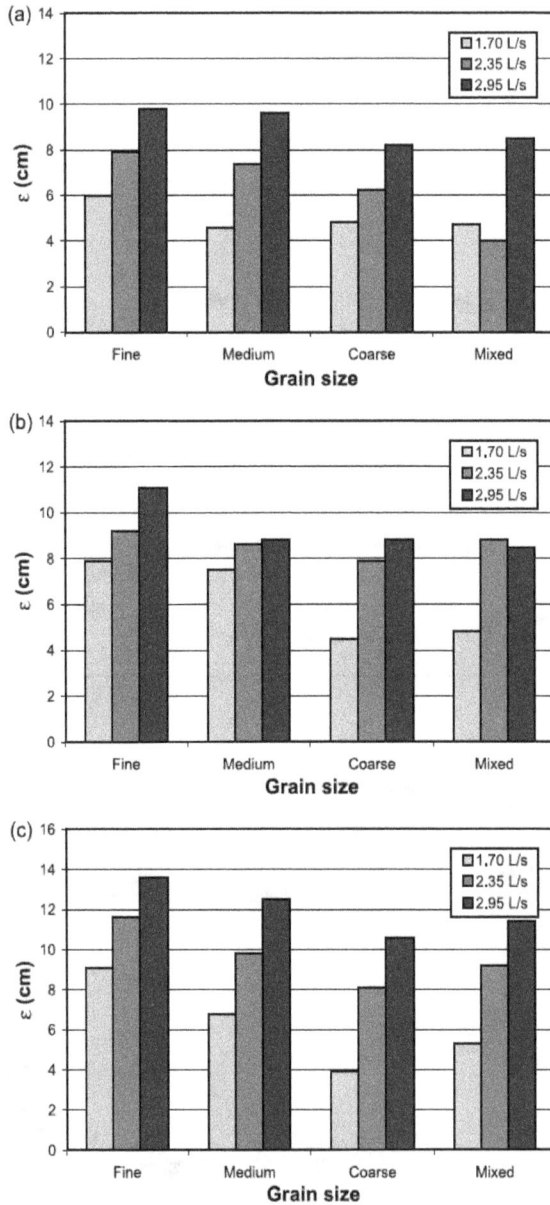

Figure 8. Effect of discharge and grain size on Phase-B maximum scour depth, ε_m, at $t = 24$h: (a) $H/b_o = 4.0$; (b) $H/b_o = 6.3$ and (c) $H/b_o = 8.8$ (Kells et al. 2001, copyright permission of NRC research press)

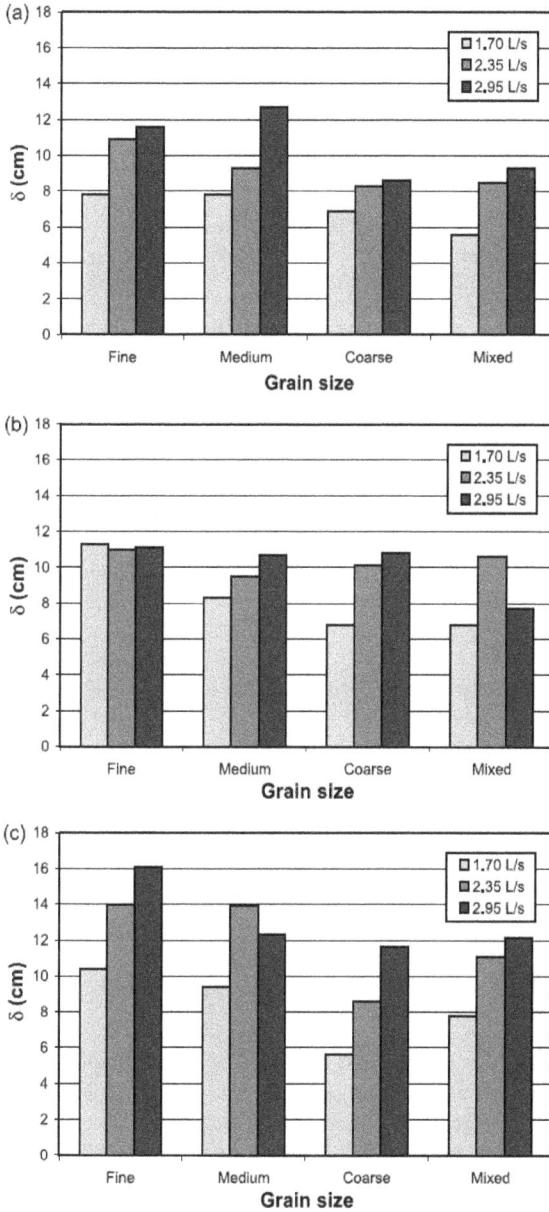

Figure 9. Effect of discharge and grain size on Phase-B location of point of maximum scour depth, x_m, at t= 24h: (a) H/b_o= 4.0; (b) H/b_o = 6.3 and (c) H/b_o= 8.8 (Kells et al. 2001, copyright permission of NRC research press)

The figures show the effect of grain size with increasing discharge and tailwater depth. Analysis of Figs. 8a, 8b, and 8c tell us that the magnitude of the maximum depth of scour (ε_m) increases with an increase in flow rate for a given grain size and maximum depth of scour decreases with increasing grain size for a given flow rate. It is also observed from Fig. 8 that maximum scour depth increases with increasing tailwater depth for a given grain size.

Figs. 9a, 9b, and 9c demonstrate that the location of maximum scour depth (δ) moves upstream with an increase in the grain size and moves downstream with increase in discharge. This is due to the fact that shear stress increases with increase in flowrate or critical shear stress decreases with decrease in grain size. The location of maximum scour depth (δ) moves downstream with increasing tailwater depth.

In summary, it was found that the grain size had a significant influence on the extent of scour, with more scour occurring with the smaller-sized material. Less scour occurred for graded sand (indicated as mixed sand in the figure) than a uniform one having a similar median grain size. Amount of scour increases with an increase in the discharge, hence the velocity of flow. For any given discharge and grain size, the greater the tailwater depth, the greater is the scour depth, extent, and volume of scour. It appears that the tailwater serves to slow the rate of jet expansion, thus increasing the length of bed exposed to high velocity, hence high shear stress conditions. It was also found that the tendency toward the dynamic alternating dig-fill cycling was lessened with a reduction in the discharge, or an increase in the grain size or the tailwater depth.

8. Effect of test startup conditions on local scour

A cursory evaluation of the test startup conditions in several of the studies mentioned in this chapter indicates that scour pattern is quite varied and depends on how the flow is initially commenced. For example, Kells et al. (2001) had the nozzle outlet plugged, whereas Balachandar et al. (2000) and Mohamed and McCorquodale (1992) had the sluice gate closed until proper head and tailwater conditions were established. Following this, the nozzle was unplugged or the sluice gate opened to a predetermined extent to generate the jet flow. This requires a certain amount of time before a steady jet discharge can be established. Rajaratnam (1981), Chatterjee et al. (1994), Mazurek (2002), and Aderibigbe and Rajaratnam (1998) created a suitable constant head difference between the downstream and upstream sections prior to generating the jet.

Johnston (1990) and Ali and Lim (1986) used a suitably sized aluminum sheet to cover the leveled bed in order to prevent it from being disrupted on commencement of the flow. The inflow was started; the flow and tailwater depths were then set to desired values, following which the sheet was slowly removed. It is thus clear that even for a seemingly simple flow emanating from nozzles or sluice gates, the flow and the corresponding scour pattern can become complex due to various influences and one among them is the startup condition.

Deshpande et al. (2007) investigated the effect of test startup conditions on plane turbulent wall jet behavior and the resulting scour profiles. The changing startup conditions also reflect practical situations where the discharge and tailwater conditions change during regular operations. Deshpande et al. (2007) study also investigated the jet behavior and the different regimes of flow for a range of submergences (both low and high) using a laser Doppler anemometer (LDA). Furthermore, the effects of three different startup conditions on scour include an instantaneous startup condition, a gradual startup condition, and a stepwise startup condition were studied. Results of Deshpande et al (2007) are discussed below.

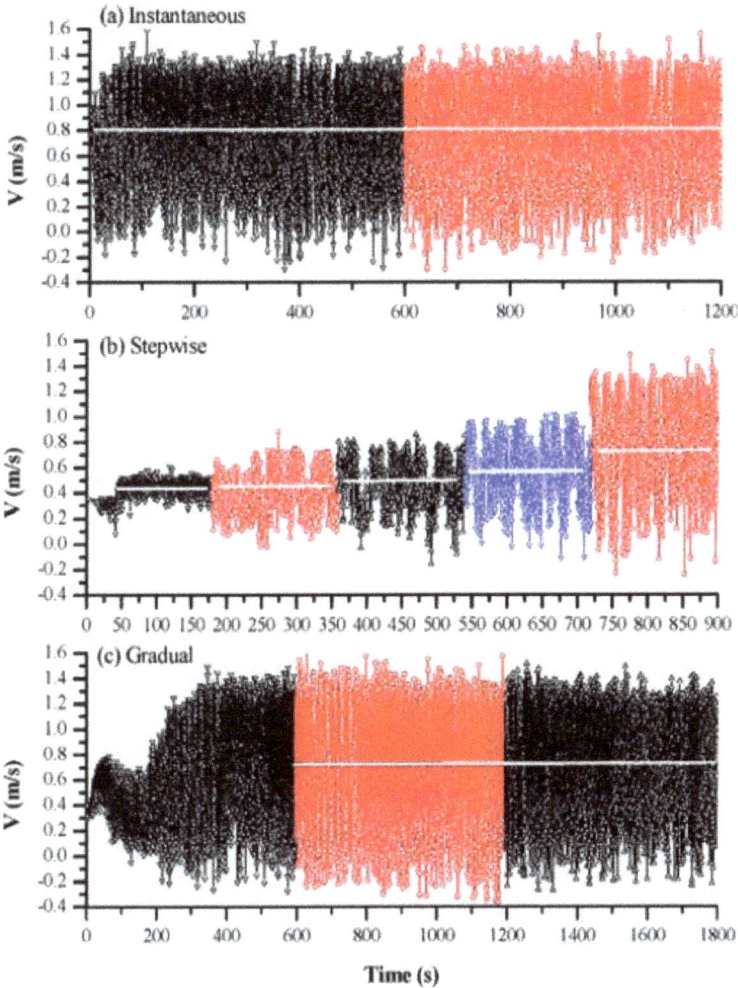

Figure 10. Effect of test startup conditions for $H/b_o = 4$ (Deshpande et al. 2007, copyright permission of Taylor and Francis)

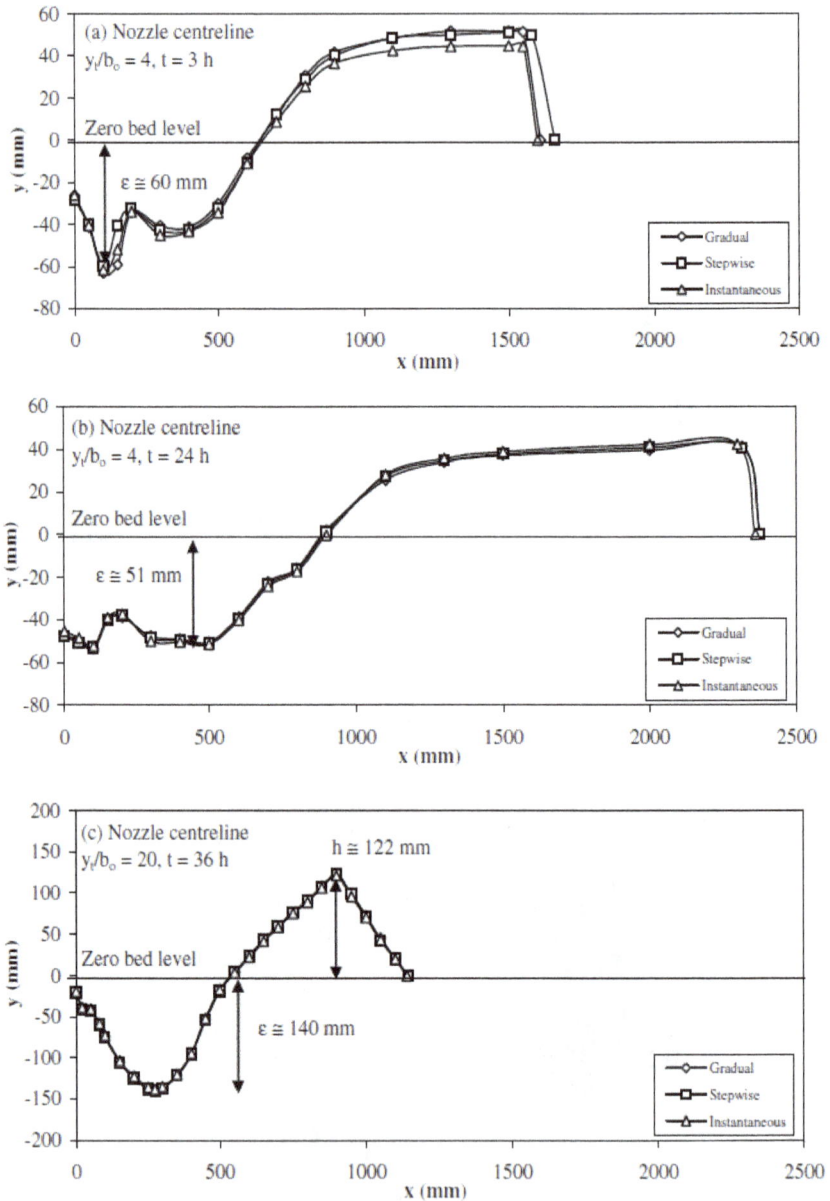

Figure 11. Effects of test startup conditions on the scour hole profiles (Deshpande et al. 2007, copyright permission of Taylor and Francis)

Fig. 10 shows the velocity-time history near the nozzle exit for three startup conditions at a low submergence ($H/b_o = 4$). Velocity data clearly indicate that there is no large scale scour upto about 150 s for stepwise and gradual change in flow. Velocity profile shows a significant dip in the mean velocity accompanied by large scale increase in turbulence after 150 s which indicates large scale scour. Further increase in valve opening increases velocity and large scale scour and causes higher turbulent intensities. For $t > 400s$, there is the presence of low frequency fluctuations. It was also found that a steady state velocity was attained after $t = 2000$ s in the case of stepwise startup condition and after $t = 600$ s in the case of gradual startup condition.

Fig. 11(a) shows the effect of the startup conditions on scour at $H/b_o = 4$ at 3h from the commencement of the flow. There are no significant differences in the profiles that can be attributed to startup conditions. However, the step wise and gradual startup conditions influence the mound to be slightly bigger and longer. It was observed that the digging phase continued for a longer time for these two startup conditions as compared to instantaneous condition. As a result, there has been more digging action and this is reflected in the profiles. This is an important aspect to note while comparing scour profiles with different startup conditions. Fig. 11(b) illustrates the profiles along the nozzle centerline for three startup conditions at 24h from the commencement of the test for $H/b_o = 4$. No significant differences can be found in the profiles and any minor differences that were noticed earlier have vanished. Figure 11(c) shows the scour profiles at the three startup conditions at $t = 36$ h at $H/b_o = 20$. Clearly the effects of test startup conditions are not evident at $t = 36$h. It was concluded that the effects of test startup conditions did not influence the long term scouring process.

9. Effect of channel width on local scour

Lim (1995) pointed out that the scour profile is not affected by the channel sidewalls when the expansion ratio is ten or greater. He also noted that the downstream channel would affect the lateral development of the scour hole if it becomes too narrow and restrict the normal diffusion of the three-dimensional jet. However, Faruque et al. (2006) have noted that scour hole dimensions are affected by the width of the downstream channel even for an expansion ratio (ER) as high as 14.5. They reported that the occurrence of the secondary ridges along the wall should be an effect of the jet expansion ratio. Faruque et al. (2006) and Sarathi et al. (2008) found that no secondary effects were observed for an ER of 41.4.

The effect of channel width on the extent of scour with varying tailwater conditions has been documented by Bey et al. (2008). They investigated the effect of jet exit velocity, tailwater depth and channel width concurrently on the scour characteristics. To this end, Bey et al. (2008) studied four groups of tests (denoted as A, B, C and D) using LDA and scour profile measurements to characterize the flow field. Tests were conducted at four widths ($w = 0.25\,W$, $0.5W$, $0.75\,W$ and W) and three different jet exit velocities ($U_o = 0.75$, 0.90 and 1.16 m/s). Here, W denotes a width of 0.4 m. In group A, for a given jet velocity and channel width, the submergence was varied in a stepwise fashion from a low tailwater ($y_t = 2b_o$) to a high tailwater condition ($y_t = 20b_o$). The tests in group B were commenced from a high submergence condition

($y_t = 20b_o$) with a stepwise reduction towards low submergence. Group C tests were chosen such that the jet would continue to flip alternately between the bed and the free surface, and yet be close to the high submergence range at a given exit velocity. Group D tests were carried out at a fixed tailwater depth ($y_t = 20b_o$) for a period of 24 hours and velocity measurements are obtained at various time intervals. The results of varying channel width on bed scour profile variation are illustrated in Figs. 12 and 13.

Figure 12(a) and (b) shows the scour profiles for group C and group D for $U_o = 1.16$ m/s at various channel widths at $t = 24$ h. It can be observed from Fig. 12 that the scour profiles are dependent on channel width. The maximum scour depth decreases as the channel width increased from $w = 0.25W$ to $0.75W$. However, the maximum scour depth at $w = 1.0$ W is more than at $w = 0.25W$ and $0.75W$. From Fig. 12(b), it is observed that the maximum scour depth decreases with decreasing channel width. For all other conditions maintained constant, at higher submergence, it can be seen from the results that as the channel width increases, the maximum scour hole depth decreases, followed by an increase in maximum scour hole depth with further increase in width, whereas, at lower submergence, the scour hole depth decreases with decreasing channel width. The elongated mound and the formation of an intermediate hump in the scour hole near the nozzle (Fig. 12b) have been also observed in previous studies of Kells et al. (2001) and Deshpande et al. (2007).

Figure 13(a) and 13(b) shows the variation of maximum depth (ε) of scour at $t = 24$ h for test groups A and C, respectively. It can be noted that maximum depth of scour is a function of jet exit velocity and channel width. Available data from literature with comparable values of densimetric Froude number (F_o) are also shown in Fig. 13. It is observed from Fig. 13 that scour depth is different for the group A and C for any given channel width.

Sui et al. (2008) studied scour caused by 3D square jets interacting with non-cohesive sand beds to further understand the effects of channel width, tailwater conditions and jet exit velocity. The tailwater ratio was varied from 2 to 12 times the nozzle width, while the channel width was 31.6 and 41.4 times the nozzle width. Three different jet exit velocities were adopted and two different bed materials fine sand ($d_{50} = 0.71$ mm) and coarse sand ($d_{50} = 2.30$ mm) was used. Fig. 14 presents the variation of the different scour parameters with varying expansion ratio and tailwater ratio for both fine (Fig. 14a-c) and coarse sand (Fig. 14d-f). As shown in Fig. 14a, it can be noted that the maximum scour width increases with increasing ER and attains a maximum value at a tailwater ratio (TWR) = 4 for the tests with ER = 41.4. Increase of TWR results in a decrease of maximum scour width. For TWR > 3, length of scour hole (L_s) decreases with increasing ER. From Fig. 14b, it can be observed that maximum ridge width increases with increasing ER. It is also observed that the maximum length of scour hole is shorter for the higher ER for TWR > 3. Figs. 14d-f show the variation of the scour geometry parameters for the coarser sand at the two expansion ratios. Fig. 14d shows that the maximum width of scour is smaller for the higher expansion ratio for $3 \leq TWR \leq 4$ and the maximum width of scour is higher for higher expansion ratio for TWR > 4.5. It is observed from Fig. 14d that the effect of ER on maximum scour width decreases with increasing sand grain size. Fig. 14e shows that there is no effect of ER on ridge width for coarse sand. Length of scour hole increases with decreasing ER (Fig. 14f). One can conclude the following from the study of Sui et al. (2008):

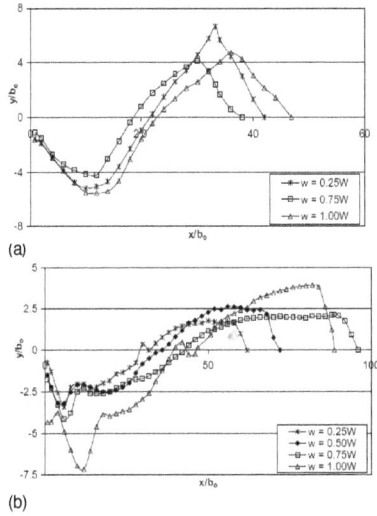

Figure 12. a) Scour profile variation with channel width with U_o = 1.16 m/s for group D, (b) Scour profile variation with channel width with U_o = 1.16 m/s for group C (Bey et al. 2008, copyright permission of Taylor and Francis

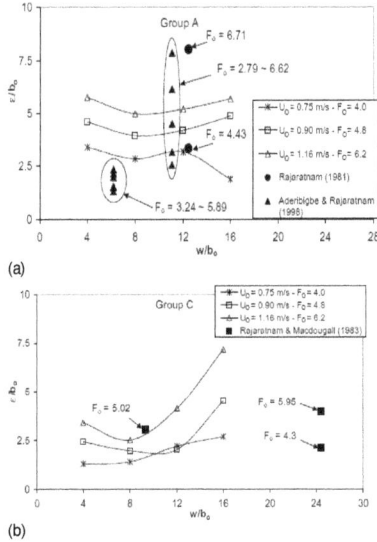

Figure 13. a) Plots of maximum scour depth with channel width for various velocities in group A, (b) Plots of maximum scour depth with channel width for various velocities in group C (Bey et al. 2008, copyright permission of Taylor and Francis)

i.   For the fine sand and for TWR > 3, scour hole was wider but shorter in length at the higher expansion ratio. For the coarse sand, a similar trend was observed for TWR > 4.

ii.   For the coarse sand and for $3 \leq$ TWR ≤ 4, scour hole is strongly dependent on the tailwater ratio, especially at the lower expansion ratio.

iii.   The effect of expansion ratio reduces as densimetric Froude number increases.

iv.   The extent of difference for different scour parameter due to expansion ratio reduces with increasing TWR irrespective of the sand grain size.

v.   Effect of ER reduces with increasing sand grain size.

vi.   Results indicate that different expansion ratios create different scour profiles even for TWR ≥ 12.

vii.   In an effort to provide useful, but a simplified scour prediction equation at asymptotic conditions, a relationship for the maximum scour depths has been proposed in terms of densimetric Froude number, tailwater ratio and expansion ratio.

viii. Proposed relationship predicted scour depths correctly for a wide range of test conditions.

10. Influence of densimetric Froude number on local scour

To better understand the scouring process, jets interacting with sand beds have been studied by many researchers and empirical relations involving densimetric Froude have been proposed to predict local scour. Rajaratnam and Berry (1977) studied the scour produced by circular wall jets and concluded that the main geometric characteristics of the scour hole are functions of the densimetric Froude number. Rajaratnam and Diebel (1981) concluded that the relative tailwater depth and width of the downstream channel only affect location of the maximum scour, whereas densimetric Froude number affects maximum scour depth. Ali and Lim (1986) indicated that the value of the critical tailwater condition increases with increasing densimetric Froude number and the effect of tailwater becomes insignificant when H/b_o is beyond 16. Chiew and Lim (1996) studied local scour by deeply submerged circular jets of both air and water and concluded that the densimetric Froude number was the characteristic parameter in describing the scour hole dimensions. Ade and Rajaratnam (1998) noted that the maximum depth of scour was larger at higher values of F_o. It was also noted that to attain an asymptotic state at higher values of F_o, a longer time was required. Faruque et al. (2004) concluded that the densimetric Froude number, tailwater depth and nozzle size-to-grain size ratio, all have an influence on the extent of scour caused by 3-D jets. They speculated about the dominance of each parameter at different flow conditions.

Sarathi et al. (2008) studied effect of the densimetric Froude number, tailwater depth and sediment grain size on scour caused by submerged square jets. Results of their study with respect to densimetric Froude number are presented in this section. Figures 15(a-c) show the

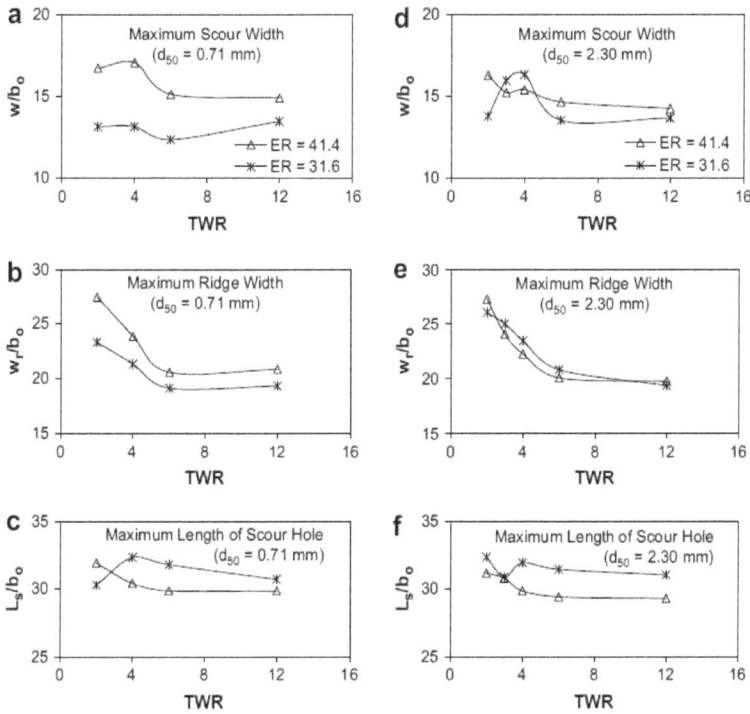

Figure 14. Variation of different scour parameters with respect to tailwater ratio (TWR) for F_o= 10, fine sand (d_{50}= 0.71 mm) and coarse sand (d_{50}= 2.30 mm) (Sui et al. 2008, copyright permission of Science Direct)

asymptotic scour profiles along the centerline of the nozzle for different F_o values. The asymptotic shape of the scour bed profiles for different F_o at H/b_o = 18 are shown in Fig. 15(a). It is observed that the maximum ridge height and its location, maximum scour depth and the distance of the maximum depth of scour hole from the nozzle (x_m) increases with increasing F_o. Figure 15(b) shows the scour profiles at three different values of F_o at lower tailwater conditions (H/b_o=2). The ridge crest is sharper at the lower value of F_o and at higher F_o the ridge is flat and directly related to the prevailing local velocity. Figure 15(c) shows the profiles at different F_o for H/b_o = 1. Ridge crests are flat and they are of constant height at different densimetric Froude numbers. It is clear that the size and shape of the ridge is clearly dependent on F_o and tailwater depth. Figure 15(d) shows the variation of the ridge height for different densimetric Froude numbers and tailwater depth. Ridge height is higher with higher densimetric Froude number. Ridge height increases with increasing tailwater depth for a given F_o however, beyond H/b_o = 6 for a given F_o, the ridge height attains a near constant value. Figure 16(a-c) shows the dependence of the normalized asymptotic scour parameters on F_o for a range of tailwater ratio. Figure 16(a) shows asymptotic maximum scour depth (ε_m) is higher for higher densimetric Froude number irrespective of the tailwater depths. Figure 16(b) and 16(c) show similar trends of width of scour hole and length of scour hole. Scour hole dimensions are higher for higher densimetric Froude numbers. Finally, Sarathi et al. (2008) concluded that the role of grain size was completely absorbed by densimetric Froude number.

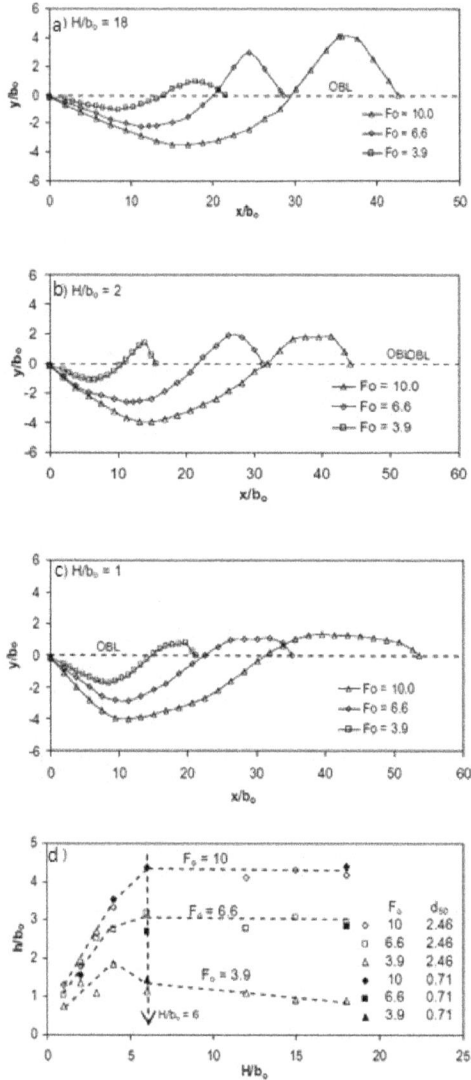

Figure 15. Scour geometry at asymptotic conditions at different densimetric Froude numbers (Sarathi et al. 2008, copyright permission of Taylor and Francis)

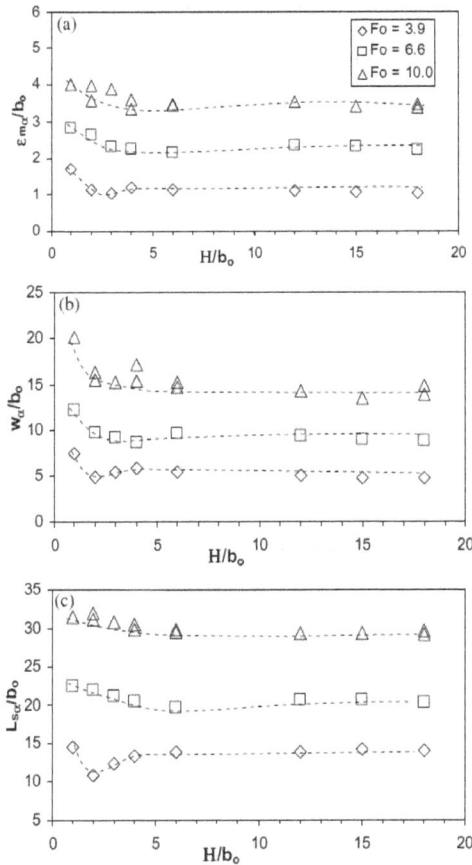

Figure 16. Variation of the asymptotic scour parameters with densimetric Froude number (Sarathi et al. 2008, copyright permission of Taylor and Francis)

11. Local scour under ice cover

Ice covers in rivers effect the resistance to flow, flow velocity and depth. Ice cover on top of the river with high sediment transport influences the development, size and shape of bedforms. This section reviews ice cover influences on the relationships between flow and local scour.

Ettema (2002) published an in-depth review on extent to which alluvial channels respond to ice-cover formation, presence, and breakup. An imposed ice cover results in an increased composite resistance to the flow. Ice cover thickness affects only a comparatively narrow region near the upper level of flow in deep flows. On the other hand, in shallow flows, an ice cover chokes the flow, which decreases the velocity resulting in the bed shear stress to be less than the critical shear stress, and thereby decreases sediment transport. The studies also confirm that sediment-transport rate increases with decreasing water temperature. The location of the maximum velocity is dependent on the relative magnitudes of the channel bed and cover underside resistance coefficients (Gogus and Tatinclaux, 1981). The bed elevation in an ice covered stream must rise locally due to decrease in bulk velocity.

Asymptotic scour in the presence of smooth and rough ice cover was investigated by Sui et al (2009) and arrived at the following conclusions:

i.   The variation of maximum scour depth with tailwater ratio under covered flow was different from that under open-water flow.

ii.   Under open-water flow conditions, maximum scour depth decreased with increasing tailwater depth, and beyond a certain tailwater depth, the values tend to increase and attain constancy.

iii.   Under covered flow, maximum scour depth increased with increasing tailwater depth, and beyond a certain tailwater depth, the values tend to decrease and attain constancy.

iv.   Ice cover conditions do not influence maximum scour depth at large values of submerged conditions the.

v.   Maximum scour depth in the fine sand bed under ice covered flow was always less than that noticed in the coarse bed material.

vi.   Impact of ice cover on scour depth was also less obvious for finer sand.

vii.   There was a negligible effect of ice cover on the scour parameters at lower densimetric Froude number.

viii.   Scour width increased with increasing tailwater depth but decrease with increasing tailwater depth beyond a critical value and finally attained an asymptotic width in free surface and with fine sand bed.

ix.   In the case of the coarse sand, scour length was larger under rough ice cover condition, but in the case of the fine sand, scour length was larger under open-flow condition.

x.   Ridge height was not affected by type of bed material and with or without ice covered flow.

12. Numerical modeling of scour caused by jets

Numerical modeling is gaining momentum since experimental investigations are time consuming and expensive. Li and Cheng (1999) proposed a mathematical model based on potential-flow theory for simulating the equilibrium scour hole underneath offshore pipelines. Karim and Ali (2000) tested effectiveness of a commercial software (FLUENT) to predict 2-D flow velocity distribution and the bed shear stress generated by a turbulent water jet impinging on rigid horizontal and scoured beds. Minimal and deeply submerged water jet simulations were carried out. The close agreement between selected various experimental and computed results were noted. The standard $k-\varepsilon$ and the RNG $k-\varepsilon$ models described the flow at the boundary better than the Reynolds stress model (RSM).

Neyshabouri et al. (2003) attempted to obtain numerical predictions of the scour hole geometry created by a free falling jet. The two-dimensional momentum equations, the continuity equation were solved using a $k-\varepsilon$ turbulence model. First, the turbulent flow due to a free falling jet was computed, then the distribution of the sand concentration was determined based on the convection-diffusion equation and the scoured bed was computed based on sediment continuity equation. The above-mentioned steps were repeated until the equilibrium scour hole was reached. Sediment transport calculations required specification of the sediment concentration near the bed at the start of simulation. Stochastic and deterministic expressions for sediment concentration near the bed proposed by van Rijn (1987) were used. The scour profiles obtained using stochastic expression was most realistic as compared to using of deterministic expression.

Adduce and Sciortino (2006) numerically and experimentally investigated local scour in clear water scour conditions downstream of a sill followed by a rigid apron. A mathematical one-dimensional model was developed which uses measured velocity fields (obtained using a ultrasonic Doppler velocimeter) to simulate the scour hole evolution. The dune profiles predicted by the model were similar to the measured profiles for large discharges, while when the discharges were smaller, the dunes predicted by the model were always longer than those measured.

Adduce and La Rocca (2006) studied different scour developments caused by a submerged jet, the surface wave jet and the oscillating jet developed downstream of a trapezoidal drop followed by a rigid apron. They have highlighted that as flow separation takes place at the edge of the rigid bed, reverse circulation motion along the longitudinal section develops near the bed downstream of the rigid bed itself. A stability analysis of the surface wave jet with both a flat and a scoured bed was performed by using modified Saint-Venant equations with correction terms accounting for the curvature of the streamlines. They concluded that the stability of the surface wave jet is weakened by the presence of the scoured bed.

Boroomand et al. (2007) mathematically modeled an offset jet entering a domain with sediment bed using ANSYS FLUENT with a two phase model in which water was primary phase and sediment bed was the secondary phase. In multiphase models, setting up of initial conditions is very important. These conditions include initial grid generation, initial phases and their properties and volume fractions of each phase. The model determines the interphase exchange coefficients, lifting force, virtual mass force, and interaction force between phases and solves the continuity and momentum equations for each phase. Scour profiles calculated by the model agreed fairly with the measurements. Also, the computed concentration profiles agreed fairly with measurements except near the bed zone.

Liu and Garcia (2008) studied turbulent wall jet scour in which bed evolution was modeled by solving the mass balance equation of the sediment. The free surface was modeled by the VOF method while the scour process was modeled by the moving mesh method. The modeling effort yielded a good agreement with measurements (velocity field, the maximum scour depths and local scour profile). Further research is needed to investigate the effect of the turbulence model for free surface waves (especially for near breaking and breaking waves) and to study the possibility of using an Eulerian approach for morphological modeling.

Abdelaziz et al. (2010) developed a bed load sediment transport module and integrated into FLOW-3D. This model was tested and validated by simulations for turbulent wall jet scour in an open channel flume. Effects of bed slope and material sliding were also taken into account. The hydrodynamic module was based on the solution of the three-dimensional Navier-Stokes equations, the continuity equation and $k - \varepsilon$ turbulence closure scheme. The rough logarithmic law of the wall equation was iterated in order to compute shear velocity u_*. The predicted local scour profile fit well with the experimental data, however the maximum scour depth was slightly under estimated and the slope downstream of the deposition dune was over estimated.

13. Scope for future research

The review of literature and our current understanding of scour by jets indicate that the following need to be considered in the future:

i.   Extend studies to include cohesive soils.

ii.   Extend studies to higher range of densimetric Froude number ($F_o > 10$).

iii.   Non-uniform sand beds need to be studied to determine the effect of different gradations.

iv.   To study the role of fluid structures with varying submergence, expansion ratio and nozzle size to grain size ratio.

v.   Developing analytical equations for three-dimensional scour volume.

vi.   To understand the effect of the removal of the mound at the asymptotic state, on scour depth, and 3-D scour hole development with time.

vii.   To investigate separation, recirculation and eddies during digging and refilling phases at various submergence ratios.

viii.   Obtain instantaneous velocity fields during scour hole development to enhance modeling efforts

ix.   To study jet scour together with sediment transport by considering continuous sediment influx through the jet and in the ambient flow.

Author details

Ram Balachandar and H. Prashanth Reddy

Department of Civil and Environmental Engineering, University of Windsor, Canada

References

[1] Abdelaziz, S, Bui, M. D, & Rutschmann, P. Numerical simulation of scour development due to submerged horizontal jet", River Flow (2010). Dittrich, Koll, Aberle and Geisenhainer (eds), 978-3-93923-000-7, 2010.

[2] Adduce, C. La Rocca, M. ((2006). Local scouring due to turbulent water jets downstream of a trapezoidal drop: Laboratory experiments and stability analysis. Water Resour. Res. 42(2), W, 02405, 1-12.

[3] Adduce, C, & Sciortino, G. (2006). Scour due to a horizontal turbulent jet: Numerical and experimental investigation. J.Hydraulic Res. 44(5), 663-673.

[4] Ade, F, & Rajaratnam, N. (1998). Generalized study of erosion by circular horizontal turbulent jets, J. Hydraulic Res. 36(4), 613-635.

[5] Aderibigbe, O, & Rajaratnam, N. (1998). Effect of sediment gradation of scour by plane turbulent wall jets." J. Hydraul. Eng. 10.1061/(ASCE)0733-9429(1998)124:10(1034), 124 (10), 1034-1042.

[6] Aderibigbe, O, & Rajaratnam, N. (1996). Erosion of loose beds by submerged circular impinging vertical turbulent jets, J. Hydraulic Res, 34(1), 19-33.

[7] Ali, K. H. M, & Lim, S. Y. (1986). Local scour caused by submerged wall jets, Proc. Ins Civil Engineers. 81(2), 607-645.

[8] Ankamuthu, S, Balachandar, R, & Wood, H. (1999). Computational Steroscopy for Three-dimensional Scour Depth Measurement in Channels. Can. J. Civ. Eng., , 26, 698-712.

[9] Balachandar, R, & Kells, J. A. (1997). Local channel scour in uniformly graded sediments: The time-scale problem." Can. J. Civ. Eng. 10.1139/cjce-24-5-799, 24 (5), 799-807.

[10] Balachandar, R, & Kells, J. A. (1998). Instantaneous water surface and bed scour profiles using video image analysis." Can. J. Civ. Eng. 10.1139/cjce-25-4-662, 25 (4), 662-667.

[11] Balachandar, R, Kells, J. A, & Thiessen, R. J. (2000). The effect of tailwater depth on the dynamics of local scour." Can. J. Civ. Eng., 27 (1), 138-150.

[12] Bey, A, Faruque, M. A. A, & Balachandar, R. (2007). Two Dimensional Scour Hole Problem: Role of Fluid Structures". J. Hydraul. Engng. 133(4), 414-430.

[13] Bey, A, Faruque, M. A. A, & Balachandar, R. (2008). Effects of varying submergence and channel width on local scour by plane turbulent wall jets". J.Hydraul. Res. 46(6), 764-776.

[14] Boroomand, M, Neyshabouri, R, & Aghajanloo, S. , A. , S. K., ((2007). Numerical simulation of sediment transport and scouring by an offset jet", Can. J. Civ. Eng., , 34

[15] Breusers, H. N. C, & Raudkivi, A. J. (1991). Scouring, Hydraulic Structure Design Manual. A. A. Balkema, Rotterdam. The Netherlands.

[16] Chatterjee, S. S, Ghosh, S. N, & Chatterjee, M. (1994). Local scour due to submerged horizontal jet." J. Hydraul. Eng., 120 (8), 973-992.

[17] Chiew, Y. M, & Lim, S. Y. (1996). Local Scour by a Deeply Submerged Horizontal Circular Jet". J. Hydraul. Engng 122(9), 529-532.

[18] Day, R. A, Liriano, S, & White, R. W. (2001). Effect of tailwater depth and model scale on scour at culvert outlets, Proceedings of the Institution of Civil Engineers Water and Maritime Engineering, 148(3), 189-198.

[19] Deshpande, N. P, Balachandar, R, & Mazurek, K. A. (2007). Effects of submergence and test startup conditions on local scour by plane turbulent wall jets Journal of Hydraulic Research, 45 (3) (2007), , 370-387.

[20] Dey, S, & Sarkar, A. (2006a). Scour downstream of an apron due to submerged horizontal jets." J. Hydraul. Eng., 132(3), 246-257.

[21] Dey, S, & Sarkar, A. (2006b). Scour response of velocity and turbulence in submerged wall jets to abrupt change from smooth to rough beds and its application to scour downstream of an apron." J. Fluid Mech., , 556, 387-419.

[22] Ettema, R. (2002). Review of alluvial-channel responses to river ice." J. Cold Reg. Eng., 16(4), 191-217.

[23] Faruque, M. A. A, Sarathi, P, & Balachandar, R. (2006). Clear Water Local Scour by Submerged Three-Dimensional Wall Jets: Effect of Tailwater Depth," J. Hydraul. Engng. 132(6), 575-580.

[24] FaruqueMd. ((2004). Transient local scour by submerged three dimensional wall jets, effect of the tailwater depth. Msc thesis, University of Windsor.

[25] Ghodsian, M, Melville, B, & Tajkarimi, D. (2006). Local scour due to free overfall jet, Ins Civil Engineers, Water Management. 159(4), 253-260.

[26] Hoffmans, G. J. C. M. (1998). Jet scour in equilibrium phase", Journal of Hydraulic Engineering, , 124(4), 430-437.

[27] Hoffmans, G. J. C. M, & Pilarczyk, K. W. (1995). Local scour downstream of hydraulic structures", Journal ofHydraulic Engineering, , 121(4), 326-340.

[28] Hogg, A. J, Huppert, H. E, & Dade, W. B. (1997). Erosion by planar turbulent wall jets." J. Fluid Mech., , 338, 317-340.

[29] Hopfinger, E. J, Kurniawan, A, Graf, W. H, & Lemmin, U. (2004). Sediment erosion by Götler vortices: The scour problem." J. Fluid Mech., , 520, 327-342.

[30] Johnston, A. J. (1990). Scourhole Developments in Shallow Tailwater". J. Hydraul. Res., IJHR 28(3), 341-354.

[31] Karim, O. A, & Ali, K. H. M. (2000). Prediction of flow patterns in local scour holes caused by turbulent water jets. J. Hydraul. Res. , 38, 279-287.

[32] Kells, J. A, Balachandar, R, & Hagel, K. P. (2001). Effect of grain size on local channel scour below a sluice gate." Can. J. Civ. Eng. 10.1139/cjce-28-3-440, 28 (3), 440-451.

[33] Li, F, & Cheng, L. (1999). A numerical model for local scour under offshore pipelines J. Hydraul. Eng., 125 (4), , 400-406.

[34] Lim, S. Y. (1995). Scour below unsubmerged full flowing culvert outlets. Proc. Ins Civil Engineers, , 112, 136-149.

[35] Liu, X, & Garcia, M. H. (2008). Three-dimensional numerical model with free water surface and mesh deformation for local sediment scour. Journal Waterway, Harbour, Coastal and Ocean Engineering, ASCE 134 (4), 203-217.

[36] Mazurek, K. A, & Rajaratnam, N. (2002). Erosion of a polystyrene bed by obliquely impinging circular turbulent air jets, J. Hydraulic Res. 40(6), 709-716.

[37] Mehraein, M, & Ghodsian, M. and Salehi Neyshaboury, S. A.A ((2010). Local scour due to an upwards inclined circular wall jet, Proc. Ins Civil Engineers, , 164, 111-122.

[38] Melville, B. W, & Chiew, Y-M. (1999). Time scale for local scour at bridge piers. ASCE Journal of Hydraulic Engineering, , 125(1), 59-65.

[39] Meulen, V, Vinje, T, & Three-dimensional, J. , J. local scour in noncohesive sediments", Proc., 161th International Association for Hydraulic Research (IAHR)-Congr., San Paulo, Brazil, , 263-270.

[40] Mih, W. C, & Kabir, J. (1983). Impingement of Water Jets on Non uniform S treambed, J. Hydraulic Eng. 109 (4), 536-548.

[41] Mohamed, M. S, & Mccorquodale, J. A. (1992). Short-term local scour", Journal of Hydraulic Research, , 30(5), 685-699.

[42] Neyshabouri, S. A. A., Ferreira Da Silva, A. M. and Barron, R. (2003). Numerical Simulation of Scour by a Free Falling Jet. J. Hydraul. Res., , 41(5), 533-539.

[43] Neyshabouri, S. A. A, & Barron, R. Ferreira da Silva, A.M. ((2001). Numerical Simulation of Scour by a Wall Jet". Water Engng Res. 2(4), 179-185.

[44] Pagliara, S, Hager, W. H, & Minor, H. E. ((2006). Hydraulics of plane plunge pool scour. J. Hydr. Engng. 132(5), 450-461.

[45] Rajaratnam, N. (1981). Erosion by plane turbulent jets, J. Hydraulic Res. 19(4), 339-358.

[46] Rajaratnam, N, & Berry, B. (1977). Erosion by circular turbulent wall jets, J. Hydraulic Res. 15(3), 277-289.

[47] Rajaratnam, N, & Mazurek, K. A. (2003). Erosion of sand by circular impinging water jets with small tailwater, J. Hydraulic Eng. 129(3), 225-229.

[48] Rajaratnam, N, Aderibigbe, O, & Pochylko, D. (1995). Erosion of sand beds by oblique plane water jets." Proc. Inst. Civ. Eng., Waters. Maritime Energ., 112(1), 31-38.

[49] Rajaratnam, N, & Diebel, M. (1981). Erosion below Culvert-Like Structure". Proceeding of the 5th Canadian Hydrotechnical Conference, 26-27 May, CSCE, , 469-484.

[50] Rajaratnam, N, & Macdougall, R. K. (1983). Erosion by Plane Wall Jets with Minimum Tailwater". J. Hydrual. Engng., 109(7), 1061-1064.

[51] Sarathi, P, Faruque, M. A. A, & Balachandar, R. (2008). Influence of tailwater depth, sediment size and densimetric Froude number on scour by submerged square wall jets, J. Hydraulic Res. 46(2), 158-175.

[52] Sui, J, Faruque, M. A. A, & Balachandar, R. (2009). Local scour caused by submerged square jets under model ice cover, J. Hydraulic Eng., 135(4), 316-319.

[53] Sui, J, Faruque, M. A. A, & Balachandar, R. (2008). Influence of channel width and tailwater depth on local scour caused by square jets, J. Hydro-environment Res, 2(1), 39-45.

[54] Van Rijn ((1987). Mathematical Modeling of Morphological Processes in the Case of Suspended sediment transport. PhD. Dissertation, Delft University of Technology, The Netherlands.

[55] Wu, S, & Rajaratnam, N. (1995). Free jumps, submerged jumps, and wall jets." J. Hydraul. Res., 33(2), 197-212.

Chapter 8

The Gravel-Bed River Reach Properties Estimation in Bank Slope Modelling

Levent Yilmaz

Additional information is available at the end of the chapter

1. Introduction

Church and Kellerhals [2] point out the difficulty of adequately characterizing a gravel bed by a single grain size distribution for a relatively long river reach. Bray [1] indicated that the initiation of motion calculations gave as a result in which the gravel bed is immobile or at least not highly mobile at flows by flooding boundary layers. The basic data for each gravel-bed river reach are directly applied to a specified equation to compute the average velocity.

The knowledge about the hydraulic geometric parameters, width, depth and area of the river at the bankful discharge are required for solving a variety of problems related to rivertraining, location of river constructions and navigation. To predict the average velocity of flow, the resistance offered to the flow by the boundary and air-water interface needs to be known. In methods for the prediction of width, depth, area and the flow velocity or resistancecoefficient the results of the analysis of the available gravel-bed river data will be given.

2. Method

The resistance characteristics and the study of hydraulic geometry for gravel-bed rivers is the main method for finding all the hydraulic characteristics. The hydraulic geometry refers to the geometrical characteristics of the cross-section such as the average width w, average depthh and area A (=wh) at the bankful discharge Q.

The basic data for each gravel-bed river reach are directly applied to a specified equation to compute the average velocity. Then for each reach the percent deviation (PDEV) of the computed average velocity from the "observed" average velocity is computed. The distribution of the percent deviations associated with a specified equation is then determined for the different gravel-bed rivers reaches. A summary of the parameters to describe the distribution of the percent deviations for each of the specified equations is given in Table. 1.

Equation (1)	Mean (2)	Standard Deviation (3)	Minimum Value (4)	Median Value (5)	Maximum Value (6)
Manning's Eq.	-3.3	29.6	-50.0	-7.0	83.2
n by modified Cowan n by Strickler	44.9	43.7	-18.6	31.8	181.9
$n = 0.41\, D^{1/6}_{50}$	37.5	40.9	-23.1	25.0	156.9
$n = 0.038\, D^{1/6}90$ n by Limerinos	2.5	28.8	-41.8	-3.1	74.4
Keulegan's Eq.	54.2	46.1	-12.7	40.4	195.3
$k_3 = D_{50}$	47.0	42.7	-17.3	35.2	169.2
$k_3 = D_{65}$	32.9	38.3	-23.9	23.0	136.4
$k_4 = D90$ Lacey's Eq.	8.6	29.4	-26.6	-0.7	116.1

Table 1. Statistics for Gravel-bed River Reaches [1]

Some of the characteristics which differentiate gravel-bed rivers from the alluvial rivers are:

a.   much steeper slope (0.001 – 0.02)

b.   resistance is higher than the alluvial rivers

There is scope of using all the available gravel-bed river data and develop non dimensional relationships for the hydraulic geometry. In the analysis of river and channel problems it must be given a relationship between the average velocity U, the depth h or the hydraulic radius R, channel slope S and some coefficient which is related to the channel boundary. This is known as the resistance relationship [3]. The work of Lacey [4] about the sand-bed rivers has shown that for such rivers depth h or hydraulic radius $R \sim (Q/f_1)^{1/3}$, width W or wetted perimeter P~ $Q^{0.50}$, Area $A \sim Q^{5/6}/ f_1^{\ 1/3}$ where f_1 is Lacey's silt factor and is given by $f_1 = 1{,}76\,(d)^{0.5}$, d being the median size of bed material in mm. As regards the gravel-bed rivers Kellerhals and Bray [5] have related W, h, A to Q and sediment size d as

$$W = 2.08\, Q^{0.528} d^{-0.70} \tag{1}$$

$$h = 0.256\, Q^{0.331} d^{-0.25} \tag{2}$$

All such equations are based on the analysis of limited amount of data and are not dimensionally homogeneous. Only the dimensionless parameters W/d, h/d and $U/(\Delta\gamma_s\ d/\varrho_f)^{0/5}$ and related them to the dimensionless discharge $Q/\ d^2\ (\Delta\gamma_s\ d/\varrho_f)^{0/5}$ are given [3]. Hence $\Delta\gamma_s$ is the difference in specific weights of sediment and water and ϱ_f is the mass density of water. Hence there is scope of using all the available gravel-bed river data and develop non-dimensional relationships for the hydraulic geometry. In the analysis of river and channelproblems we need also a relationship between the average velocity U, the depth h or the hydraulic radius R, channel slope S and some coefficient which is related to the channelboundary. This is known as the resistance relationship [3]. The resistance relationship isexpressed in dimensionless form as [3],

$$\text{in Manning's Equation} : \frac{U}{\sqrt{ghS}} = \frac{h^{1/6}}{n\sqrt{g}} \tag{3}$$

$$\text{in Chezy's Equation} : \frac{U}{\sqrt{ghS}} = \frac{C}{\sqrt{g}} \tag{4}$$

$$\text{in Darcy-Weisbach Equation}: \frac{U}{\sqrt{ghS}} = \sqrt{\frac{8}{f}} \tag{5}$$

$$\frac{U}{\sqrt{ghS}} = F\left(\frac{h}{d}\right) \tag{6}$$

In the above equations n is Manning's roughness coefficient, C is Chezy's discharge coefficient, f is Darcy-Weisbach resistance coefficient and F is a function. These coefficientsdepend on the resistance, offered to the flow by the channel boundary and air-water interface [3]. The available data in Turkey at East Black Sea Basin have been analysed in a unified manner to obtain dimensionally homogeneous relationships for W, h, A and U.

3. Data

A summary of data were classified as bankful discharge and variable discharge. The bankful discharge data were used to study the hydraulic geometry. The variable discharge data pertainto discharges other than the bankful in any stream. In order to study the effect of bed condition, each set of data were further subdivided into those with mobile bed, and those withpaved bed. There is no need to subdivide the data, because both sets of data behaved in similar manner.

4. Analysis of hydraulic geometry

The dependent variables can be any two of the four variables average width W, average depth h, area of flow A = Wh and the average velocity U. The independent variable related to the flow is bankful discharge Q. The sediment representing the bed and the banks will be described by the median size of the bed material d, its geometric standard deviation σ_g andthe difference in the specific weights of sediment and water $\Delta\gamma_s$ [3]. It is known that for a given stream the channel slope is related to the bankful discharge Q, the slope decreasingas Q increases in the downstream direction [3]. If we deal with the data from different basins,S and Q will not be related and hence S should be taken as an independent variable. If we ignore Q_B, because the gravel-bed rivers carry a small amount of sediment load, we cananalyse as [3],

$$W, h, A, U = F\ (Q,\ d,\sigma_g,\Delta\gamma_s,\rho_f,\ \mu,\ S) \tag{7}$$

With simplifications Garde [3] gave the dimensionless relationship for hydraulic geometry of different river basins as,

$$\frac{W}{d},\frac{h}{d},\frac{A}{d^2},\frac{U}{\sqrt{\dfrac{\Delta\gamma_s}{\rho_f}d}} = F\left[\frac{Q}{d^2\sqrt{\dfrac{\Delta\gamma_s}{\rho_f}d}},S\right] \tag{8}$$

If studied regime types of relations, we must plot W, h, and A against Q on log-log scale which yielded straight lines giving equations by [3],

$$W = 4.547\ Q^{0.507} \tag{9}$$

$$h = 0.293\ Q^{0.332} \tag{10}$$

$$A = 1.330\ Q^{0.839} \tag{11}$$

By comparing this equations, also the North Anatolian River Reaches will be investigated. The exponents of Q obtained in Eqs. [9], [10], [11] are very close to those obtained by Lacey [4]. Similar investigation was carried out using W/d, h/d and A/d² and determining their variation with $\dfrac{W}{d},\dfrac{h}{d},\dfrac{A}{d^2}, = F\left[\dfrac{Q}{d^2\sqrt{\dfrac{\Delta\gamma_s}{\rho_f}d}}\right]$ by plotting on log-log scales [3].

The relationships given by Garde [3] are,

$$W/h = 7.675 \, Q^{0.448} \tag{12}$$

$$h/d = 0.504 \, Q^{0.373} \tag{13}$$

$$A/d^2 = 3.872 \, Q^{0.821} \tag{14}$$

5. Method

From www.terrasol.com the program for landslides can be estimated by TALREN 4 which is ideal for checking the stability of natural slopes, cut or fill slopes, earth dams and dikes. It takes into account various types of reinforcements, such as: anchors and soil nails, piles and micropiles, geotextiles and geogrids, steel and polymer strips. There is another new user-friendly graphical interface with:

a. In the program, definition of the profile using a mouse, rulers and a grid. Other features include pop up menus and choice of soil colours.

b. Ability to load background drawings (.jpg and.gif formats) and adjust to scale.

c. Several construction stages and calculation alternatives can be handled in the same file.

d. Tables illustrating main soil, load and reinforcement data.

e. Various output options for graphical display and tables (shadings, forces in reinforcements, detailed results for each failure surface, etc.)

f. Wizards and databases to help produce the best model and choice of input data (partial safety factors).

New calculation functionalities:

a. Automatic search option for circular failure surfaces (no need to define a manual grid).

b. No limit on the number of elements you can define (points, layers, reinforcements, hydraulic mesh, etc.)

c. Future upgrade option for TALREN 4 users: calculation method based on limit analysis theory.

d. TALREN still benefits from extensively used methods as limit equilibrium calculation along potential failure surfaces using the Fellenius, Bishop or perturbations methods.

e. Ability to take into account hydraulic conditions.

f.   Seismic loads are taken into account by the pseudostatic method.

6. Data uncertainties

For estimation of landslides condition we require precipitation, streamflow, evapotranspiration and watershed morphology. The effects of data uncertainties must be considered in different ways:

1.   whether the model parameters are determined from calibration or from physical measurements and principles,

2.   whether the model is used to estimate real events (landslide forecasting), or to estimate synthetic events (design storms and generation of synthetic flows which reasoned the landslides. These issues are considered separately.

In Turkey the landslides can be seen in the Karst environment. Karst is a term applied to topography fdrmed in regions of limestone or dolomite bedrock by the vigorous solution work of groundwater. One recognizes karst topography by the presence of large numbers of sinkholes, solution valleys, disappearing streams, and landslides. The development of karst topography is enhanced by the presence of well-jointed carbonates or evaporites near the surface. It is also enhanced by rainfall. And sufficient relief to insure continuous movement of groundwater that will carry away dissolved matter. The term karst comes from a limestone plateau in Yugoslavia where solution features are well developed. Similar topography can be found in Turkey, Kentucky, Tennessee, Indiana, northern Florida, and Puerto Rico.Types of mass wasting Earth materials on slopes shows the movement where it shows as the result of landslides.

	Rate of Movement	Amount of Water Present
Flow (movement distributed throughout material)		
Creep	Slow	Water not necessary
Rock glacier, rock, stream	Slow	Water not necessary
Solifluction	Fast	Water-saturated
Mudflow, debris flow	Slow or fast	Much water
Earthflow	Slow or fast	Much water
Slide (movement as one Mass on a slip surface)		
Debris avalanche	Fast	Wet or dry
Slump	Fast	Wet or dry
Landslide, rockslide	Fast	Wet or dry
Fall (free fall of rock or soil)		

Table 2. Types of mass movement (Levin, 1986)

Creep is a small form of land movement where the amount of water is not necessary and its measure is only a few centimeters in one year (Watkins et all., 1975). Creep can decrease if we can follow it through the earth surface and is a form of small earth flow. There are two types of creep, soil and rock creep which can be observed.

Landslide prevention

Simple engineering techniques have been used to prevent the landslide, for example, by flattening the cut-slope angle the landslide movement of erosion can improve in an easy way by construction of infrastructure (Levin, 1986). Meandering environment shows us another way of landslides as an example, polygonal ground on the flood plain of the Kogosukruk River, Alaska. Scale of air photograph 1: 20.000 (Courtesy of U.S. Geological Survey)

7. Relation for meander tortuosity

The relationship between the tortuosity ratio and other parameters can be expressed as,

$$LR/LV= f\left(W, D, S, m\right) \tag{15}$$

This can be reduced to the dimensionless equation,

$$LR/LV=f\left(W/D, S, m/D\right) \tag{16}$$

If mean velocity and discharge per length of channel width are assumed as two more relevant parameters, Eq.(3) can further be modified as,

$$LR/LV=f\left(W/D, S, m/D, R, F\right) \tag{17}$$

in which $R=q/v=$ the Reynolds number; $F= V/ (gD)^{1/2}=$ the Froude number; $q=$ discharge per unit length width; $v=$ kinematic viscosity; $V=$ mean velocity; and $g=$ gravitational acceleration.

To investigate the actual relationship and its validity, river data or laboratory data for meandering flumes were needed for all the parameters involved. The study of the effect of parameters W/D, S and m/D individually on meander tortuosity, plots of LR/LV against these three parameters indicate that channels with low tortuosity ratio, i.e., more or less straight channels, have wide and shallow cross sections, steeper slopes, and relatively coarser bed material. A value of LR/LV equal to one indicates straight channels. With gradual reduction in the value of all three parameters W/D, S, and m/D, the tortuosity ratio increases, indicating that meanders become more and more acute.

Flow curvature creates superelevation and transverse flow across the section of a channel bend. The strength of the transverse current depends on boundary friction. In wide and shallow channels the ratio of roughness elements to flow depth is higher than in deep channels because of coarser bed material as well as shallower depth, the higher roughness ratio results in more frictional resistance and hence weaker transverse flow than in narrow and deep channels. In considering river patterns, channels can logically be divided into two main groups, single channel streams and multichannel streams, with a transition range between the two. Single channel streams can be further subdivided into meandering channels and straight channels with a transition between them. Meanders can be classified as regular or irregular, simple or compound, acute or flat, and sine, parabolic, circular or sine-generated curves.

Meandering channels are formed if the flow dynamics corresponds with the channel morphology. Braided channels occur if flow dynamics and channel morphology are incompatible. Alluvial channels are unstable because the stability criteria for the channel bed and for the channel banks are different.

Meander flow takes place in one single channel which oscillates more or less regularly with meandering river amplitudes that tend to increase with time. Meanders are found in beds of fine sediment with gentle slopes.

x/L (Distance)	Elevation from bottom(mm)	Run.1 (τ_{max}=100mN/m²)	Run. 2(τ_{max}=239mN/m²)	Run. 3(τ_{max}=300mN/m²)	Run. 4(τ_{max}=390mN/m²)
0	-5.00	0.63	1.06	1.27	1.5
0.05	-4.00	0.64	1.17	1.32	1.58
0.1	-3.00	0.655	1.30	1.40	1.77
0.15	-2.00	0.84	1.40	1.50	1.89
0.20	-1.00	0.92	1.5	1.67	1.96
0.25	0.00	0.97	1.58	1.78	2.06
0.30	1.00	1.09	1.70	1.93	2.2
0.35	2.00	1.19	1.80	2.00	2.3
0.40	3.00	1.25	1.85	2.08	2.3
0.45	4.00	1.25	1.92	2.1	2.4
0.50	5.00	1.27	1.94	2.1	2.4
0.55	4.00	1.27	1.95	2.1	2.4
0.60	3.00	1.14	1.87	2.0	2.4
0.65	2.00	1.00	1.76	2.0	2.36
0.70	1.00	0.79	1.67	1.97	2.29
0.75	0.00	0.71	1.50	1.87	2.2
0.80	-1.00	1.0	1.58	1.80	2.1
0.85	-2.00	1.0	1.40	1.58	1.9
0.90	-3.00	0.72	1.25	1.48	1.7
0.95	-4.00	0.62	1.10	1.20	1.5
1.00	-5.00	0.556	1.10	1.19	1.4

Table 3. Shear Stress Distribution by landslides at mendering channels experimental set-up

8. Model application

The results of model applications were carried out for the same situations as the mathematical model at the Technical University of Berlin, Institute Wasserbau and Wasserwirtschaft (Yilmaz, 1990), started with flat bed, continued until $\partial z/\partial t=0$. Then the beds were solidified, and precise measurements of the bed configuration and the velocity were performed. Plan geometries of runs consist of a sine-generated curve and an asymmetrical meander loop, respectively. The latter is derived by a Fourier series analysis on several typical bends. The meso-scale bed configuration in alluvial streams is highly dependent on the width-depth ratio of the channel. The velocity measurements were made with small mechanical current meters fixed to a 1m high frame that rested on the bottom while measuring the lower points on the profile. The frame was suspended at different levels above the bottom to collect the data represented by the higher points. Velocity profiles are plotted semi-logarithmically with the dots representing field data and the smooth lines showing model predictions. The mean velocity was calculated from a fit to the entire data set, not for each profile. Smaller dunes (0.50 m high, 2 m long) were superimposed on the large sand wave. Smith and McLean (1977) estimated the roughness parameter for both the skin friction and the form drag due to the smaller dunes to be 0.141 cm^3. Three different perturbations are recognized:

1. **Alternating bars:** The bed configuration reached an equilibrium state after one hour, and the quantitative and qualitative agreements are given. Sensitivity analyses of each term in Eqs. 2 and 3 into the development of alternating bars were also carried out. The term in the Eq. (3) $\partial v/\partial s$ was found to play the most important role in developing alternating bars.

2. **Braided bars :** The calculated velocity vectors and bed configurations were given after one hour. Divergence and convergence of flow streamlines in a wide straight channel and the meso-scale bed configuration of braided bars can be clearly seen.

3. **No bars:** The calculated velocity vector and bed configuration were given after one hour. The velocity distributions are almost uniform and the bed configuration is two-dimensional with less scour and fill than in the case of alternating bars and braided bars.

Numerical calculations are performed using the hydraulic conditions as listed in Table.4.

9. Observations

If the sediment transport behaves as bed load, the sediment surface at meandering channels will deform into transverse waves. These bed forms can have a variety of scales ranging from ripples through small dunes to fully developed dunes or sand waves. Smith (1970) gave that, under pure bed load transport, a flat sand bed is unstable at all wavelengths to small perturbations in boundary topography so that with sufficient time all infinitesimal undulations will grow in height. His analysis predicts that, for bed features of finite wave number, a fastest growing wave exists only when there is a lag between the boundary shear stress and the

Run Number (1)	Width Of Channel B(m) (2)	Size of Bed Material $Dx10^{-2}$ (3)	Average Bed Slope (4)	Flow rate Q x 10^3 (m³/s) (5)	Average Water Depth h (m) (6)	Froude number (7)
10	0.50	0.30	1/75	1.00	0.70	"/>1
8	0.50	0.30	1/100	1.25	0.80	"
5	0.50	0.30	1/120	1.15	0.69	"
9	0.50	0.30	1/60	1.20	0.75	"
7	0.50	0.30	1/80	1.00	0.70	"
9	0.50	0.30	1/90	1.10	0.73	"
11	0.50	0.30	1/110	1.20	0.75	"
5	0.50	0.30	1/70	1.00	0.70	"
8	0.50	0.30	1/50	1.10	0.74	"

Table 4. Experimental Condition for Alternating Bars

sediment transport rate; this is the ripple instability. The tendency for larger bed forms to have a seemingly discrete wavelength distribution, and a wavelength associated with fastest growth, is not explained by such a primitive stability model, so Smith (1970) suggested that wake affects also had to be taken into account.

It was showed that once perturbations are of finite amplitude, the larger stresses at the crests cause the crests to propagate faster than the troughs, thus imparting asymmetrical shapes to the waves. When the asymmetry is strong enough, the flow will separate, which creates a momentum deficit downstream of the wave crest much like that found in the wake of a circular cylinder. At the point of reattachment, the near-bottom velocity and stress are both zero. Downstream from this point, an internal boundary layer must develop beneath the momentum defect, or wake, region. The internal boundary layer adjusts to the velocity of the wake region above it, which increases downstream due to the flux of momentum into the wake from the interior. This produces two competing processes that are critical to determination of the boundary shear stress. They are : accelerating effect of an outward diffusing velocity defect; and the decelerating effect of a thickening boundary layer. In the near-field, spatial acceleration of the fluid in the wake dominates the decelerative effects of the internal boundary layer, but in the mid-field the opposite is true, and the net result is a decrease in the near-bed velocity in this region. In the far-field the boundary layer ultimately engulfs the wake entirely, and the boundary shear stress asymptotically approaches equilibrium. The essential features of this response to separation are preserved over an upsloping surface such as the stoss side of a bed form. Consequently, the resulting maximum in the stress profile has important consequences for both bed deformation and bed-form growth.

10. Conclusions

The flow resistance in a meander bend is considerably increased due to the form resistance of the patterns about which much is not known. It depends on a number of factors including grain friction, form resistance of two- and three dimensional patterns, skin friction of the non-separated oscillatory component and the sediment transport rate.

Author details

Levent Yilmaz

Address all correspondence to: lyilmaz@itu.edu.tr

Hydraulic Division, Civil Engineering Department, Technical University of Istanbul, Maslak, Istanbul, Turkey

References

[1] Bray, D. I. *Generalized Regime-Type Analysis of Alberta Rivers*, thesis presented to the University of Alberta, at Edmonton, Canada, in partial fulfilment of the requirements for the degree of Doctor of Philosophy, (1972).

[2] Church, M, & Kellerhals, R. On the Statistics of Grain Size Variation Along a gravel River, *Canadian Journal of Earth Sciences*, 15(7), (1978).

[3] Garde, R. J. Hydraulic Geometry and Resistance of Gravel-Bed Rivers, *Proceedings of the Indian National Science Academy*, 67 (A), (2001). , 2001, 597-609.

[4] Lacey, G. A General Theory of Flow in Alluvium, *Journal of the Institution of Civil Engineers*, 27 (5518), (1947).

[5] Levin, H. L. Contemporary Physical Geology, Saunders College Publishing, New York, (1986). , 558.

[6] Kellerhals, R, & Bray, D. I. Sampling Procedures for Coarse Fluvial Sediments, *Journal of the Hydraulics Division*, 97 (HY8), (1971).

[7] Watkins, J. S, Bottino, M. L, & Morisawa, M. Our Geological Environment, Philadelphia, Saunders College Publishing, (1975). , 335.

Numerical Modelling of Sediment Transport

Chapter 9

Quasi-3D Modeling of Sediment Transport for Coastal Morphodynamics

Yun-Chih Chiang and Sung-Shan Hsiao

Additional information is available at the end of the chapter

1. Introduction

Sand transport plays a very important role in many aspects of coastal and marine engineering. The balance of moving sands influences the construction of harbours, coastal defence, offshore wind turbine and oil rig, offshore platform and pipeline and many other engineering. Coastal sands may be carried by currents (such as tidal currents, wind-driven-currents, wave-driven-currents, storm surge driven currents), or by waves (monsoon waves or typhoon waves), or influenced by bedform changes, or all of them acting together and interacting in general sea state.

We can easily consider a sediment budget for a coastal area where a control section or a control volume is selected. The changing rate of net accretion or erosion of the coastal area of sea bed-level depends on the net transport rates at which sediments are entering or leaving the control section or the control volume. If the sum of the inflow sediment transport rates is larger than that of the outflows, the bed-level will tend to accrete; if the sum of the inflow sediment transport rates is smaller than that of outflow rates, the bedform will erode.

Consequently, accurate prediction of sediment transport rates is an important element in coastal engineering, foundations of offshore structures and morphological studies for the coastal environment. The procedure of the coastal morphological modeling system is shown as Fig. 1. The prediction of net sediment transport rates is a subject of great importance to coastal engineers and morphological modellers concerned with mediumand long-term shoreline changes. The aim of this chapter is to provide models for calculating the hydrodynamics and dynamic quantities of sediment transports in coastal zone, especially for the applications in surf zone.

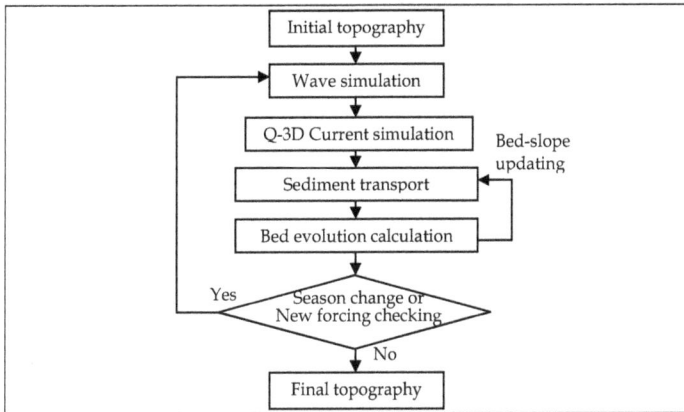

Figure 1. The procedure of the coastal morphological modeling system

1.1. Threshold of sediment motion

In order to estimate the changes of sediment budget, a quantitative evaluation of the net sediment transport rates are required. The sediment transport rate is defined as the amount of sediment per unit time entering or leaving a control volume, which is the vertical plane of unit width perpendicular to the sea mean water-level. The mechanics of sediment motion is depended on the effects of sediment dynamics while the frictions exerted on the sea bed by the hydrodynamics forcing agents, such as waves and currents.

While the friction exerted on the sea bed by waves and currents, the sediment 'entrainment' is the sand grains are carried up from bed. The 'bed load' sediment transport is the entraining sand grains rolling, hopping and sliding along the bedform, and is dominant with the inertial force and drag force on grains less than gravity force for slow flow or large grains. Suspended load sediment transport is the portion of the entraining sediment that is carried by the larger flow (or the wave large enough) which settles slowly enough and moves with the stream.

To estimate the sediment transport rate is more difficult, it may be divided into two components by cross-shore and longshore direction. The cross-shore sediment transport is mainly carried by the skewness and asymmetry wave orbital motion and cycle-mean water level around the surf zone. The longshore sediment transport is primarily dominated with wave-driven currents. In order to quantitate the sediment transport rate, included both bed load and suspended load, in each direction, an empirical relationship has been derived between 'transport rate' and the energetic-based components, such as bed shear-stress, wave-driven currents, wave orbit velocity….. and so on. The agreement between measurements (or experiments in hydraulic laboratory) and calculations associated with the relationship has been widely applied and predicts accurately sediment transport along long, straight-like beaches.

The total load sediment transport models are widely applied with coastal engineering in last three decades. Although the models are relatively simple and easy to use, but there are some weaknesses when applied around the surf zone. We will discuss in next section.

1.2. Interaction between sediment motion and bed features for morphodynamics

Nearshore sandbars are the important and popular feature of natural beaches morphody-namics. The cross-shore location of sandbars changes by the interactions between the sandbar and the sediment transport fluxes from waves and wave-driven-currents. Hoefel and Elgar (2003) indicated the mechanics of wave-induced sediment transport and sand-bar migration: large waves breaking on the sandbars caused offshore mean currents, which maximum near the sandbar crest, will lead sandbars moved offshore; small waves pitching forward on the sandbars made the onshore acceleration skewness of wave orbi-tal velocities, which maximum near the sandbar crest, will lead the sandbar moved on-shore.

The nearshore sandbars could protect shorelines from wave attack by dissipating wave en-ergy offshore through sandbar-crest-induced wave breaking. In general coasts, the dynam-ic behavior of nearshore sandbars are similar to quasi-cycle for storm and seasons waves alternated. However, the formation and evolution of sandbars are very important to coast-al planners and engineers. Prediction of the dynamic behaviour of nearshore sandbar sys-tems could be of great importance, there are many studies about the evolutions and migrations of nearshore sandbars by numerical simulations in last decade (Hsu et al., 2006; Long et al., 2006; Ruessink et al., 2007; Drønen and Deigaard, 2007; Houser and Greenwood, 2007; Ruessink and Kuriyama, 2008; Ruessink et al., 2009; Pape et al., 2010; Almar et al., 2010).

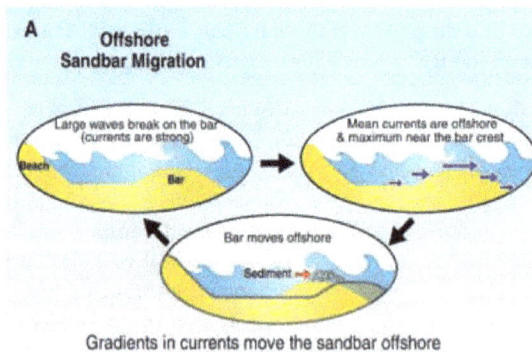

Figure 2. The mechanics of offshore sandbar migration. (Hoefel and Elgar,2003)

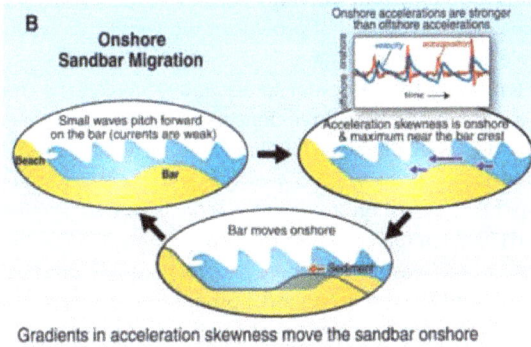

Figure 3. The mechanics of onshore sandbar migration. (Hoefel and Elgar,2003)

According to Hoefel and Elgar (2003), two mechanisms are commonly used in the explanation of morphodynamics of sandbars migration. The first mechanism type is the migration of offshore sandbars. The offshore sandbars migrated seaward observed during storms were driven primarily by a maximum in the offshore mean current (Under highly energetic storm conditions, breaking waves cause near bottom seaward flows, also called "undertow") near the sandbar crest. Offshore sandbar migration during storms results from feedback and interaction between breaking waves driven the "undertow" and bathymetric evolution (Elgar et al., 2001).

The second mechanism associates the migration of onshore sandbars. There are many studies have suggested mechanisms that could drive sandbars migration shoreward. Trowbridge and Young (1989) and Trowbridge and Madsen (1984) demonstrated that nonlinear wave boundary layer processes might play a role. Onshore sandbars migration might also be derived by the systematic changes in wave kinematics when passing over nearshore sandbars. As waves shoaling, their shapes are often described as "skewed" and "asymmetric" (Elgar, 1987), the mean water elevation depressed (the "Wave set-down") leads mean currents been weak. The non-breaking wave caused sediment transport over sandbars is driven predominately by wave asymmetric orbital velocities. Under the steep skewed and asymmetric waves, the water particle velocity is accelerated strongly as the asymmetric orbital velocity rapidly changes from maximum offshore to maximum onshore (e.g., Elgar et al., 1988).

In order to describe morphodynamics sandbars well, a key parameter for cross-shore sediment transport under breaking and near-breaking waves is well performed the near-bed skewed and asymmetric wave orbital velocity. Therefore, the three dimensionality of the hydrodynamic system should be considerable and must be taken into account. Most sediment transport models are based on phase-averaged wave models, depth-integrated hydraulic models (currents and wave driven currents) and sediment transport formula. In the hydraulic models, nearshore currents have previously been predicted by using two-dimensional models in the horizontal plane (2DH model). However, in the surf zone, the direction of current vectors near the water surface is different from that at the sea-bottom because of the effect of undertow velocities. Nearshore currents have spiral profiles in the vertical direction. Undertows also play an important role in the morphodynamical changes on a littoral beach such as the cross-shore migration of longshore bars. In order to accurately predict the changes of sandbars migration, it is very important that the three-dimensional distribution of nearshore currents is determined. Therefore, well predicted nonlinear wave dynamics, vertical current and sediment transport models can be a good tool for nearshore sandbar morphodynamics, especially for cases where cross-shore transport mechanisms over sandbars are important. Drønen and Deigaard (2007) compared 2D horizontal depth-integrated approach and quasi-3D numerical model with formations of alongshore bars on gradual slope beach by normal and oblique incident waves, and the quasi-3D model produces a crescentic bar while the depth integrated model predicts almost straight sections of the bar interrupted by rip channels. Consequently, considering the accuracy and efficiency, a nonlinear waves model with quasi-3D sediment transport model can be applied as morphodynamics sandbar models.

1.3. The importance of quasi-3D sediment transport modeling

The numerical simulation of hydrodynamic and sediment transport processes form a powerful tool in the description and prediction of morphological changes and sediment budgets in the coastal zone. One of the key elements in a morphodynamics model is the correct quantification of local sand transport. Most sediment transport models are based on phase-averaged wave models, depth-integrated hydraulic models (currents and wave driven currents) and sediment transport formula. In the hydraulic models, nearshore currents have previously been predicted by using two-dimensional models in the horizontal plane (2DH model). However, in the surf zone, the direction of current vectors near the water surface is different from that at the sea-bottom because of the effect of undertow velocities. Nearshore currents have spiral profiles in the vertical direction. Undertows also play an important role in the morphological changes on a littoral beach such as the cross-shore migration of longshore bars. In order to accurately predict the changes of beach profile, it is very important that the three-dimensional distribution of nearshore currents is determined. Therefore, well predicted vertical current and sediment transport models can be a good tool for coastal area morphological modelling, especially for cases where cross-shore transport mechanisms are important. Considering the accuracy and efficiency, a quasi-3D model can be applied as a coastal profile model or a coastal area model.

Some models for determining the vertical distribution of nearshore currents have previously been proposed. de Vriend et al. (1987) presented a semi-analytical model and suggested that a 3D model is required when the sediment transport in the cross-shore direction becomes important; and then Svendsen and Lorenz (1989) proposed an analytical model composed of cross-shore and longshore current velocities. In recent years, many quasi-3D numerical models have been developed by extending 2DH model with one-dimensional velocity profile model defined in the vertical direction (1DV model), have also been proposed (Sanchez et al., 1992; Briand and Kamphuis, 1993; Okayasu et al., 1994; Elfrink et al., 1996; Rakha, 1998; Kuroiwa et al,. 1998; Drønen and Deigaard, 2000; Davis and Thorne, 2002; Fernando and Pan, 2005; Drønen and Deigaard, 2007; Li et al., 2007). In these models, the mean flow is determined by the 2DH model, and the velocity profiles across water column in the vertical direction are resolved by using a 1DV model. While surface wave field and 3D flow field have been analyzed, sediment transport vectors in the horizontal plane can be calculated with the sediment transport profile across water column in the vertical direction.

The aim of this chapter is to develop an accurate model for estimation of local sediment transport rate of the nearshore both inside and outside of the surf zone. A two-dimensional 2D fully nonlinear Boussinesq wave module is combined with a quasi-3D hydrodynamic module (2DH and extended 1DV module). The 1DV hydrodynamic modules similar to those described by Elfrink et al. (1996) with surface-roller concept and a one-equation turbulence model are developed. The calculation of sediment transport rates is based on the formula with wave asymmetric and ripple-bed effects developed by Lin et al. (2009). The quasi-3D hydrodynamic modules are validated and compared, for regular waves over fixed beds. The local sediment transport rates is also calculated and validated with experimental data.

2. 2DH waves and nearshore currents models

In this section, the two-dimensional wave and nearshore current models are described as below:

2.1. Wave model

The wave model is based on the fully nonlinear Boussinesq equation developed by Wei et al. (1995); the equation is expressed by velocity with an arbitrary water depth. Bottom friction, wave breaking and subgrid lateral turbulent mixing as proposed by Kennedy et al. (2000), are also expressed by equations. The governing equations are shown as below:

$$\frac{\partial \eta}{\partial t} + \nabla \cdot [(h+\eta)u_\alpha] + \nabla \cdot \{(\frac{z_\alpha^2}{2} - \frac{h^2}{6})h\nabla(\nabla \cdot u_\alpha) + (z_\alpha + \frac{h}{2})h\nabla[\nabla \cdot (hu_\alpha)]\} = 0 \tag{1}$$

$$\frac{\partial u_\alpha}{\partial t} + g\nabla\eta + (u_\alpha \cdot \nabla u_\alpha) + z_\alpha \{\frac{z_\alpha}{2}\nabla(\nabla \cdot \frac{\partial u_\alpha}{\partial t}) + \nabla[\nabla \cdot (h\frac{\partial u_\alpha}{\partial t})]\} + R_f - R_b - R_s = 0 \tag{2}$$

In the above equations, $x=(x,y)$ are the horizontal coordinates coincident with the still water surface and z is the vertical coordinate; t is the time; ∇ is the horizontal gradient operator, defined as $(\partial/\partial x, \partial/\partial y)$; $\eta(x,t)$ is water surface elevation; g is the gravitational acceleration. $h=h(x)$ represents the water depth, $u_\alpha=(u_p,v_p)$ is the particle velocity vector at $z=z_\alpha$. R_f, R_b and R_s are the effects of bottom friction, wave breaking and subgrid lateral turbulent mixing, respectively. The detail mathematical operations are shown below,

$$R_f = \frac{K}{h+\eta} u_\alpha \mid u_\alpha \mid \tag{3}$$

K is the friction coefficient. The wave breaking term, $R_b = (R_{bx}, R_{by})$ is represented as

$$R_{bx} = \frac{1}{h+\eta}\{[\nu((h+\eta)u_\alpha)_x]_x + \frac{1}{2}[\nu((h+\eta)u_\alpha)_y + \nu((h+\eta)v_\alpha)_x]_y\} \tag{4}$$

$$R_{by} = \frac{1}{h+\eta}\{[\nu((h+\eta)v_\alpha)_y]_y + \frac{1}{2}[\nu((h+\eta)u_\alpha)_y + \nu((h+\eta)v_\alpha)_x]_x\} \tag{5}$$

The eddy viscosity (ν) is defined as

$$\nu = B\delta_b^2 (h+\eta)\eta_t \tag{6}$$

Kennedy et al. (2000) proposed the mixing length, δ_b, is 1.2. The parameter B controls the occurrence of energy dissipation is defined as

$$B = \begin{cases} 1, & \eta_t \geq 2\eta_t^* \\ \dfrac{\eta_t}{\eta_t^*} - 1, & \eta_t^* < \eta_t \leq 2\eta_t^* \\ 0, & \eta_t \leq \eta_t^* \end{cases} \tag{7}$$

The onset and cessation of wave breaking using the parameter, η_t^*, is represented as

$$\eta_t^* = \begin{cases} \eta_t^{(F)}, & t - t_0 > T^* \\ \eta_t^{(I)} + \dfrac{t-t_0}{T^*}(\eta_t^{(F)} - \eta_t^{(I)}), & 0 \leq t - t_0 < T^* \end{cases} \tag{8}$$

$$\eta_t^{(I)} = 0.65\sqrt{gh}, \eta_t^{(F)} = 0.15\sqrt{gh}, T^* = 5\sqrt{h/g} \tag{9}$$

where T^* is the transition time, t_0 is the time when wave breaking occurs, and $t-t_0$ is the age of the breaking event. The subgrid lateral mixing terms $R_s = (R_{sx}, R_{sy})$ is displayed as follow

$$R_{sx} = \frac{1}{h+\eta}\{[v_s((h+\eta)u_\alpha)_x]_x + \frac{1}{2}[v_s((h+\eta)u_\alpha)_y + v_s((h+\eta)v_\alpha)_x]_y\}$$ (10)

$$R_{sy} = \frac{1}{h+\eta}\{[v_s((h+\eta)v_\alpha)_y]_y + \frac{1}{2}[v_s((h+\eta)u_\alpha)_y + v_s((h+\eta)v_\alpha)_x]_x\}$$ (11)

The parameter v_s is the eddy viscosity due to the subgrid turbulence.It can be calculated by:

$$v_s = c_m \Delta x \Delta y [(U_x)^2 + (V_y)^2 + \frac{1}{2}(U_y + V_x)^2]^{1/2}$$ (12)

where c_m is the mixed coefficient.

2.2. 2DH nearshore current model

Based on computed characteristics of wave fields, the radiation stress terms can then be found and input into the depth integrated (2DH) nearshore current module, which solves the depth-and-wave-period averaged continuity and momentum equations at each local point on horizontal plane, for calculating wave driven current:

$$\frac{\partial \eta}{\partial t} + \frac{\partial}{\partial x}[U(h+\eta)] + \frac{\partial}{\partial y}[V(h+\eta)=0]$$ (13)

$$\frac{\partial U}{\partial t} + U\frac{\partial U}{\partial x} + V\frac{\partial U}{\partial y} = fV - g\frac{\partial \eta}{\partial x} + \frac{1}{\rho}\left(\frac{\partial \tau_{xx}}{\partial x} + \frac{\partial \tau_{yx}}{\partial y}\right) + \frac{1}{\rho(h+\eta)}(\tau_{sx}-\tau_{bx}) - \frac{1}{\rho(h+\eta)}\left(\frac{\partial S_{xx}}{\partial x} + \frac{\partial S_{yx}}{\partial y}\right)$$ (14)

$$\frac{\partial V}{\partial t} + U\frac{\partial V}{\partial x} + V\frac{\partial V}{\partial y} = -fV - g\frac{\partial \eta}{\partial y} + \frac{1}{\rho}\left(\frac{\partial \tau_{xy}}{\partial x} + \frac{\partial \tau_{yy}}{\partial y}\right) + \frac{1}{\rho(h+\eta)}(\tau_{sy}-\tau_{by}) - \frac{1}{\rho(h+\eta)}\left(\frac{\partial S_{xy}}{\partial x} + \frac{\partial S_{yy}}{\partial y}\right)$$ (15)

where U and V are depth-integrated nearshore current velocities in x and y direction respectively, S_{xx}, S_{xy} and S_{yy} are radiation stress tensor, g is acceleration due to gravity, ϱ is water density, h is water depth, η is water surface elevation, τ_{xx}, τ_{xy} and τ_{yy} are Reynolds stress tensor, τ_s and τ_b are shear stress on surface and bottom. The friction factor for combined wave-current flow in bottom shear stresses and the mixing coefficient in Reynolds stresses are suggested by Chiang et al. (2010).

3. Quasi-3D extended: 1DV velocity model

Fig. 4 depicts the coordination of the quasi-3D hydraulic system. The x coordinate is defined in the cross-shore direction towards shore. The y coordinate denotes the long shore direction. The Z coordinate is toward from sea bed to surface in depth-direction.

3.1. Numerical formulation

The distributions of the velocity profiles in the long-shore and cross-shore direction at each local point along the vertical water column are found through the following momentum equations:

$$\frac{\partial u}{\partial t} = -\frac{1}{\rho}\frac{\partial p}{\partial x} + \frac{\partial}{\partial z}\left(v_t\frac{\partial u}{\partial z}\right) + \frac{1}{\rho}\frac{\partial \overline{\tau^x}}{\partial z} \tag{16}$$

$$\frac{\partial v}{\partial t} = -\frac{1}{\rho}\frac{\partial p}{\partial y} + \frac{\partial}{\partial z}\left(v_t\frac{\partial v}{\partial z}\right) + \frac{1}{\rho}\frac{\partial \overline{\tau^y}}{\partial z} \tag{17}$$

where u and v are wave-period-averaged velocities across the water column in the cross-shore and long-shore direction respectively, p is the pressure, v_t is the turbulence viscosity, $\overline{\tau^x}$ and $\overline{\tau^y}$ are wave-period-averaged wave-induced shear stress through water column. In equation (16) and (17), the convection terms in the left hand side of the equations have been neglected which follows the conventional 1DV type of models by assuming the spatial gradient at the interested site is small. This is due to the fact that the convection terms have been counted in the 2DH nearshore current model and the 1DV model is only used to resolve the vertical profile of the flow velocities and sediment concentration in suspension.

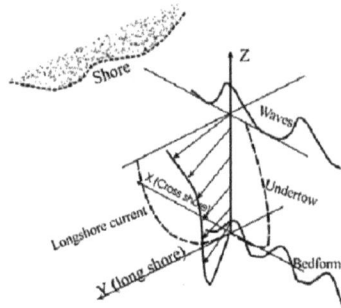

Figure 4. The coordination of the quasi-3D hydraulic system

The pressure gradient term may be divided into two components by inside/outside boundary layer. The pressure gradient term can be easily calculated from the variation of cycle-mean water free surface. Within the boundary layer, we can calculate from the time differential of velocity at boundary layer:

$$\frac{\partial p}{\partial x} = \frac{dU_{w0}^x}{dt}, \frac{\partial p}{\partial y} = \frac{dU_{w0}^y}{dt} \tag{18}$$

where U_{w0}^x and U_{w0}^y is the wave orbit velocity outside the boundary layer.

The cycle-mean wave-induced shear stress under waves is consist of wave motion component (τ_w), wave breaking surface roller component (τ_r), boundary layer streaming component (τ_b) and mean water surface changed (wave set-up/set-down) component (τ_{su}), as shown in Fig. 5. The Boundary layer streaming components can be neglected, because it is small than the others. The total shear stress is defined as

$$\bar{\tau} = \tau_w + \tau_r + \tau_{su} \tag{19}$$

The shear stress distribution due to wave motion (τ_w) is in accordance with the derivations of Deigaard and Fredsoe (1989):

$$\tau_w = \frac{-1}{c} \frac{dE_f}{dx} \left(1 + \frac{h-y}{2h} \right) \tag{20}$$

where E_f is cycle-mean energy flux due to wave motion (Svendsen, 1984):

$$E_f = \rho g c \bar{\eta}^2 = B \rho g c H^2 \tag{21}$$

The coefficient B is 1/12 while wave breaking and 1/8 in general.

The shear stress due to wave breaking (τ_w) can be calculated by the concept of surface roller. According to Svendsen (1984) and Deigaard et al. (1986), the shear stress of surface roller is assumed to be constant by experiment:

$$\tau_r = -\frac{\rho}{T} \frac{d(Ac)}{dx} \tag{22}$$

where c is wave celerity, T is wave period, and A is the area of surface roller (as shown in Fig. 7) can be easily calculated by $A=0.09H$ (Deigaard et al., 1986).

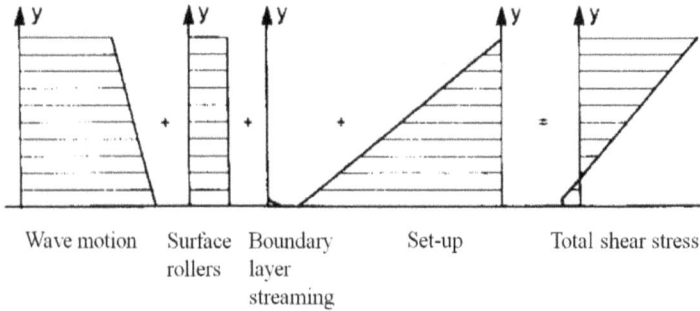

| Wave motion | Surface rollers | Boundary layer streaming | Set-up | Total shear stress |

Figure 5. Total shear stress and its component from surface to bed in vertical direction under wave

The shear stress due to variation of mean water level (τ_{su}) can be easily calculated by

$$\tau_{su} = -\rho g h \left(1 - \frac{z}{h}\right)\frac{\partial \overline{\eta}}{\partial x} \tag{23}$$

The turbulence viscosity should be specified through certain turbulence models. In the present study, the one-equation k-closure is adopted as follows:

$$\frac{\partial k}{\partial t} = \frac{\partial}{\partial z}\left(\frac{v_{tb}}{\sigma_k}\frac{\partial k}{\partial z}\right) + v_{tb}\left[\left(\frac{\partial u}{\partial z}\right)^2 + \left(\frac{\partial v}{\partial z}\right)^2\right] + \frac{P_r}{\rho} - \varepsilon \tag{24}$$

where k is the turbulent kinetic energy, constant σ_k=1.0, P_r is the turbulent production due to surface wave breaking by surface-roller concept and ε is dissipation rates which is given as:

$$\varepsilon = c_1 k^{3/2} / l \tag{25}$$

in which l is the turbulence length scale computed as (Deigaard et al., 1991):

$$l = \begin{cases} c_1^{1/4}\kappa z & , z \leq l_{max} / \left(\kappa c_1^{1/4}\right) \\ l_{max} & , z > l_{max} / \left(\kappa c_1^{1/4}\right) \end{cases} \tag{26}$$

where l_{max}=0.1h, c_1=0.09.

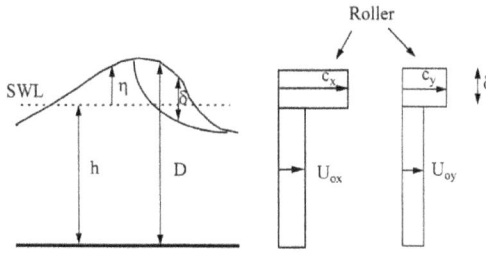

Figure 6. The concept of surface roller due to wave breaking (Deigaard et al., 1986)

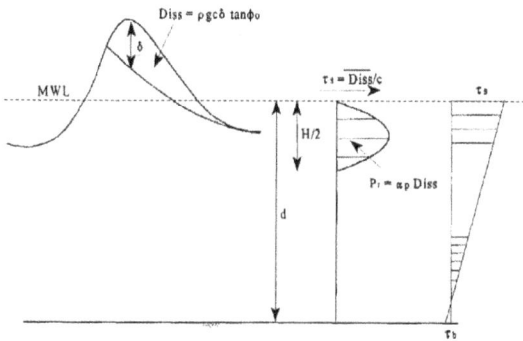

Figure 7. The turbulence generated and energy dissipation due to wave breaking (Deigaard et al., 1986)

The turbulence generated and energy dissipation at the water surface due to wave breaking is computed following Deigaard et al. (1991), as shown in Fig. 7 :

$$P_r = \alpha_p \text{DISS} \tag{27}$$

where constant $\alpha_p = 0.33$, the energy dissipation term DISS is suggested by Deigaard (1989):

$$\text{DISS} = \rho g c \delta_r \tan \phi_0 \tag{28}$$

where c is wave celerity, φ_0 is wave initial breaking angle as 10 degree, δ_r is thickness of surface roller head suggested by Deigaard et al. (1986):

$$\delta_r = \frac{2(h + H/2)}{L} \tag{29}$$

3.2. Boundary condition and key parameters

3.2.1. Eddy viscosity

According to Brøker et al. (1991), the eddy viscosity is calculated by assuming the total kinetic energy could be the sum of three contributions: the oscillatory near bed boundary layer (v_{tw}), wave breaking (v_{tb}), and the time-averaged currents (v_{tU}) respectively. The eddy viscosity outside and inside the wave boundary layer is calculated by eq. (30), and eq. (31).

$$v_t = \sqrt{v_{tU}^2 + v_{tb}^2} \tag{30}$$

$$v_t = \sqrt{v_{tw}^2 + v_{tb}^2} \tag{31}$$

The component of oscillatory near bed boundary layer (v_{tw}) is calculated from (Okayasu et al., 1988)

$$v_{tw} = z\kappa U_f \tag{32}$$

where $\kappa=0.4$ is von karman constant, $U_f = \sqrt{\tau_b/\rho}$ is the frictional velocity under wave.

The component of wave breaking (v_{tb}) is calculated from (Rakha, 1998):

$$v_{tb} = l\sqrt{k} \tag{33}$$

where k is turbulence kinetic energy, and l is turbulence length scale.

The contribution of the time-averaged currents (v_{tU}) is calculated also from (Rakha, 1998):

$$v_{tU} = l_{max}^2 \sqrt{\left(\frac{du}{dz}\right)^2 + \left(\frac{dv}{dz}\right)^2} \tag{34}$$

where l_{max} is the same as eq. (26).

3.2.2. Thickness of wave boundary layer

The thickness of wave boundary layer (δ_w) is calculated from (Soulsby et al., 1993):

$$\delta_w = \frac{u_{*\max} T}{2\pi} \tag{35}$$

where $u_{*\max} = \sqrt{\tau_{\max}/\rho}$, $\tau_{\max}$ is the maximum shear stress by waves and currents.

4. Sediment transport formula

The total sediment transport is consisted of bed load and suspended load suggested by Chiang et al. (2011):

$$q_{total} = q_b + q_s \tag{36}$$

where q_b is the bed load sediment transport rate, and the q_s is the suspended load sediment transport rate.

4.1. Bed load sediment transport

Following Chiang et al. (2011), the instantaneous bed load transport rate due to wave asymmetric and ripple-bed effects is given as:

$$q_b = \overline{\Phi_b} \left[g(s-1) d_{50}^3 \right]^{1/2} \tag{37}$$

$$\overline{\Phi_b} = \left(\overline{\Phi_{bx}}^2 + \overline{\Phi_{by}}^2 \right)^{1/2} \tag{38}$$

$$\overline{\Phi_{bx}} = 11 \left[t_c \left(\theta_{cx} - \theta_c \right)^{1.65} - t_t \left(\theta_{tx} - \theta_c \right)^{1.65} \right] \tag{39}$$

$$\overline{\Phi_{by}} = 11 \left(\theta_y - \theta_c \right)^{1.65} \tag{40}$$

where g being the gravity acceleration, s the relative density ($s = \rho_s/\rho$, with ρ_s the density of sediment), d_{50} is median diameter, Φ_b is dimensionless sediment transport rates, subscript x, y indicates along wave propagating and perpendicular direction. The wave-crest-half period, wave-trough-half period, and perpendicular wave-period averaged equivalent Shields stress are suggested by Chiang et al. (2011) as following:

$$\theta_{cx} = \frac{1}{2} \frac{(f_{cw})_{cx} u_{cx}^2}{(s-1)gd_{50}} \tag{41}$$

$$\theta_{tx} = \frac{1}{2} \frac{(f_{cw})_{tx} u_{tx}^2}{(s-1)gd_{50}} \tag{42}$$

$$\theta_y = \frac{1}{2} \frac{f_c U_0^2 \sin^2(\phi)}{(s-1)gd_{50}} \tag{43}$$

where f_w, f_c, and f_{cw} are the friction coefficient of wave, current and wave-current interaction, u_{cx} and u_{tx} are equivalent phase-averaged near-bed velocity under crest and trough half period.

$$f_w = \begin{cases} 0.00251 \times \exp\left[5.21 \times \exp\left(\dfrac{a_w}{K_s}\right)^{-0.19}\right], & \dfrac{a_w}{K_s} > 1.57 \\[2ex] 0.3, & \dfrac{a_w}{K_s} \le 1.57 \end{cases} \tag{44}$$

$$f_c = 2\left[\frac{0.4}{\ln(h/z_0)-1}\right]^2 \tag{45}$$

$$(f_{cw})_i = \varepsilon_i f_c + (1-\varepsilon_i) f_{wi} \tag{46}$$

where $i = c$(crest) or t(trough), and weigh coefficient given as:

$$\varepsilon_c = \frac{U_0}{U_{w\,max} + U_0} \tag{47}$$

$$\varepsilon_t = \frac{U_0}{|U_{w\,min}| + U_0} \tag{48}$$

In eq. (44), a_w is half wave orbit closure:

$$a_w = \frac{U_w T}{2\pi} \tag{49}$$

According to Soulsby (1997), the total roughness (K_s) of sea bed during wave passing is consisted of grain related component (K_{ss}), form drag component (K_{sf}) and sediment transport component (K_{st}):

$$K_s = K_{ss} + K_{sf} + K_{st} \tag{50}$$

In eq. (37), the grain related component is given as (Nielson, 1992 and Soulsby, 1997),

$$K_{ss} = 2.5 \times d_{50} \tag{51}$$

The form drag component associated with sandy ripples is defined (Davis and Villaret, 2003):

$$K_{sf} = 25\eta_r \left(\eta_r / \lambda_r \right) \tag{52}$$

where the λ_r and η_r is the wave length and wave height of full-developped sandy ripples, the empirical relationship formula is shown as Nielson (1992). The sediment transport component of friction roughness (K_{st}) is given as (Wilson, 1989):

$$K_{st} = 50\theta_s d_{50} \tag{53}$$

In eq. (40), d_{50} is the sediment median grain size, and θ_s is the entraining bed shear-stress.

4.2. Suspended load sediment transport

The suspended sediment transport can be calculated by integrated sediment concentration (C) of vertical water column from bottom to surface. The sediment concentration C in the water column is found through the mass conservation equation:

$$\frac{\partial C}{\partial t} = w_s \frac{\partial C}{\partial z} + \frac{\partial}{\partial z} \left(\varepsilon_{sd} \frac{\partial C}{\partial z} \right) \tag{54}$$

where w_s is settling velocity, ε_{sd} is the coefficient of sediment diffusion. According to Fredsøe and Deigaard (1992), ε_{sd} is equivalent to turbulent eddy viscosity v_t. The settling velocity is defined as (Soulsby, 1997):

$$w_s = \frac{v}{d_{50}} \left[\left(10.36^2 + 1.049 D_*^3 \right)^{1/2} - 10.36 \right] \tag{55}$$

where D_* is the dimensionless grain diameter:

$$D_* = \left[\frac{g(s-1)}{v^2} \right]^{1/3} d_{50} \tag{56}$$

After the suspended sediment concentration and vertical velocity profiles obtained, the instantaneous suspended transport rate can be evaluated as:

$$q_s = \int_{z_a}^{h} \overline{c}(z)\overline{u}(z)dz \tag{57}$$

where z_a is the The reference height a is specified as $2.5d_{50}$ (Soulsby, 1997), h is the water depth.

5. Model validation and discussion

The quasi-3D sediment tranport model is validated against wave flume tests and existed numerical models.

5.1. Model validation with Cox and Kobayashi (1996)

The wave and quasi-3D model system described above was firstly tested by regular wave with uniform sloped bed, and compared with experiment (Cox and Kobayashi, 1996). The test conditions are given: wave height 13.2cm, wave period 2.2 sec, the length of wave flume is 14.0m, width is 1.5m, depth is 30.0cm, slope-1:35; sand medium diameter is 1.0mm, and bed roughness height k_s=1mm. The numerical results of wave height and layouts of wave flume is shown as Fig. 8. In Fig. 8, the black triangles indicate experimental data by Cox and Kobayashi (1996), the red line is the results of present wave model, and the black line is the numerical results from Rakha (1998). The wave nonlinear effects and wave breaking and regenerating can be observed well, it's shown good agreement with experiment.

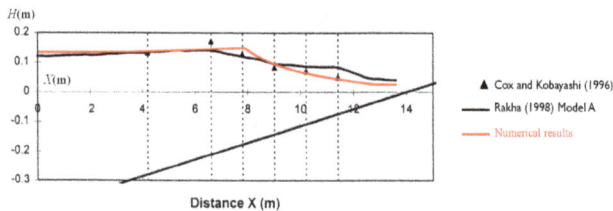

Figure 8. The validation of wave height and the layout of wave flume with results from Cox and Kobayashi (1996)

The comparisons of the vertical velocity profile at various position used by Cox and Kobayashi (1996) are also adopted here to assess the present model's accuracy for wave propagating at a slope in wave flume. The velocity profiles of various position are shown as Fig. 9 ~ Fig. 14. The black triangles also indicate experimental data by Cox and Kobayashi (1996), the blue line is the results of present 1DV model, and the black line is the numerical results from Rakha (1998). Compared with experiments, they are shown good performance for the validation of distribution of the vertical velocity profile before and after wave breaking.

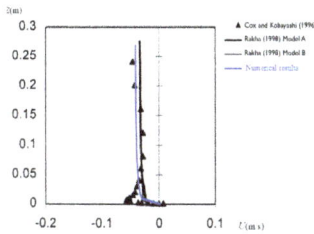

Figure 9. The validation of vertical velocity profile at wave flume x=4.2m from Cox and Kobayashi (1996)

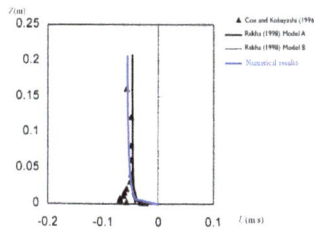

Figure 10. The validation of vertical velocity profile at wave flume x= 6.6m from Cox and Kobayashi (1996)

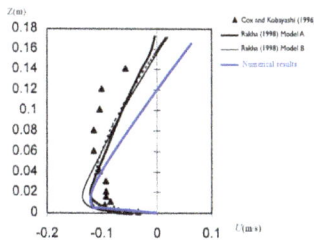

Figure 11. The validation of vertical velocity profile at wave flume x=7.8m from Cox and Kobayashi (1996)

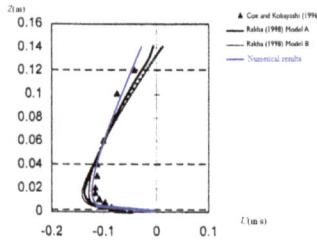

Figure 12. The validation of vertical velocity profile at wave flume x= 9.0m from Cox and Kobayashi (1996)

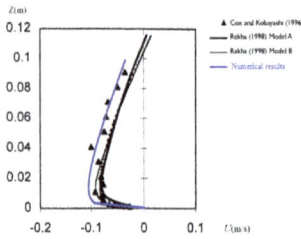

Figure 13. The validation of vertical velocity profile at wave flume x= 10.2m from Cox and Kobayashi (1996)

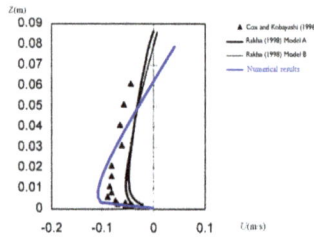

Figure 14. The validation of vertical velocity profile at wave flume x= 11.4m from Cox and Kobayashi (1996)

5.2. Model validation with Ting and Kirby (1994)

The Boussinesq wave model and quasi-3D nearshore current model system described above was tested by regular wave with uniform sloped bed, and compared with experiment (Ting and Kirby, 1994, test 1). The test conditions are given: wave height 12.5cm, wave period 2.0 sec, the length of wave flume is 13.0m, width is 1.5m, depth is 30.0cm, slope-1:35; sand medium diameter is 1.0mm, and bed roughness height ks=1mm. The numerical results of wave height and layouts of wave flume is shown as Fig. 15. In Fig. 15, the black triangles indicate experimental data by Ting and Kirby (1994), the red line is the results of present wave model, and the black line is the numerical results from Rakha (1998). The wave nonlinear effects and wave breaking and regenerating can be observed well, it's also shown good agreement with experiments.

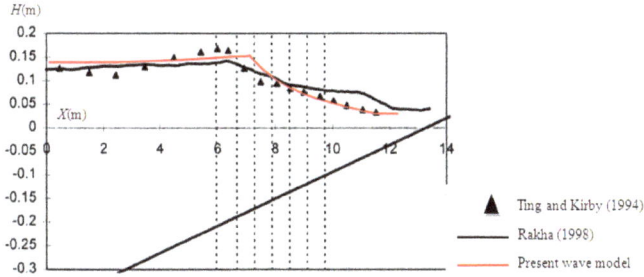

Figure 15. The validation of wave height and the layout of wave flume with results from Ting and Kirby (1994), test 1

The comparisons of the vertical velocity profile at various position used by Ting and Kirby (1994) are also adopted here to assess the present model's accuracy for wave propagating at a slope in wave flume. The velocity profiles of various position are shown as Fig. 16 ~ Fig. 21. The black triangles also indicate experimental data by Ting and Kirby (1994), the blue line is the results of present 1DV model, and the black line and dotted line are the numerical results from Rakha (1998) model A and model B. Compared with experiments, they are shown good performance for the validation of distribution of the vertical velocity profile before and after wave breaking. Because of full-nonlinear Bossinesq wave model and accuracy wave breaking dissipation terms in 1DV model, the present model performed well than Rakha (1998).

5.3. Validation for sediment transport calculations

To assess the present quasi-3D sediment transport model's ability of prediction for local sediment transport under combined waves and currents for a range conditions, a series of

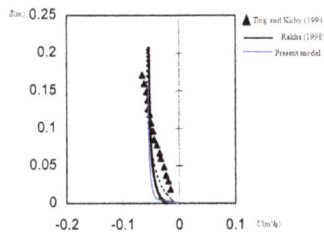

Figure 16. The validation of vertical velocity profile at wave flume x=5.945m with results from Ting and Kirby (1994)

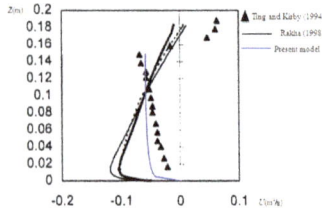

Figure 17. The validation of vertical velocity profile at wave flume x=6.665m with results from Ting and Kirby (1994)

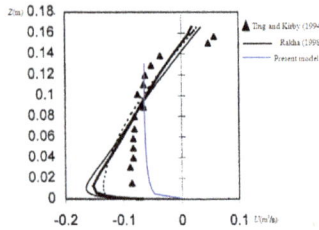

Figure 18. The validation of vertical velocity profile at wave flume x=7.275m with results from Ting and Kirby (1994)

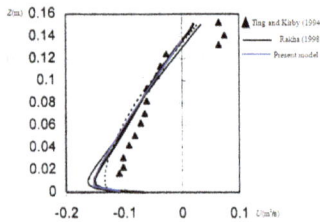

Figure 19. The validation of vertical velocity profile at wave flume x=7.885m with results from Ting and Kirby (1994)

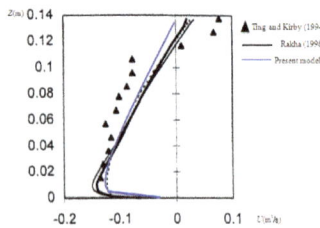

Figure 20. The validation of vertical velocity profile at wave flume x=8.495m with results from Ting and Kirby (1994)

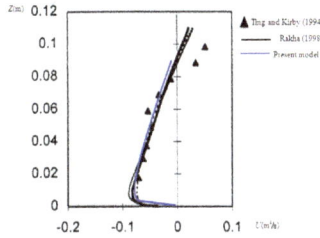

Figure 21. The validation of vertical velocity profile at wave flume x=9.110m with results from Ting and Kirby (1994)

experiments in the wave flume have been observed by Dibajnia and Watanabe (1998), Dibajnia and Kioka (2000), and Dibajnia et al. (2001). All of experimental data are shown as Table 1.

The comparison has been carried out between the period-averaged net sediment transport and experimental results for different wave-current conditions and the results are reasonable accurate within a factor of 2, as shown in Fig. 22. There are 66% numerical results greater than experiments, and 8% numerical results out of compared accurate factor of 2. They are shown good performance to predict the local sediment transport rates in nearshore region.

$D_{50}(cm)$	$2U_w(cm/s)$	$T(sec)$	$u_{max}/2U_w$	$2T_{pc}/T$	$U_0(cm/s)$	$q(cm^2/s)$
0.02	163.8	3.9	0.67	0.29	0	0.133
0.02	148.2	3.9	0.68	0.29	0	0.094
0.02	136.1	3.9	0.68	0.29	0	0.079
0.02	126.9	3.9	0.67	0.29	0	0.042
0.02	103.9	3.7	0.59	0.41	0	0.014
0.02	119.4	3.7	0.59	0.41	0	0.055
0.02	128.5	3.7	0.59	0.41	0	0.048
0.02	136.5	3.8	0.58	0.42	0	0.051
0.02	140.5	3.5	0.68	0.32	0	0.08
0.02	129	3.5	0.68	0.32	0	0.049
0.02	121.2	3.6	0.68	0.31	0	0.061
0.02	108.4	3.6	0.6	0.39	0	0.031
0.02	144.8	3.8	0.58	0.42	0	0.072
0.02	122.8	3.6	0.68	0.31	0	0.041
0.02	117.1	3.5	0.59	0.41	0	0.032
0.02	119.8	3.5	0.59	0.41	0	0.046
0.02	114.6	3.6	0.60	0.4	0	0.044
0.02	109.5	3.9	0.67	0.29	0	0.062

$D_{50}(cm)$	$2U_w(cm/s)$	$T(sec)$	$u_{max}/2U_w$	$2T_{pc}/T$	$U_0(cm/s)$	$q(cm^2/s)$
0.02	126.3	3.5	0.59	0.41	0	0.032
0.02	105.6	3.6	0.68	0.31	0	0.043
0.02	137.3	3.5	0.67	0.33	0	0.086
0.02	122.9	3.9	0.67	0.29	0	0.066
0.02	112.7	3.7	0.59	0.41	0	0.046
0.02	135.9	3.5	0.59	0.41	0	0.069
0.02	119.2	3.9	0.68	0.3	16.3	0.117
0.02	114.7	3.8	0.58	0.42	14.3	0.095
0.02	116.1	3.7	0.59	0.41	5.6	0.063
0.02	118.8	3.9	0.68	0.29	11	0.1
0.02	189.5	4	0.65	0.31	0	0.21
0.02	184.2	3.8	0.64	0.35	0	0.171
0.02	167.5	3.8	0.65	0.33	0	0.134
0.02	151.7	3.8	0.67	0.31	0	0.109
0.02	183.2	3.6	0.65	0.34	0	0.142
0.02	183.2	3.6	0.65	0.35	0	0.139
0.02	179.3	3.6	0.65	0.34	0	0.123
0.02	175.6	3.6	0.65	0.34	0	0.103
0.02	165.5	3.6	0.65	0.33	0	0.125
0.02	163.4	3.6	0.65	0.34	0	0.119
0.02	155.6	3.6	0.66	0.33	0	0.082
0.02	146.4	3.6	0.65	0.33	0	0.075
0.055	275.3	4.2	0.57	0.43	0	1.184
0.055	265	4.1	0.59	0.41	0	0.773
0.055	239.1	4	0.62	0.37	0	0.683
0.055	208.3	4	0.64	0.36	0	0.439
0.055	264.8	3.6	0.59	0.41	0	0.634
0.055	260.1	3.6	0.61	0.39	0	0.534
0.055	250.1	3.6	0.62	0.38	0	0.56
0.055	225.3	3.6	0.63	0.37	0	0.459
0.08	280.5	4.2	0.57	0.43	0	1.373
0.08	276.3	4.1	0.57	0.43	0	1.44
0.08	270.2	3.6	0.57	0.43	0	1.137
0.08	264.2	4	0.58	0.42	0	0.8

Table 1. Experiments from Dibajnia and Watanabe (1998), Dibajnia and Kioka(2000), and Dibajnia et al. (2001).

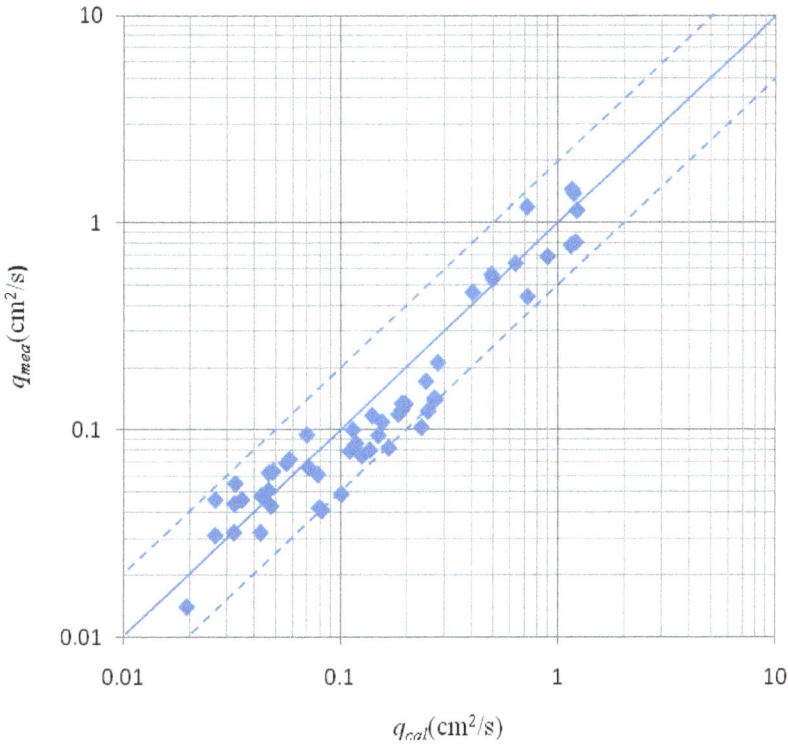

Figure 22. Measured against calculated sediment transport rates with present model and experimental results from Dibajnia and Watanabe (1998), Dibajnia and Kioka(2000), and Dibajnia et al. (2001).

6. Conclusion

In this chapter, a quasi-3D numerical model has been developed to predict sand transport in the coastal region. The whole model system is consisted of a fully nonlinear Bossinesq wave module, 2HD depth-integral wave-driven current module, 1DV velocity profile module, and the sediment transport formula. Numerical results indicate that the wave breaking and regenerating are good agreement with experiments. The phenomenon of undertow for wave breaking or not is performed well, and the vertical velocity profiles are shown good accuracy with experiments. The numerical results of sediment transport have been compared with experiment and obtained are reasonably accurate within a factor of 2. The quasi-3D sediment transport model system was then used to simulate several laboratory studies to test its ability to reproduce the important nearshore morphodynamic processes.

Acknowledgements

The work was partially financially supported by the National Science Council (Taiwan, R.O.C.) Project: NSC99-2218-E-320- 002-MY3.

Author details

Yun-Chih Chiang[1] and Sung-Shan Hsiao[2]

1 Tzu Chi University, Taiwan

2 National Taiwan Ocean University, Taiwan

References

[1] Almar, R, Castelle, B, Ruessink, B. G, Sénéchal, N, Bonneton, P, & Marieu, V. (2010). Two- and three-dimensional double-sandbar system behaviour under intense wave forcing and a meso-macro tidal range", Continental Shelf Research, , 30, 781-792.

[2] Briand, M, & Kamphuis, J. W. (1993). Waves and currents on natural beaches: a quasi 3d numerical model," Coastal Eng. 20, , 101-134.

[3] Brøker, I, & Stive, M. J. F. (1991). *Validation of hydrological models, phase 1*, Danish Hydraulic Institute, Horsholm, Denmark.

[4] Brøker, I, Deigaard, R, & Fredsøe, J. (1991). Onroffshore sediment transport and morphological modelling of coastal profiles," *Proc. ASCE Specialty Conf. Coastal Sediments'* 91, Seattle, WA, , 643-657.

[5] Chiang, Y. C, Hsiao, S. S, & Lin, M. C. D Sediment Transport Modeling for Coastal Morphodynamics", Proc. of the 21 (2011) International Offshore and Polar Eng. Conference, Maui, Hawaii, USA, June 19-24, , 1053-1058.

[6] Cox, D, & Kobayashi, N. (1996). Undertow profiles in the bottom boundary layer under breaking waves." 25th *Int. Conf. Coastal Eng.*, ASCE , 3, 3194-3206.

[7] Davies, A, & Thorne, P. (2002). DV modelling of sand transport by waves and currents in the rippled bed regime," In: *Coastal Eng. Conf.* 2002 (ICCE). ASCE, New York, , 2600-2611.

[8] Davis, A. G, & Villaret, C. (2003). Sediment Transport Modelling for Coastal Morphodynamics," *Proc. of the International Conference on Coastal Sediments,* 2003, Florida, USA, , 1-14.

[9]   Deigaard, D, Fredsøe, J, & Hedegaard, I. B. (1986). Suspended sediment in the surf zone," *J. of Waterway, Port, Coastal and Ocean Eng.*, ASCE, , 112, 115-128.

[10] Deigaard, R, & Fredsøe, J. (1989). Shear stress distribution in dissipative water waves," *Coastal Eng.*, , 13, 357-378.

[11] Deigaard, R, Justesen, P, & Fredsøe, J. (1991). Modelling of undertow by a one-equation turbulence model." *Coastal Eng.*, , 15, 431-458.

[12] Deigaard, R. (1993). A note on the three-dimensional shear stress distribution in a surf zone." *Coastal Eng.*, 20, 157-171.

[13] De Vriend, H. J, & Stive, M. J. F. (1987). Quasi-3D Modelling of Nearshore Currents," *Coastal Eng.*, , 11, 565-600.

[14] Dibajnia, M, & Watanabe, A. (1996). A transport rate formula for mixed-size sands," *Proc. 25th Int. Conf. on Coastal Eng.*, , 3791-3804.

[15] Dibajnia, M, & Watanabe, A. (1998). Transport rate under irregular sheet flow conditions," *Coastal Eng.*, 35, 167-183.

[16] Dibajnia, M, & Kioka, W. (2000). Long waves and the change in cross-shore sediment transport rates on a sheet flow dominated beach," *Coastal Eng. Jpn.*, JSCE & World Scientific, , 42-1, 87-110.

[17] Dibajnia, M, & Watanabe, A. (2001). A representative wave model for estimation of nearshore local transport rate," *Coastal Eng. Jpn.*, 43, 1-38.

[18] Drønen, N, & Deigaard, R. (2000). Three dimensional near-shore bar morphology." *Proc. 27th Int. Conf. on Coastal Eng.* ASCE, Sydney, , 3205-3217.

[19] Drønen, N, & Deiggard, R. (2007). Quasi-three-dimensional modelling of the morphology of longshore bars," *Coastal Eng.*, 54, 197-215.

[20] Elfrink, B, Broker, I, Deigaard, D, Hansen, E, & Justesen, P. (1996). Modelling of q3d sediment transport in the surf zone," In: Edge, B.L. (Ed.), *Coastal Eng. Conf.* ASCE, Orlando, Florida, USA, , 3805-3817.

[21] Elgar, S. (1987). Relationships involving third moments and bispectra of a harmonic process," *IEEE Trans. on Acoustics, Speech, and Signal Processing*, 35, , 1725-1726.

[22] Elgar, S, Guza, R. T, & Freilich, M. H. (1988). Eulerian measurements of horizontal accelerations in shoaling gravity waves," *J. Geophys. Res.*, , 93, 9261-9269.

[23] Elgar, S, Gallagher, E. L, & Guza, R. T. (2001). Nearshore sandbar migration," *J. of Geophysical Research*, , 106(C6), 11623-11627.

[24] Fernando, P. T, & Pan, S. (2005). Modelling wave of hydrodynamics around a scheme of detached leaky breakwaters," In: Smith, J.M. (Ed.), *Proceeding of the 29th International Conference on Coastal Eng.*. World Scientific, Lisbon, Portugal, , 830-841.

[25] Fredsøe, J, & Deigaard, R. (1992). *Mechanics of Coastal Sediment Transport*, World Scientific, Singapore.

[26] Hoefel, F, & Elgar, S. (2003). Wave-induced sediment transport and sandbar migration", Science, , 299, 1885-1887.

[27] Houser, C, & Greenwood, B. (2007). Onshore Migration of a Swash Bar During a Storm", J. of Coastal Research, , 23, 1-14.

[28] Hsu, T, Elgar, J, Guza, S, & Wave-induced, R. T. sediment transport and onshore sandbar migration", Coastal Eng., , 53, 817-824.

[29] Kuroiwa, M, Noda, H, & Matsubara, Y. (1998). Applicability of a quasi-three dimensional numerical model to nearshore currents," *Proc. of 26th ICCE*, , 815-828.

[30] Kennedy, A. B, Chen, Q, Kirby, J. T, & Dalrymple, R. A. (2000). Boussinesq Modeling of Wave Transformation, Breaking, and Runup. I: 1D," *Journal of Waterway, Port, Coastal and Ocean Eng.*, ASCE, , 126(1), 39-47.

[31] Li, M, Fernando, P. T, Pan, S, Connor, O, & Chen, B. A. D., ((2007). Development of a quasi-3d numerical model for sediment transport prediction in the coastal region." *Journal of Hydro-environment Research*, , 1, 143-156.

[32] Lin, M. C, Kuo, J. C, Chiang, Y. C, & Liou, J. Y. (1996). Numerical Modeling of Topography Changes in Sea Region," *Proc. 18th Conf. on Ocean Engineering in Republic of China*, Nov. 1996, , 627-637.

[33] Lin, M. C, Chiang, Y. C, & Hsiao, S. S. (2009). A study of the sediment transport formula under combined wave and current conditions." *J. of Coastal and Ocean Eng.*, , 9(2), 177-205.

[34] Long, W, Kirby, J. T, & Hsu, T. J. ((2006). Cross shore sandbar migration predicted by a time domain Boussinesq model incorporating undertow", *Coastal Eng. Conf.*, , 2655-2667.

[35] Nielsen, P. (1992). *Coastal Bottom Boundary Layers and Sediment Transport*, Advanced Series on Ocean Engineering. Singapore: World Scientific Publication., 4

[36] Nwogu, O. (1993). An Alternative Form of the Boussinesq Equations for Modeling the Propagation of Waves from Deep to Shallow Water," *J. of Waterway, Port, Coastal and Ocean Eng.*, ASCE, , 119(6), 618-638.

[37] Okayasu, A, Shibayama, T, & Horikawa, K. (1988). Vertical variation of undertow in the surf zone," *Proc. 21st Int. Conf. on Coastal Eng.*, ASCE, Malaga, , 478-491.

[38] Okayasu, A, & Katayama, H. (1992). Distribution of undertow and long-wave component velocity due to random waves," *Proc. of 23rd Int. Conf. on Coastal Eng.*, ASCE, , 883-893.

[39] Okayasu, A, Hara, K, & Shibayama, T. (1994). Laboratory experiments on 3-D nearshore currents and a model with momentum flux by breaking wave," *Proc. 24th Int. Conf. Coastal Eng.*, , 2461-2475.

[40] Pape, L, Plant, N. G, & Ruessink, B. G. (2010). On cross-shore migration and equilibrium states of nearshore sandbars," *J. of Geophysical Research*, , 115, 1-15.

[41] Rakha, K. A. d phase resolving hydrodynamic and sediment transport model." *Coastal Eng.* , 34, 277-311.

[42] Roelvink, J. A, & Reniers, A. J. H. M. (1995). LipllD Delta Flume Experiments, a profile dataset for profile model validation. Report H2130, January 1995, Delft Hydraulics, The Netherlands.

[43] Ruessink, B. G, Kuriyama, Y, Reniers, A. J. H. M, Roelvink, J. A, & Walstra, D. J. R. (2007). Modeling cross- shore sandbar behavior on the timescale of weeks", *J. Geophys. Res.*, 112, F03010.

[44] Ruessink, B. G, & Kuriyama, Y. (2008). Numerical predictability experiments of cross-shore sandbar migration", *Geo. Research Letters*, , 35, 16-35.

[45] Ruessink, B. G, Pape, L, & Turner, I. L. (2009). Daily to interannual cross-shore sandbar migration: Observations from a multiple sandbar system", *Continental Shelf Research*, , 29, 1663-1677.

[46] Sanchez-arcilla, A, Collado, F, & Rodriguez, A. (1992). Vertically varying velocity field in Q-3D nearshore circulation," *Proc. of Int. Coastal Eng. Conference*, , 2811-2824.

[47] Soulsby, R. L. (1997). Dynamics of marine sands, Thomas Telford Publications, London.

[48] Soulsby, R. L, Hamm, L, Klopman, G, Myrhaug, D, Simons, R, & Thomas, G. (1993). Wave-current interaction within and outside the bottom boundary layer," *Coastal Eng.* , 21, 41-69.

[49] Splinter, K. D, Holman, R. A, & Plant, N. G. dynamic model for sandbar migration and 2DH evolution", J. of Geophysical Research, , 116, 10-20.

[50] Svendsen, I. A. (1984). Wave heights and set-up in a surf zone." *Coastal Eng.* , 8, 303-329.

[51] Svendsen, I. A, & Lorenz, R. S. (1989). Velocities in combined undertow and longshore currents," *Coastal Eng.*, , 13, 55-79.

[52] Ting, F. C. K, & Kirby, J. T. (1994). Observation of undertow and turbulence in a laboratory surf zone." *Coastal Eng.* , 24, 51-80.

[53] Trowbridge, J, & Madsen, O. S. (1984). Turbulent Wave Boundary Layers 1. Model Formulation and First- Order Solution," *J. Geophys. Res.*, C5), , 89, 7989-7997.

[54] Trowbridge, J, & Young, D. (1989). Sand transport by unbroken waves under sheet flow conditions," *J. Geophys. Res.*, , 94, 10.

[55] Wai, O, Chen, Y, & Li, Y. driven coastal sediment transport model." *Coastal Eng. Journal,* , 46, 385-424.

[56] Wei, G, Kirby, J. T, Grilli, S. T, & Subramanya, R. (1995). A Fully Nonlinear Boussinesq Model for Surface Waves. Part 1. Highly Nonlinear Unsteady Waves," *J. Fluid Mech.,* , 294, 71-92.

[57] Wilson, K. C. (1989). Friction of wave induced sheet flow," *Coastal Eng.* , 12, 371-379.

Chapter 10

Derivation of Sediment Transport Models for Sand Bed Rivers from Data-Driven Techniques

Vasileios Kitsikoudis,
Epaminondas Sidiropoulos and Vlassios Hrissanthou

Additional information is available at the end of the chapter

1. Introduction

Hydraulic engineers and geologists have studied sediment transport in natural streams and rivers for centuries due to its importance in understanding river hydraulics. Erosion and deposition of sediment alters the hydraulic geometry of the channel and may cause increase of flood frequency as well as navigation problems from excessive deposition. Moreover, discharge of industrial and agricultural residuals sets the sediment particles to be the primary transporters of toxic substances that contaminate aquatic systems. High sediment discharge peaks may be destructive for fish habitats and ecosystems, and long-term sediment yield affects the design and function of constructions such as dams and reservoirs, as well as the coastal erosion at the basin outlet.

Sediment transport in sand bed rivers and natural streams is a complex process. For its quantification, numerous sediment transport functions have been introduced in the past years based on different concepts. There are four basic approaches used in the derivation of sediment transport formulae (Yang, 1977): 1) The deterministic approach, which obeys the laws of physics and usually is based on an independent variable like slope, shear stress, stream power, unit stream power etc. 2) The regression approach, which has emerged from the thought that sediment transport is such a complex phenomenon that cannot be described by a single dominant variable. 3) The pioneering probabilistic approach of Einstein (1942), which highlighted the complexity and the stochastic nature of the sediment transport in a rather laborious way for common usage in engineering, and 4) The regime approach, which was developed as a result of long-term measurements in equilibrium conditions.

The emerging results from all these concepts usually differ drastically from each other and from the measured data. Consequently, none of the published sediment transport equations has gained universal acceptance in confidently predicting sediment transport rates, especially in rivers. An alternative approach may be the usage of data-driven modeling, which is especially attractive for modeling processes, in which knowledge of the physics of the problem is inadequate. The scope of this chapter is the utilization of some widely used data-driven techniques, namely artificial neural networks (ANNs) and symbolic regression based on genetic programming (GP) in order to determine the dominant dimensionless variables that can be used as inputs in such schemes and generate sediment transport models for natural streams and rivers that are based solely on the data without presuming anything about their structure and their degree of nonlinearity.

For the proper training of a data-driven scheme, data of good quality are needed. Since field measurements accommodate the peculiarities of the considered streams and the inclusion of noise in the measurement process is inevitable, the training data comprise solely laboratory flume measurements. The testing data, however, comprise exclusively field measurements in order to implement the models in actual applications. Based on this concept, the approach of the basic trend of the function is feasible and the derived model will be applicable to the data range for which it will be trained. Regarding the efficiency of scaling in the sediment transport context, model-prototype comparisons have shown that correspondence of behavior is often well beyond expectations, as has been attested by the successful operation of many structures designed from model tests (Pugh, 2008). This study exhibits the potential of machine learning in capturing functions with physical meaning since the training and testing sets have significant differences in their statistical distributions. The determination of the input variables that best define the problem is accomplished by the assessment of some common independent dimensionless variables based on their correlation with the sediment concentration and the aid of ANNs on the basis of a tentative trial-and-error procedure. Subsequently, ANNs and symbolic regression are utilized in order to derive equations from the selected input combinations.

2. Data mining and data-driven techniques in the context of sediment transport

The recorded observations of a system can be further analyzed in the search for the information they encode. Such automated search for models accurately describing data constitutes a direction that can be identified as that of data mining. Data mining and knowledge discovery aim at providing tools to facilitate the conversion of data into a number of forms, such as equations. The latter provide a better understanding of the process generating or producing these data. These models combined with the already available understanding of the physical processes result in an improved understanding and novel formulations of physical laws and improved predictive capability (Babovic, 2000).

Data-driven modeling (DDM) and machine learning techniques used for predictions are essentially modernized regression schemes with the significant advantage over the classical regression schemes that they do not have to presume the structure of the nonlinear model, which they attempt to fit. They are based on simple ideas, usually inspired from the way nature works, and their only prerequisite is a good, although usually large, data set. The data are usually divided into three sets, namely the training, validation and testing set. The training set trains the scheme on the basis of a minimization criterion and the validation set is used as a stopping criterion for training to avoid overfitting to the data used for training. The test set is used to evaluate the generated model. The minimization criterion, on the basis of which the training process takes place, is usually a sum of errors between the computed outputs and the actual measured data. The optimization model that is used for the minimization depends on the data-driven scheme and may be deterministic as well as stochastic.

Inferring models from data is an activity of deducing a closed-form explanation based solely on observations. These observations, however, represent a limited source of information. The question emerges as to how this, a limited flow of information from a physical system to the observer, can result in the formation of a model that is complete in the sense that it can account for the entire range of phenomena encountered within the physical system in question and describe even the data outside the range of previously encountered observations. The present efforts are characterized by the search for a model that is capable of acquiring semantics from syntax. Clearly, every model has its own syntax. Artificial neural networks have the syntax of a network of interconnected neurons, whereas genetic programming has the syntax of treelike networks of symbolic expressions in reverse Polish notation. The question is whether such a syntax can capture the semantics of the system it attempts to model (Babovic, 2000). Witten et al. (2011) argued that the universal learner is an idealistic fantasy since experience has shown that no single machine learning scheme is appropriate to all data mining problems. Certain classes of model syntax may be inappropriate as a representation of a physical system. One may choose the model whose representation is complete, in the sense that a sufficiently large model can capture the data's properties to a degree of error that decreases with an increase in the model size. For example, one may decide to expand Taylor or Fourier series and decrease the error by adding terms in a series. However, in these cases, semantics almost certainly would not be caught (Babovic, 2000).

2.1. Artificial neural networks

ANN is the most widely used data-driven method. Since abundant information on ANNs is available in the literature [e.g. Haykin (2009)], only a brief description of ANNs is provided, with regard only to the methodology applied herein. ANN is a broad term covering a large variety of network architectures and structures. The most common of them, and the one utilized herein, is the multilayer feedforward network. This type of network is a parallel distributed information processing system that consists of the input layer, the hidden layer(s), and the output layer, and the information goes only in a forward direction. Each layer comprises a number of neurons, each one of which is connected with those in the successive layer with synaptic weights that determine the strength of the connections. The hidden and output layer neurons have an inherent activation function, which accommodates the nonlinear transformation of the input data to the targets. In this study, the neurons of the hidden layer(s) will have the hyperbolic tangent activation function, which squashes the data between (-1, 1), and the single neuron of the output layer will have the linear activation function, which simply returns the value that is passed to it. The input data are scaled to the range (-0.9, 0.9) because, if the values are scaled to the extreme limits of the transfer function, the size of the weight updates is extremely small and flat-spots in training are likely to occur (Maier and Dandy, 2000).

The training process of an ANN may be viewed as a "curve fitting" problem and the network itself may be considered simply as a nonlinear input-output mapping (Haykin, 2009). Supposing that a deterministic relation between sediment load concentration and some specific independent variables exists, a multilayer feedforward ANN is able to approximate this function, if it includes at least one hidden layer with a sufficient number of neurons (Hornik et al., 1989). However, this universal approximation theorem does not specify if a single hidden layer is optimal in the sense of learning time, ease of implementation, or (more importantly) generalization (Haykin, 2009). As a result, several network architectures are tested in order to determine the optimal one.

Although the implementation of ANNs is extensive and successful in water resources applications [e.g. Maier and Dandy (2000)] and in the prediction of daily suspended sediment data [e.g Cigizoglu (2004)], it is quite sparse in the prediction of sediment concentration from other independent hydraulic variables. Nagy et al. (2002) reviewed some widely used sediment discharge equations and selected some of the dominant dimensionless variables of the problem as input neurons for an ANN that was trained and tested with field data. Bhattacharya et al. (2005) used dimensionless parameters obtained from the Engelund and Hansen (1967) formula in order to train and test an ANN with a mixture of flume and field data, whilst in similar studies Bhattacharya et al. (2004, 2007) scrutinized further the possible input parameters based on the same data. Yang et al. (2009) chose as input variables combinations of dimensional quantities and applied them to field data. All of these works used the back-propagation algorithm (Rumelhart et al., 1986) for training the ANNs and compared the results with some of the most popular sediment transport formulae. For all the cases, the ANNs generated superior results.

2.2. Symbolic regression based on genetic programming

Many seemingly different problems in artificial intelligence, symbolic processing and machine learning can be viewed as requiring discovery of a computer program that produces some desired outputs for particular inputs. The process of solving these problems can be reformulated as a search for a highly fit individual computer program in the space of possible ones. GP extends the concept of genetic algorithms and provides a way to search for this fittest individual computer program (Koza, 1992).

GP works by randomly generating a population of computer programs (represented by tree structures) and each individual program in the population is measured in terms of how well it performs in the particular problem environment. This measure is called the fitness measure (Koza, 1992) and usually is a sum of errors between the outputs predicted by the program and the actual ones. Initially, the generated computer programs will have exceedingly poor fitness. Nonetheless, some individuals in the population will turn out to be somewhat fitter than others. These differences in performance are subsequently exploited. The Darwinian principle of reproduction and survival of the fittest and the genetic operations of sexual recombination (crossover) and mutation are used to create a new offspring population of individual programs from the current population. The reproduction principle involves the selection, in proportion to fitness, of a computer program from the current population that survives from the generation by being copied into the new population. The genetic process of sexual recombination is used to create new offspring programs from two parental programs selected in proportion to fitness. The parental programs are typically of different sizes and shapes. The offspring programs are composed of subexpressions from their parents and are, typically, of different sizes and shapes as well. Intuitively, if two programs are somewhat effective in solving a problem, then some of their parts probably have some merit. By recombining randomly chosen parts of somewhat effective programs, the result may be the production of new programs that are even fitter in solving the problem (Koza, 1992). Mutation serves the potentially important role of restoring lost diversity in a population by replacing random subtrees of variable length with other random ones. Its purpose is to prevent premature convergence to unsatisfactory solutions. After the operations of reproduction, crossover and mutation are performed on the current population, the offspring population replaces the old one. Each individual in the new population of programs is then measured for fitness and the process is iterated for a predetermined number of generations. This algorithm will produce populations of programs, which over many generations tend to exhibit increasing average fitness in dealing with their environment. The individual computer program that performs best in the evolved generations is considered to be the fittest.

A multigene individual consists of multiple genes, each of which is a GP evolved tree. In multigene symbolic regression, each prediction $\hat{y}$ of the output variable y is formed linearly by the weighted output of each of the genes plus a bias term (Searson, 2009). Each tree is a function of the input variables. Mathematically, a multigene regression model can be written as:

$$\hat{y} = d_0 + d_1 \times tree1 + ... + d_M \times treeM \tag{1}$$

where d_0=bias (offset) term; d_1, ..., d_M are the gene weights and M is the number of genes comprising the current individual. The gene weights are automatically determined by a least squares procedure for each multigene individual. The number and structure of the trees is evolved automatically during a run (subject to user defined constraints) using the training data. Hence, multigene symbolic regression combines the power of classical linear regression with the ability to capture nonlinear behavior without needing to pre-specify the structure of the nonlinear model. During a run, genes are acquired and deleted using a tree crossover operator called two-point high level crossover. This allows the exchange of genes between individuals and it is used in addition to the "standard" GP recombination operators (Searson et al., 2010).

GP has been implemented in hydraulic engineering in the last years with very good results. Babovic and Abbott (1997) applied GP to some representative problems, while Babovic and Keijzer (2000) highlighted the usage of GP as a data mining tool in which the
human expert interprets models suggested by the computer, aiming at knowledge discovery. Minns (2000) suggests that the symbolic expressions obtained from GP may be less
accurate than the ANN in mapping the experimental data. However, these expressions
may be more easily examined in order to provide insights into the processes that created
the data. In the context of sediment transport, Zakaria et al. (2010) applied gene-expres-
sion programming, which is similar to multigene symbolic regression, to predict the total
bed material load for rivers using dimensional quantities from field data, and outper-
formed some of the traditional sediment load formulae. Azamathulla et al. (2010) utilized
GP in order to predict the scour depth at bridge piers and obtained results superior to
those of ANNs and regression equations.

3. Sediment transport

Sediment load is the material being transported, and it can be divided into wash load and
bed material load. The wash load is the fine material of sizes, which are not found in appreciable quantities on the bed, and is not considered to be dependent on the local hydraulics of
the flow, but instead is dependent on the upstream supply. As a practical definition, the
wash load is considered to be the fraction of the sediment load finer than 0.062 mm. The bed
material load is the material of sizes, which are found in appreciable quantities on the bed
and it can be conceptually divided into the bed load (the portion of the load that moves near
the bed) and the suspended load (the portion of the load that moves in suspension), although the division is not precise. The consequent difficulty, however, to separate bed load
from turbulence dominated suspended load leads to a total load definition for the quantification of sediment transport in sand bed rivers. A dimensionless, commonly used measure
for sediment quantification is concentration by weight in parts per million (ppm), which is
the ratio of the sediment discharge to the discharge of the water-sediment mixture, both expressed in terms of mass per unit time, here called C_t. This can be given as

$$C_t = 10^6 \frac{\rho_s Q_{st}}{\rho Q + \rho_s Q_{st}} \tag{2}$$

For practical reasons, the density of the water-sediment mixture is taken to be approximately equivalent to the density of water. This approximation will cause errors of less than one
percent for concentrations less than 16000 ppm (Brownlie, 1981a).

The parameters governing a sediment transport process can be described by (Yalin, 1977)

$$q_t = f(V, D, d, S, g, \rho_s, \rho, v) \tag{3}$$

Since the data-driven schemes are trained and validated with flume data but tested with field data and in order to ensure dimensional consistency in the derived models, the input and output variables should be dimensionless. Instead of applying dimensional analysis and Buckingham's π theorem, the independent variables of Eq. (3) will be introduced by some common and well-known dimensionless variables that have physical meaning and have been utilized for the creation of various sediment transport formulae. These variables are directly related to quantities the engineer can readily visualize and measure; they are listed as follows and summarized in Table 1.

Froude number, which gives a measure of the ratio of inertial forces to gravitational forces of the flow. For the flume data, the depth will be the hydraulic radius of the bed which is equivalent to the mean depth of an infinitely wide channel with the same slope, velocity and bed friction as the flume, and is calculated according to the sidewall correction of Vanoni and Brooks (1957). This elaboration is due to the fact that in flume experiments the sand covered bed will generally be much rougher than the flume walls, and thus will be subjected to higher shear stresses.

$$Fr = \frac{V}{\sqrt{gD}} \tag{4}$$

Reynolds number, which gives a measure of the ratio of inertial forces to viscous forces of the flow

$$Re = \frac{VD}{v} \tag{5}$$

Shear Reynolds number, the physical meaning of which, is the ratio of particle size to the thickness of the viscous sublayer δ, because δ is proportional to v/U_*.

$$Re^* = \frac{U_* d_{50}}{v} \tag{6}$$

Dimensionless shear stress or Shields number

$$\tau^* = \frac{\tau}{(\gamma_s - \gamma)d_{50}} \tag{7}$$

Dimensionless grain diameter. It is a dimensionless expression for grain diameter that can be derived by eliminating shear stress from the two Shields parameters (Shields, 1936); or from the drag coefficient and Reynolds number of a settling particle, by eliminating the settling velocity; or dimensionally, with immersed weight of an individual grain, fluid density, and viscosity as the variables (Ackers and White, 1973). The dimensionless grain diameter is, therefore, generally applicable to coarse, transitional, and fine sediments and is the cube root of the ratio of immersed weight to viscous forces. Thus

$$d_{gr} = d_{50} \left[\frac{g(\gamma_s/\gamma - 1)}{v^2} \right]^{1/3}$$

(8)

Dimensionless stream power. The power equation appears first to have been applied to sediment transport by Rubey (1933) and later by Velikanov (1955). It was again suggested by Knapp (1938), and was later introduced by Bagnold (1956) in a paper wherein the flowing fluid was regarded as a transporting machine. The available power supply, or time rate of energy supply, to unit length of a stream is the time rate of liberation in kinetic form of the liquid's potential energy as it descends the gravity slope S. Denoting this power by Ω, Bagnold (1966) derived the formula

$$\Omega = \rho g Q S$$

(9)

The mean available power supply to the column of fluid over unit bed area, to be denoted by ω, is therefore

$$\omega = \frac{\Omega}{W} = \frac{\rho g Q S}{W} = \rho g D S V = \tau V$$

(10)

In order to define a dimensionless transport parameter that encapsulates Bagnold's view of sediment transport as a stream power related phenomenon, Eaton and Church (2011) developed the following formula

$$\omega^* = \frac{\omega}{\rho \left[g(\gamma_s/\gamma - 1)d \right]^{3/2}}$$

(11)

Dimensionless unit stream power. Yang (1972) reviewed the basic assumptions used in the derivation of conventional sediment transport equations. He concluded that the assumption that sediment transport rate could be determined from water discharge, average flow velocity, energy slope, or shear stress is questionable. Consequently, the generality and applicability of any equation derived from one of these assumptions is also questionable. The rate of energy per unit weight of water, available for transporting water and sediment in an open channel with reach length x and total drop of Y, is

$$\frac{dY}{dt} = \frac{dx}{dt} \frac{dY}{dx} = VS$$

(12)

Yang (1972) defines the unit stream power as the velocity-slope product and argues that the rate of work being done by a unit weight of water in transporting sediment must be directly related to the rate of work available to a unit weight of water. Thus, total sediment concentration or total bed material load must be directly related to unit stream power. While Bagnold (1966) emphasized the power that applies to a unit bed area, Yang (1972, 1973) emphasized the power available per unit weight of fluid to transport sediments. The fact that sediment discharge or concentration is dominated by the unit stream power has been confirmed by Vanoni (1978) as well. While Yang divided unit stream power VS by fall velocity ω_s to obtain a dimensionless variable, Vanoni (1978) divided the product VS by $(gd_{50})^{1/2}$. Both d_{50} and ω_s are commonly used for describing the size of sediment particles. However, d_{50} can only reflect the physical size of sediment particles, while ω_s can also reflect the interaction between sediment particles and water, which is affected by particle shape, water viscosity and temperature. On the other hand, the computation of fall velocity is problematic and a common source of errors. The emerging variables expressing dimensionless unit stream power according to Yang and Vanoni are, respectively, the following

$$\frac{VS}{\omega_s} \tag{13}$$

and

$$\frac{VS}{\sqrt{gd_{50}}} \tag{14}$$

No	Dimensionless variables
1	Froude number, Fr
2	Reynolds number, Re
3	Shear Reynolds number, Re^*
4	Dimensionless shear stress, τ^*
5	Dimensionless grain diameter, d_{gr}
6	Dimensionless stream power, ω^*
7	Yang's dimensionless unit stream power, VS/ω_s
8	Vanoni's dimensionless unit stream power, $VS/(gd_{50})^{1/2}$

Table 1. Dimensionless variables assessed for the determination of the dominant ones

Yang (1977, 2003) argued that total sediment discharge correlates best with unit stream power based on the plots of Figure 1. Nonetheless, equations based on the other hydraulic variables have been used successfully as well.

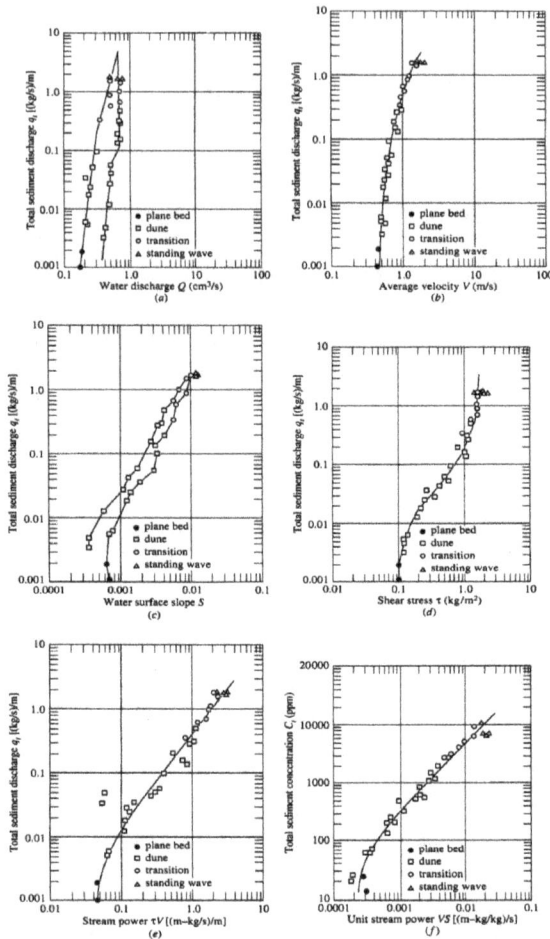

Figure 1. Relationships between total sediment discharge and (a) water discharge, (b) velocity, (c) slope, (d) shear stress, (e) stream power, and (f) unit stream power, for 0.93 mm sand in an 8 ft wide flume [obtained from Yang (2003)]

4. Data preparation and determination of the inputs

Since data-driven techniques require a large number of quality data that represent a wide spectrum of the considered problem in order to be trained efficiently, the database assembled by Brownlie (1981b) is utilized. Brownlie's (1981b) database contains 7027 records (5263 laboratory records and 1764 field records) in 77 data files. These data were subjected to a screening process similar to the one Brownlie (1981a) used for the derivation of his formula. Firstly, the measurements that were not verified by Brownlie, were incorrect or incomplete, were removed. Secondly, because only flows with sand beds were considered, median particle sizes were limited to values between 0.062 mm and 2.0 mm. To avoid samples with large amounts of gravel or fine, cohesive material, geometric standard deviations were restricted to values smaller than 5, and some other constraints were imposed in order to reduce sidewall effects, eliminate shallow water effects, and overcome accuracy problems associated with low sediment concentration. In addition to these, only flume measurements with uniform flows were considered and supercritical flows were removed due to the subcritical flows that usually prevail in nature, in sand bed rivers. Finally, the measurements with specific gravity outside the quartz density range were neglected as well as measurements that had extreme temperature values. Wherever the temperature was missing, a value of 15 °C was used for the calculation of kinematic viscosity. For the laboratory data, the sidewall correction of Vanoni and Brooks (1957) was utilized to adjust the hydraulic radius to eliminate the effects of the flume walls. If sediment concentration is correlated with velocity, however, the sidewall correction will be of little use. These restrictions are shown in Table 2.

Restriction	Reason
0.062 mm $\leq d_{50} \leq$ 2.0 mm	Sand only
$\sigma_g \leq 5$	Eliminate bimodal distributions
$W/D > 4$	Reduce sidewall effects (only for laboratory data)
$R/d_{50} > 100$	Eliminate shallow water effects
$C > 10$ ppm	Accuracy problems associated with low concentration
$Fr < 1$	Subcritical flows
$2.57 \leq$ Specific gravity ≤ 2.68	Natural sediments

Table 2. Restrictions imposed on data

Since measurements in natural streams and rivers are notoriously difficult, and sometimes inaccurate, and the inclusion of field data to the training set would result in a model applicable only to rivers similar to those the data were obtained from, field data are excluded from the training set. Consequently, the training set consists solely of laboratory flume data so that the noise embedded in the training set is minimized. The testing set, however, comprises exclusively field data in order to test the derived mathematical models in actual problems that occur in nature. With this technique, the generated models will have general applicability to the data range for which they are trained. The final database consists of 984 laboratory records and 600 field records that lie within the range of the laboratory records that constitute the training set, due to the data sensitive nature of DDM.

Further pruning of the outliers in the training dataset and the subsequent increase of data homogeneity would be beneficial for the training procedure, however, this would be at the expense of the amount of training data, which are already significantly reduced from the screening process. Since most DDM methods perform well when the data has a distribution that is close to normal (Bhattacharya et al., 2005), a log-transformation of the input and out-

put variables of all datasets was applied so that the distributions of the transformed variables were closer to normal. Figure 2 depicts the distribution of the flume sediment
concentrations for the original and the log-transformed values.

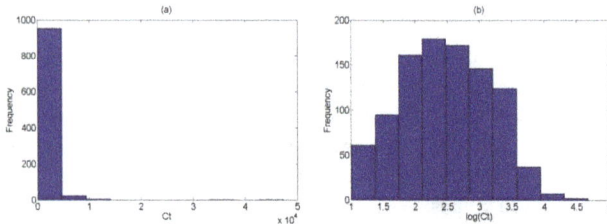

Figure 2. Distribution of flume sediment concentration in ppm (a) before log-transformation and (b) after log-trans-formation

For the creation of training and validation sets the available 984 laboratory measurements
were placed in descending order with respect to sediment concentration and for every three
successive measurements that were picked for the training set, the fourth one was selected
for the validation set. This procedure was iterated for all the laboratory data and the
emerged training and validation sets comprise 739 and 245 measurements, respectively. The
600 field measurements constitute the test set. Table 3 shows some statistical measures of the
potential variables of these sets. Table 4 shows the datasets from which the data used in this
study were obtained and some representative values of each set. The abbreviations used in
Table 4 are the same with those Brownlie (1981b) used in his data compilation; consequently, all the references to the original datasets may be obtained from that study.

		Statistical measures	d_{gr}	Fr	Re^*	VS/ω_s	$VS/(gd_{50})^{0.5}$	ω^*	τ^*	C_t
Laboratory data	train set	Minimum value	2.405	0.166	2.468	0.0010	0.0014	0.093	0.0323	10.2
		Maximum value	38.800	0.999	158.292	0.5350	0.2818	134.6	5.828	47300
		Mean value	10.069	0.490	18.356	0.0297	0.0183	5.065	0.450	977.9
		Standard deviation	6.337	0.188	17.476	0.0425	0.0198	9.263	0.412	2575.8
		Skewness coefficient	1.456	0.943	3.219	5.127	5.633	6.598	4.618	11.758
	validation set	Minimum value	2.508	0.202	2.938	0.0017	0.0016	0.227	0.0578	11.3
		Maximum value	38.505	0.968	146.416	0.2984	0.1150	81.56	2.555	10630
		Mean value	9.708	0.482	17.220	0.0292	0.0173	5.397	0.459	861.4
		Standard deviation	6.760	0.173	17.467	0.0337	0.0153	9.128	0.374	1412.2
		Skewness coefficient	1.590	0.912	3.579	3.200	2.429	4.212	1.671	3.372
Field data	test set	Minimum value	2.531	0.166	6.165	0.0017	0.0015	0.365	0.094	11
		Maximum value	33.931	0.992	131.127	0.1528	0.0857	118.7	5.805	11400
		Mean value	10.654	0.375	27.910	0.0202	0.0156	14.13	0.943	1239.3
		Standard deviation	6.968	0.150	18.994	0.0194	0.0121	16.73	0.648	1472.2
		Skewness coefficient	1.705	1.027	2.105	2.676	1.893	2.350	1.726	2.554

Table 3. Statistical measures of the train, validation and test set

		Range of field variables							
Code	No.	Velocity (m/s)		Depth (m)		Slope (‰)		C_t (ppm)	
		min	max	min	max	min	max	min	max
BAL	25	0.226	1.093	0.091	0.256	0.44	2.1	19	3776
BEN	1	0.205	0.205	0.038	0.038	0.5	0.5	10.2	10.2
BRO	6	0.372	0.616	0.047	0.060	2.4	3.5	1200	5300
CHY	7	0.423	0.586	0.066	0.101	1.11	2	99.4	345
COS	12	0.403	0.503	0.140	0.156	0.45	1.01	10.954	102.08
DAV	69	0.244	0.792	0.076	0.305	0.248	2.67	11.3	1760
EPA	16	0.440	0.706	0.088	0.300	0.6	3.68	32	1017
EPB	19	0.265	0.762	0.148	0.304	0.262	1.6	45	1810
FOL	6	0.388	0.599	0.036	0.047	3.74	4.02	845	1848
FRA	11	0.361	0.450	0.129	0.161	0.938	1.693	39.979	166.34
GKA	27	0.302	0.635	0.032	0.124	1.8	6.401	205	3160
GUY	145	0.225	1.321	0.058	0.405	0.37	9.5	12	47300
JOR	7	0.401	0.557	0.070	0.105	1.12	1.67	95.8	306.7
KEN	6	0.412	0.799	0.047	0.109	1.7	4.2	550	2070
KNB	9	0.277	0.674	0.070	0.168	0.56	2.5	14	1740
LAU	10	0.326	0.671	0.076	0.221	0.8	2.1	550	4240
MCD	11	0.480	0.660	0.082	0.146	1.11	1.67	151.2	615.8
MPR	15	0.426	0.835	0.112	0.490	0.42	4.066	14.357	1091.1
MUT	17	0.131	0.505	0.029	0.102	0.5	7.5	11	10630
NOR	27	0.524	1.802	0.256	0.585	0.47	5.77	33	8870
OBR	45	0.214	0.953	0.088	0.165	0.57	3.23	17	1332.5
OJK	14	0.338	0.586	0.075	0.135	1.09	2.67	66.791	3355.7
PRA	25	0.254	0.701	0.076	0.305	0.282	2.87	11.63	560
SAT	1	0.332	0.332	0.193	0.193	0.44	0.44	66.877	66.877
SIN	58	0.277	0.597	0.066	0.117	1	4	35.7	1105
SON	1	0.465	0.465	0.043	0.043	6.7	6.7	6300	6300
STE	27	0.514	1.364	0.091	0.302	2.01	4.03	640	4615
STR	15	0.345	0.835	0.047	0.223	0.950	4.62	417	6300
TAY	11	0.348	0.878	0.077	0.160	0.89	2.09	13.979	2269.7
VAB	12	0.234	0.772	0.071	0.169	0.7	2.8	37	2500
VAH	6	0.319	0.558	0.176	0.238	0.642	1.303	31	1490
WLM	5	0.538	0.669	0.204	0.223	0.912	2.14	31.125	196.1
WLS	61	0.358	1.360	0.110	0.302	0.269	1.98	102	11700
WSA	195	0.165	0.555	0.034	0.170	1	2	11.3	587.19
WSB	36	0.444	0.578	0.108	0.176	1	2	55.8	379
WSS	13	0.377	0.388	0.073	0.075	1	1	53.8	94.6
ZNA	13	0.224	0.783	0.05	0.783	1.66	4.7	150	1975
Total	984	0.131	1.802	0.029	0.585	0.248	9.5	10.2	47300

(a)

		Range of field variables						
Code	No.	Velocity (m/s)		Depth (m)		Slope (‰)		C_t (ppm)	
		min	max	min	max	min	max	min	max
AMC	5	0.473	0.739	0.796	1.009	0.237	0.33	52	448
ATC	6	1.739	2.028	10.881	14.112	0.038	0.0513	102.374	567.343
CHO	26	0.846	1.597	2.103	3.414	0.115	0.254	149.826	1316.9
COL	58	0.617	1.266	1.134	3.371	0.107	0.407	35.6	768.7
HII	34	0.186	0.930	0.025	0.732	0.84	10.7	116.311	5638.6
MID	35	0.593	1.125	0.247	0.412	0.928	1.572	437.760	2269.2
MIS	5	1.756	2.423	11.400	17.282	0.082	0.134	178.001	511.707
MOU	91	0.366	1.350	0.040	0.438	1.36	3.15	26.763	2600.6
NIO	40	0.625	1.271	0.398	0.588	1.136	1.799	392	2750
RGC	8	0.805	1.518	0.923	1.512	0.53	0.8	674	2695
RGR	254	0.295	2.384	0.159	2.326	0.69	2.31	11	11400
RIO	38	0.624	2.384	0.332	1.463	0.74	0.89	463.65	4544.38
Total	600	0.186	2.423	0.025	17.282	0.038	10.7	11	11400

(b)

Table 4. (a) Range of laboratory variables, (b) Range of field variables

Data-driven techniques can be used for data mining since the only prerequisite for their function is the determination of the input parameters without the need to predefine the structure of the model and the degree of nonlinearity. The determination of the input parameters for the data-driven schemes will be made with a tentative assessment through a trial-and-error procedure. The correlation coefficient r has been employed in order to reveal any existing linear dependence in log-log plots between sediment concentration and any of the variables listed in Table 1

$$r = \frac{\sum_{i=1}^{N}(Y_i - \bar{Y})(X_i - \bar{X})}{\sqrt{\sum_{i=1}^{N}(Y_i - \bar{Y})^2 \sum_{i=1}^{N}(X_i - \bar{X})^2}} \qquad (15)$$

where Y denotes sediment concentration and X denotes the independent variable. Table 5 shows the correlation coefficient for log-log plots for the flume and field data of Tables 4a and 4b. From the techniques proposed, the trial-and-error process will be accomplished with the aid of ANNs, due to their speed, and after the determination of the most promising combinations that may serve as an input layer, the other data-driven techniques will be implemented as well.

	VS/ω	$VS/(gd_{50})^{1/2}$	ω^*	τ^*	Fr	Re	Re^*	d_{gr}
Flume data r	0.862	0.885	0.754	0.681	0.759	0.463	-0.014	-0.314
Field data r	0.730	0.687	0.601	0.587	0.492	0.148	-0.144	-0.405

Table 5. Correlation between sediment concentration and independent dimensionless variables of the flume and field data of Table 4 in log-log plots

The findings shown in Table 5 partially agree with the diagrams depicted in Figure 1, since sediment discharge is best correlated with unit stream power and stream power both for laboratory and for field data.

After the tentative assessment based on ANNs, of several input combinations, the most potent ones, which will be applied to the data-driven schemes, seem to be those listed in Table 6. These combinations include the independent variables of Eq. (3) and others that are relatively easily measured and commonly used in engineering. It is noteworthy that all combinations comprise dimensionless grain diameter and Froude number among others. Whilst Froude number gives a measure of the ratio of inertial forces to gravitational forces of the flow and is a commonly used variable in hydraulic engineering, the potential usage of dimensionless grain diameter is twofold. Firstly, it introduces kinematic viscosity and median grain diameter and secondly provides homogeneity in the input data. The necessity for the provided homogeneity can be seen from combination (a) where shear Reynolds number, which essentially includes dimensionless grain diameter, is included as well. The absence of any of these two terms in combination (a) has detrimental effects in the predictive capability of the generated model. The other variables for the combinations examined herein are those that most sediment transport formulae rely heavily on, namely dimensionless unit stream power, dimensionless stream power and dimensionless shear stress, and are best correlated with sediment concentration as shown in Table 5. For combination (a) Yang's dimensionless unit stream power was preferred to Vanoni's because, despite the fact that the calculation of fall velocity may be problematic, it reduced significantly the sum of errors between calculated and observed values. The other two combinations (b) and (c) comprise just three variables because shear is embedded in dimensionless stream power and dimensionless shear stress, respectively. Furthermore, it seems that there is no other potential input combination, besides those listed in Table 6, since any other combination tested gave results that declined by orders of magnitude.

Input combinations
a $d_{gr}, Fr, Re^*, VS/\omega_s$
b d_{gr}, Fr, ω^*
c d_{gr}, Fr, τ^*

Table 6. Input combinations that will be applied to the data-driven schemes

5. Applications and results

The potential of training a DDM scheme solely with flume data and subsequently applying it to a test set comprising exclusively field data has been shown in Kitsikoudis et al. (2012a, 2012b) where ANNs and symbolic regression were utilized, respectively, for the prediction of sediment concentration in sand bed rivers. In these studies, however, the data were not subjected to elaboration and screening, in order to demonstrate the potential modeling ability of this technique with crude data. As a result, input data were kept in large numbers, and the generated models yielded very good results, better than those obtained from the common sediment transport formulae. However, it is known that the incorporation of knowledge can be proved beneficial to the predictive capability of DDM schemes as long as this is accomplished by transformation and elaboration of the fundamentals. Sediment transport and open channel hydraulics rely heavily on empirical equations and ideal flows; therefore, data transformation based on such assumptions does not guarantee the enhancement of the predictive capabilities of the DDM scheme. Nevertheless, the sidewall correction of Vanoni and Brooks (1957) was applied for the proper calculation of the shear stress in flume measurements and additionally the restrictions of Table 2 were imposed to the data for the removal of various biases resulting to a significantly reduced data amount. On the contrary, a criterion for the initiation of motion has been omitted, due to the stochastic character of turbulence, and was left up to the DDM scheme to define the effective portion of the flow that quantifies the transport rate.

Since every data-driven technique has its own syntax, the three possible input combinations of Table 6 are tested individually with the aid of both ANNs and symbolic regression. The evaluation of the modeled results P_i with respect to the observed ones O_i will be made on the basis of the root mean square error (*RMSE*),

$$RMSE = \sqrt{\frac{\sum_{i=1}^{N}(O_i - P_i)^2}{N}} \tag{16}$$

coefficient of determination (R^2) or Nash-Sutcliffe model efficiency coefficient (E) (Nash and Sutcliffe, 1970),

$$R^2 = 1 - \frac{\sum_{i=1}^{N}(O_i - P_i)^2}{\sum_{i=1}^{N}(O_i - \bar{O})^2} \tag{17}$$

and discrepancy ratio (*DR*). The latter is the percentage of calculated concentrations that lie between one half and two times the respective measured concentrations.

5.1. ANNs application

This study was implemented in MATLAB with the aid of the neural network toolbox (Demuth et al., 2009). Since the usage of Levenberg-Marquardt training function gave the best results in a similar study in Kitsikoudis et al. (2012a), it was utilized for training in this application as well. Due to the importance of the initial values of the synaptic weights in the search for local minima of the error function, which is the mean square error between calculated and observed values, a MATLAB code was written, which determines the most efficient ANN within 5000 training executions, for each network architecture, with random initial weights for every repetition. The most efficient ANN is taken to be the one that yields only positive sediment concentrations, in order for the results to have physical meaning, and after the training provides the highest DR in the test set. For this evaluation, DR is preferred over $RMSE$, because the latter emphasizes on large concentrations. Models that derived slightly worse results than others, but had much simpler structure were preferred due to the principle of parsimony. Figures 3-5 depict the scatter plots of the best derived models, for each input combination of Table 6, for the field data of the test set. These models that perform best are described in Table 7. Table 8 shows the best models and their performance measures for the training, validation and test sets. Finally, Table 10 shows a comparison between the ANN induced models and some of the commonly used sediment transport functions for the rivers data constituting the test set. It should be mentioned that several of these formulae are calibrated with part of the data (especially the Brownlie formula) that are used for the comparison and despite that significant advantage they still generate inferior results to those of the ANNs.

	Input combination from Table 6	Network architecture (neurons in input-hidden-output layers)
ANN (a)	a	4-5-1
ANN (b)	b	3-6-1
ANN (c)	c	3-11-2-1

Table 7. Best performing models for each possible input combination

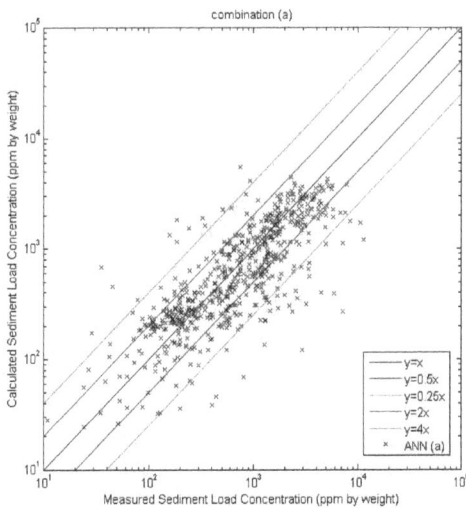

Figure 3. Scatter plot for the field data of the test set, of measured sediment concentration and computed from ANN, based on input combination (a)

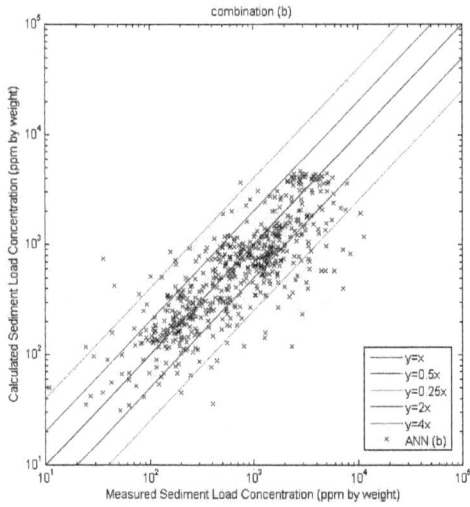

Figure 4. Scatter plot for the field data of the test set, of measured sediment concentration and computed from ANN, based on input combination (b)

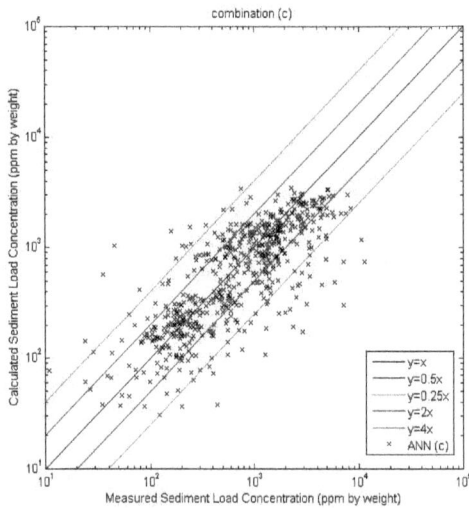

Figure 5. Scatter plot for the field data of the test set, of measured sediment concentration and computed from ANN, based on input combination (c)

		DR 0.5-2 (%)	DR 0.25-4 (%)	RMSE	R^2
ANN (a)	Training set	85.12	98.51	722.20	0.9213
	Validation set	82.86	97.55	936.72	0.5582
	Test set	72.50	92.33	1168.76	0.3687
ANN (b)	Training set	82.81	97.56	734.13	0.9187
	Validation set	85.71	95.92	884.41	0.6062
	Test set	71.67	93.17	1202.69	0.3315
ANN(c)	Training set	84.84	98.24	509.42	0.9608
	Validation set	84.08	98.37	872.36	0.6169
	Test set	70.67	91.33	1221.72	0.3102

Table 8. Performance measures of the optima ANNs

From Table 8 can be inferred that any of the three combinations listed in Table 6 has its own merit and that sediment transport can be quantified by physical quantities that can be either vectors or scalars.

5.2. Symbolic regression application

The basic computation tool for the implementation of symbolic regression is provided by GPTIPS (Searson, 2009), which is an open source MATLAB toolbox. Since every problem has its own peculiarities, proper adjustments must be made to the GPTIPS parameters in order to obtain good results. The most important parameters are the population size, the number of generations, the using functions, the maximum number of genes and the maximum tree depth. Searson et al. (2010) have found that enforcing stringent tree depth restrictions often allows the evolution of relatively compact models that are linear combinations of low order nonlinear transformations of the input variables. After several runs, only input combination (b) gave results superior to those of the classical formulae. The GPTIPS derived formula for this combination is the following

$$C_t = 1542\left(\omega^* Fr^3\right)^{0.4} - 0.4794 \frac{\sqrt{d_{gr}Fr}\left(d_{gr} + Fr - \omega^*\right) + \dfrac{\omega^*}{Fr}}{\left(\dfrac{6.218}{\omega^*} - Fr + 1\right)^3} - 313.6 \tag{18}$$

Figure 6 depicts the scatter plot of measured and calculated from Eq. (18) sediment concentrations for the field data of the test set, whilst Table 9 and Table 10 show its performance for the training, validation and testing set, and the comparison with other formulae, respectively.

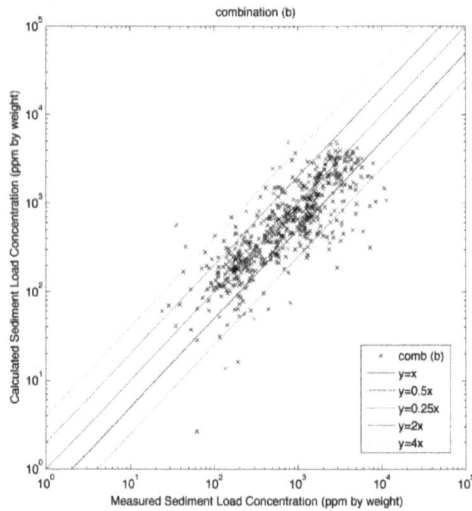

Figure 6. Scatter plot for the field data of the test set, of measured sediment concentration and computed from symbolic regression, based on input combination (b)

	DR 0.5-2 (%)	DR 0.25-4 (%)	RMSE	R^2
Training set	68.20	84.84	632.21	0.9397
Validation set	66.53	85.31	964.70	0.5315
Test set	71.00	91.33	1218.12	0.3143

Table 9. Performance measures of symbolic regression, based on combination (b)

	DR 0.5-2 (%)	DR 0.25-4 (%)	RMSE	R^2
ANN (a)	72.50	92.33	1168.76	0.3687
ANN (b)	71.67	93.17	1202.69	0.3315
ANN (c)	70.67	91.33	1221.72	0.3102
Symb. regression (b)	71.00	91.33	1218.12	0.3143
Ackers & White	58.33	88.67	1405.38	0.0872
Brownlie	68.33	91.00	1274.44	0.2494
Engelund & Hansen	69.67	92.33	1244.83	0.2838
Karim & Kennedy	64.83	91.50	1341.34	0.1685
Molinas & Wu	52.83	84.83	1423.06	0.0641
Yang	49.50	83.33	1403.07	0.0902

Table 10. Comparison of ANNs of the input combinations (a), (b) and (c) and Eq. (18), derived from symbolic regression for the combination (b), with sediment transport formulae based on the river data of the test set

The results obtained from ANNs for all the combinations are superior to those of the classi-cal sediment transport formulae in terms of DR, $RMSE$ and R^2. Combination (a) performed best in all evaluation measures, besides the second DR criterion in the range 0.25-4 where combination (b) gave better results. The third combination came up third with respect to all evaluation measures. However, these results by no means can be considered conclusive, since it is essentially unknown whether they are the best results derived from the ANN or just results obtained from the trapping in a local minimum of the minimization process in the network's training algorithm. From the results generated from symbolic regression, only combination (b) managed to surpass the classical sediment transport functions. The other two combinations gave results inferior to those of Engelund and Hansen and Brownlie for-mulae, but superior to those of the others. In addition, symbolic regression derived its best results without utilizing the log-transformation of the input data. Regarding the other sedi-ment transport functions, the formula of Engelund and Hansen performed best. The small values of the coefficient of determination R^2 in Table 10 reflect the difficulty of predicting sediment transport rates in natural streams and rivers, due to random turbulent bursts that accentuate the stochastic nature and exacerbate the complexity of the problem.

Although these results cannot be considered conclusive, it seems that the ANNs yield better results. GPTIPS sometimes (usually when only a few input variables are involved) lags be-hind a neural network model in terms of raw predictive performance, but the equivalent GP models are often simpler, shorter and may be open to physical interpretation (Searson, 2009). This is partially due to the fact that ANNs are much faster than the time consuming GP and for given time they can run multiple times comparing to GP. Moreover, since the testing set comes from a database with different statistical distributions than the one from which the training set originates, the exploration of as many as possible local minima of the training function may prove beneficial to the training process. ANNs have this property, whilst GP is based on a stochastic concept seeking the global minimum. This may be one reason for the superiority of ANNs in this study, where the training data comprise flume measurements, whilst the testing data consists of field measurements.

6. Conclusions

This study utilized two widely used data-driven techniques, namely ANNs and symbolic regression, in a novel way since the data used for training and those used for testing came from datasets with different statistical distributions. This difference is owned to the fact that the training and validation set comprises exclusively laboratory flume data, while the test-ing set consists solely of field data. Based on this concept, the inclusion of noise emanated from the field measurements will not be embedded in the training data and additionally the generated models will have general applicability since the inclusion of field data in the training set would confine them to the specific streams from which the data were obtained. The determination of the input parameters was accomplished by a tentative assessment of some of the widely used dimensionless parameters in sediment transport and open channel hydraulics. This assessment showed that three combinations had the potential to serve as inputs and were involved in this application, in which they all yielded very good results, better than those obtained from the commonly used formulae on the basis of root mean square error and the ratio of computed to measured transport rates. Unit stream power, stream power, and shear stress were the dominant independent variables of the three combinations, respectively, and the results have shown that each one, of these widely used variables in the context of sediment transport, has its own merit. The results generated from the ANNs were better from those obtained from symbolic regression; however, the explicit equation that was derived from the latter can be more easily interpreted. Finally, the results obtained in this study may enhance the confidence in using data-driven techniques, despite their black-box nature, because, in order to perform well in a dataset from a different system from the one they were trained, the induced equations must have physical meaning.

Notation

The following symbols are used in this chapter:

C_t = sediment concentration by weight in parts per million (ppm)

D = mean flow depth (m)

Q = water discharge (m³/s)

Q_{st} = sediment discharge (m³/s)

R = hydraulic radius (m)

S = energy slope

T = water temperature (°C)

u_* = shear velocity (m/s)

u_{*c} = critical shear velocity (m/s)

V = mean flow velocity (m/s)

W = channel width (m)

d = grain diameter (m)

d_{50} = median grain diameter (m)

f = friction factor

g = gravitational acceleration (m/s²)

γ = specific weight of water (N/m³)

γ_s = specific weight of sediment (N/m³)

ν = kinematic viscosity of water (m²/s)

ϱ = density of water (kg/m³)

ϱ_s = density of sediment (kg/m³)

σ_g = geometric standard deviation of bed particles [$(d_{84}/d_{50} + d_{50}/d_{16})/2$]

τ = shear stress [kg/(m.s²)]

τ^* = dimensionless shear stress

ω = stream power (kg/s³)

ω^* = dimensionless stream power

ω_s = settling or fall velocity of sediment (m/s)

Appendix A

In flume experiments, the sand covered bed will generally be much rougher than the flume walls, and thus will be subjected to higher shear stress. Separation of the shear force exerted on the bed from that on the lateral boundaries was first proposed by Einstein (1950). The line of analysis pursued as follows is that proposed by Johnson (1942) and modified by Vanoni and Brooks (1957). The principal assumption is that the cross-sectional area can be divided into two parts, A_b and A_w, in which the streamwise component of the gravity force is resisted by the shear force exerted in the bed and walls, respectively. It is further assumed that the mean velocity and energy gradient are the same for A_b and A_w, and that the Darcy-Weisbach relation can be applied to each part of the cross section as well as to the whole, i.e.

$$\frac{V^2}{S} = \frac{8gA}{fp} = \frac{8gA_b}{f_b p_b} = \frac{8gA_w}{f_w p_w} \tag{19}$$

in which, p = the wetted perimeter; and the subscripts b and w refer to the bed and wall sections, respectively. For a rectangular channel p=2D+W; p_w=2D; p_b=W. Introducing the geometrical requirement A=A_b+A_w into Eq. (19) results in

$$f_b = f + \frac{2D}{W}(f - f_w) \tag{20}$$

The wall friction factor f_w is further related to the ratio of Re/f, where Re=4VR/ν and f can be calculated from the experimental data. This relationship, which was originally given as a graph of f_w against Re/f by Vanoni and Brooks (1957), can also be described by the function

$$f_w = \left[20(\text{Re}/f)^{0.1} - 39\right]^{-1} \tag{21}$$

which is obtained by curve fitting (Cheng and Chua, 2005). Finally, f_b is calculated from Eq. (20) and $R_b = A_b/p_b$ from Eq. (19). R_b is consequently used for the calculation of the bed shear velocity and bed shear stress.

Despite its several obvious deficiencies (division of the cross section into two noninteracting parts, determination of friction factors for section components on the basis of a pipe friction diagram, use of the same mean velocity for each subsection, etc.), the side-wall correction procedure appears to yield fairly reliable estimates of the friction factors for flow over sand beds with no flume walls present (Vanoni, 2006).

Appendix B

For the calculation of particle fall velocity in a clear, still fluid, van Rijn (1984) suggested the use of the Stokes law for sediment particles smaller than 0.1 mm

$$\omega_s = \frac{1}{18} \frac{\rho_s - \rho}{\rho} g \frac{d^2}{\nu} \tag{22}$$

For suspended sand particles in the range 0.1 to 1 mm, the following type of equation, as proposed by Zanke (1977), can be used

$$\omega_s = 10\frac{\nu}{d}\left\{\left[1+0.01\left(\frac{\rho_s}{\rho}-1\right)\frac{gd^3}{\nu^2}\right]^{1/2} -1\right\} \tag{23}$$

For particles larger than about 1 mm, the following simple equation can be used (van Rijn, 1982)

$$\omega_s = 1.1\left[\left(\frac{\rho_s}{\rho}-1\right)gd\right]^{1/2} \tag{24}$$

Appendix C

Ackers and White formula: Ackers and White (1973) applied dimensional analysis to express the mobility and transport rate of sediment in terms of some dimensionless parameters. It has been shown that the transport of fine materials is best related to gross shear, shear velocity being the representative variable, and that the transport of coarse materials is best related to the net grain shear, mean velocity being the representative variable. The following equations do not necessarily apply in an upper phase of transport. However, it was shown that the following relationships are not sensitive to bed form; they apply to plain, rippled, and duned configurations. Their mobility number for sediment is

$$F_{gr} = \frac{U_*^n}{\sqrt{gd(\gamma_s/\gamma - 1)}} \left[\frac{V}{\sqrt{32}\log(10D/d)} \right]^{1-n} \tag{25}$$

Coefficients C, A, m and n are related to the dimensionless grain diameter d_{gr} based on best-fit curves of laboratory data with sediment sizes greater than 0.04 mm and Froude numbers less than 0.8. They are shown in Table 11.

$$d_{gr} = d \left[\frac{g(\gamma_s/\gamma - 1)}{v^2} \right]^{1/3} \tag{26}$$

Finally, they related the bed material load to the mobility number as follows

$$G_{gr} = \frac{XD}{d\gamma_s/\gamma} \left(\frac{U_*}{V} \right)^n = C \left(\frac{F_{gr}}{A} - 1 \right)^m \tag{27}$$

where X = rate of sediment transport in terms of mass flux per unit mass flow rate

$d_{gr} \geq 60$	$1 < d_{gr} < 60$
$n = 0.00$	$n = 1.00 - 0.56\log d_{gr}$
$A = 0.17$	$A = 0.23 d_{gr}^{-0.5} + 0.14$
$m = 1.50$	$m = 9.66 d_{gr}^{-1} + 1.34$
$C = 0.025$	$\log C = -3.53 + 2.86\log d_{gr} - (\log d_{gr})^2$

Table 11. Coefficients of the Ackers and White formula

Brownlie formula: The Brownlie (1981a) relations are based on regressions of over 1000 experimental and field data points. For normal or quasi-normal flow, the transport relation takes the form

$$C_t = 7115 c_f \left(F_g - F_{go} \right)^{1.978} S^{0.6601} \left(\frac{R}{d_{50}} \right)^{-0.3301} \tag{28}$$

where

$$F_g = \frac{V}{\sqrt{(\rho_s/\rho - 1)gd_{50}}} \tag{29}$$

$$F_{go} = 4.596\left(\tau_c^*\right)^{0.5293} S^{-0.1405} \sigma_g^{-0.1606} \tag{30}$$

$$\tau_c^* = 0.22Y + 0.06 \cdot 10^{-7.7Y} \tag{31}$$

$$Y = \left(\frac{\sqrt{(\rho_s/\rho - 1)gd_{50}}\, d_{50}}{v}\right)^{-0.6} \tag{32}$$

$c_f = 1$ for laboratory flumes and 1.268 for field channels.

Engelund and Hansen formula: Using Bagnold's stream power concept and the similarity principle, Engelund and Hansen (1967) established the following sediment transport formula

$$f'\phi = 0.1\theta^{5/2} \tag{33}$$

where

$$f' = \frac{2gDS}{V^2} \tag{34}$$

$$\phi = \frac{q_t}{\rho_s\sqrt{(\rho_s/\rho - 1)gd_{50}^3}} \tag{35}$$

$$\theta = \frac{\tau}{(\gamma_s - \gamma)d_{50}} \tag{36}$$

where q_t = total sediment discharge by weight per unit width. Strictly speaking, the Engelund and Hansen formula should be applied to those flows with dune beds in accordance with the similarity principle. However, many tests have shown that it can be applied to the upper flow regime with particle size greater than 0.15 mm without serious deviation from the theory.

Karim and Kennedy formula: Karim and Kennedy (1990) applied nonlinear multiple regression analysis to derive relations among flow velocity, sediment discharge, bed form geometry, and friction factor of alluvial rivers. A database comprising 339 river flows and 608 flume flows was used in their analysis. The obtained sediment load predictor is given by

$$\begin{aligned}
\log\frac{q_s}{\sqrt{(\gamma_s/\gamma - 1)gd_{50}^3}} &= -2.279 + 2.972\log\frac{V}{\sqrt{(\gamma_s/\gamma - 1)gd_{50}}} \\
&+ 1.060\log\frac{V}{\sqrt{(\gamma_s/\gamma - 1)gd_{50}}}\log\frac{U_* - U_{*_c}}{\sqrt{(\gamma_s/\gamma - 1)gd_{50}}} + 0.299\log\frac{D}{d_{50}}\log\frac{U_* - U_{*_c}}{\sqrt{(\gamma_s/\gamma - 1)gd_{50}}}
\end{aligned} \tag{37}$$

where q_s = volumetric total sediment discharge per unit width.

Molinas and Wu formula: This empirical relation is based on Velikanov's gravitational power theory, which assumes that the power available in flowing water is equal to the sum of the power required to overcome flow resistance and the power required to keep sediment in suspension against gravitational forces. Molinas and Wu (2001) argued that the predictors of Ackers and White, Engelund and Hansen, and Yang have been developed with flume experiments representative of shallow flows and cannot be applied to large rivers having deep flow conditions. Motivated by the need for having a total bed material load predictor for application to large sand bed rivers, they used stream power and energy considerations together with data from large rivers (e.g., Amazon, Atchafalaya, Mississippi, Red River), to obtain an empirical fit for the total bed material load concentration in ppm

$$C_t = \frac{1430\left(0.86 + \sqrt{\Psi}\right)\Psi^{1.5}}{0.016 + \Psi} \tag{38}$$

where Ψ = universal stream power, which is defined as

$$\Psi = \frac{V^3}{g\left(\rho_s/\rho - 1\right)D\omega_s\left[\log\left(D/d_{50}\right)\right]^2} \tag{39}$$

One advantage of this approximation is that the energy slope does not have to be measured directly, which is always a challenge in large alluvial rivers. On the other hand, since Molinas and Wu (2001) do not mention how the wash load was separated from the bed material load and the same large river data were used both to develop and to test their formulation, Eq. (38) might overestimate bed material load concentrations when applied to other large rivers not included in the calibration (Garcia, 2008).

Yang formula: To determine total sediment concentration, Yang (1973) used Buckingham's π theorem and the concept of unit stream power, which is given by the product of mean flow velocity and energy slope. The coefficients in Yang's equation were determined by running a multiple regression analysis for 463 sets of laboratory data. The equation obtained is

$$\begin{aligned}
\log C_t = 5.435 - 0.286\log\frac{\omega_s d_{50}}{v} - 0.457\log\frac{U_*}{\omega_s} \\
+ \left(1.799 - 0.409\log\frac{\omega_s d_{50}}{v} - 0.314\log\frac{U_*}{\omega_s}\right)\log\left(\frac{VS}{\omega_s} - \frac{V_{cr}S}{\omega_s}\right)
\end{aligned} \tag{40}$$

The critical dimensionless unit stream power $V_{cr}S/\omega$ is the product of the dimensionless critical velocity V_{cr}/ω and the energy slope S, where

$$\frac{V_{cr}}{\omega_s} = \begin{cases} \dfrac{2.5}{\log\left(\dfrac{U_* d_{50}}{v}\right) - 0.06} + 0.66 & for \quad 1.2 < \dfrac{U_* d_{50}}{v} < 70 \\ \\ 2.05 & for \quad 70 \le \dfrac{U_* d_{50}}{v} \end{cases} \tag{41}$$

Author details

Vasileios Kitsikoudis[1], Epaminondas Sidiropoulos[2] and Vlassios Hrissanthou[1]

1 Department of Civil Engineering, Democritus University of Thrace, Xanthi, Greece

2 Department of Rural and Surveying Engineering, Aristotle University of Thessaloniki, Thessaloniki, Greece

References

[1] Ackers P., White W. R. (1973). Sediment Transport: New Approach and Analysis. Journal of the Hydraulics Division, ASCE; 99(HY11) 2041-2060.

[2] Azamathulla H. Md, Ab Ghani A., Zakaria N. A., Guven A. (2010). Genetic Programming to Predict Bridge Pier Scour. Journal of Hydraulic Engineering, ASCE; 136(3) 165-169.

[3] Babovic V. (2000). Data Mining and Knowledge Discovery in Sediment Transport. Computer Aided Civil and Infrastructure Engineering; 15(5) 383-389.

[4] Babovic V., Abbott M. B. (1997). The Evolution of Equations from Hydraulic Data. Part II: Applications. Journal of Hydraulic Research, IAHR; 35(3) 411-430.

[5] Babovic V., Keijzer M. (2000). Genetic Programming as a Model Induction Engine. Journal of Hydroinformatics; 2(1) 35-60.

[6] Bagnold R. A. (1956). Flow of Cohesionless Grains in Fluids. Philosophical Transactions of the Royal Society [London], Series A; 249 235-297.

[7] Bagnold R. A. (1966). An Approach to the Sediment Transport Problem from General Physics. Prof. Paper 422-I. U.S. Geological Survey.

[8] Bhattacharya B., Price R. K., Solomatine D. P. (2004). A Data Mining Approach to Modelling Sediment Transport. In: Liong S., Phoon K., Babovic V. Proceedings of the 6th International Conference on Hydroinformatics, 21-24 June 2004, Singapore. World Scientific.

[9]   Bhattacharya B., Price R. K., Solomatine D. P. (2005). Data Driven Modeling in the Context of Sediment Transport. Physics and Chemistry of the Earth; 30(4-5) 297-302.

[10] Bhattacharya B., Price R. K., Solomatine D. P. (2007). Machine Learning Approach to Modeling Sediment Transport. Journal of Hydraulic Engineering, ASCE; 133(4) 440-450.

[11] Brownlie W. R. (1981a). Prediction of Flow Depth and Sediment Discharge in Open Channels. Report No. KH-R-43A. Pasadena, California: W. M. Keck Laboratory of Hydraulics and Water Resources, California Institute of Technology.

[12] Brownlie W. R. (1981b). Compilation of Alluvial Channel Data: Laboratory and Field. Report No. KH-R-43B. Pasadena, California: W. M. Keck Laboratory of Hydraulics and Water Resources, California Institute of Technology.

[13] Cheng N. S., Chua L. H. C. (2005). Comparisons of Sidewall Correction of Bed Shear Stress in Open-Channel Flows. Journal of Hydraulic Engineering, ASCE; 131(7) 605-609.

[14] Cigizoglu H. K. (2004). Estimation and Forecasting of Daily Suspended Sediment Data by Multi-Layer Perceptrons. Advances in Water Resources; 27(2) 185-195.

[15] Demuth H. B., Beale M. H., Hagan M. T. (2009). Neural Network Toolbox: For Use With MATLAB. The Mathworks Inc.; 2009.

[16] Eaton B. C., Church M. (2011). A Rational Sediment Transport Scaling Relation Based on Dimensionless Stream Power. Earth Surface Processes and Landforms; 36(7) 901-910.

[17] Einstein H. A. (1942). Formulas for the transportation of Bed Load. Transactions, ASCE; 107 (Paper No. 2140) 561-573.

[18] Einstein H. A. (1950). The Bedload Function for Sediment Transportation in Open Channel Flows. Technical Bulletin No. 1026. Washington D.C.: U.S. Department of Agriculture, Soil Conservation Service.

[19] Engelund F., Hansen E. (1967). A Monograph on Sediment Transport in Alluvial Streams. Copenhagen: Teknisk Vorlag.

[20] Garcia M. H. (2008). Sediment Transport and Morphodynamics. In: Garcia M. H. (ed.) ASCE Manuals and Reports on Engineering Practice No. 110, Sedimentation Engineering: Processes, Measurements, Modeling, and Practice. Virginia, U.S.A.: ASCE, p21-163.

[21] Haykin S. (2009). Neural Networks and Learning Machines, 3rd Edition. New Jersey: Prentice Hall.

[22] Hornik K., Stinchcombe M., White H. (1989). Multilayer Feedforward Networks are Universal Approximators. Neural Networks; 2(5) 359-366.

[23] Johnson J. W. (1942). The Importance of Considering Side-Wall Friction in Bed-Load Investigations. Civil Engineering, ASCE; 12(6) 329-331.

[24] Karim M. F., Kennedy J. F. (1990). Menu of Coupled Velocity and Sediment-Discharge Relations for Rivers. Journal of Hydraulic Engineering, ASCE; 116(8) 978-996.

[25] Kitsikoudis V., Sidiropoulos E., Hrissanthou V. (2012a). A New Approach for ANN Modeling of Sediment Transport in Sand Bed Rivers. Proceedings of the 10th International Conference on Hydroinformatics, CD-ROM format, 14-18 July 2012, Hamburg, Germany.

[26] Kitsikoudis V., Sidiropoulos E., Hrissanthou V. (2012b). Implementation of Multigene Symbolic Regression in the Sediment Transport Quantification Problem for Sand Bed Rivers. Proceedings of the 9th International Symposium on Ecohydraulics, CD-ROM format, 17-21 September 2012, Vienna, Austria.

[27] Knapp R. T. (1938). Energy Balance in Stream Flows Carrying Suspended Load. Transactions, American Geophysical Union; 501-505.

[28] Koza J. (1992). Genetic Programming: On the Programming of Computers by Means of Natural Selection. Cambridge MA: MIT Press.

[29] Maier H. R., Dandy G. C. (2000). Neural Networks for the Prediction and Forecasting of Water Resources Variables: A Review of Modeling Issues and Applications. Environmental Modeling and Software; 15(1) 101-124.

[30] Minns A. W. (2000). Subsymbolic Methods for Data Mining in Hydraulic Engineering. Journal of Hydroinformatics; 2(1) 3-13.

[31] Molinas A., Wu B. (2001). Transport of Sediment in Large Sand-Bed Rivers. Journal of Hydraulic Research, IAHR; 39(2) 135-146.

[32] Nagy H. M., Watanabe K., Hirano M. (2002). Prediction of Sediment Load Concentration in Rivers Using Artificial Neural Network Model. Journal of Hydraulic Engineering, ASCE; 128(6) 588-595.

[33] Nash J. E., Sutcliffe J. V. (1970). River Flow Forecasting Through Conceptual Models, Part I – A Discussion of Principles. Journal of Hydrology: 10(3) 282-290.

[34] Pugh C. A. (2008). Sediment Transport Scaling for Physical Models. In: Garcia M. H. (ed.) ASCE Manuals and Reports on Engineering Practice No. 110, Sedimentation Engineering: Processes, Measurements, Modeling, and Practice. Virginia, U.S.A.: ASCE, p1057-1066.

[35] Rubey W. W. (1933). Equilibrium Conditions in Debris-Laden Streams. Transactions, American Geophysical Union 14th Ann. Mtg; 497-505.

[36] Rumelhart D. E., Hinton G. E., Williams R. J. (1986). Learning Representations of Back-Propagation Errors. Nature (London); 323(9) 533-536.

[37] Searson D. (2009). GPTIPS: Genetic Programming & Symbolic Regression for MAT-LAB User Guide.

[38] Searson D. P., Leahy D. E., Willis M. J. (2010). GPTIPS: An Open Source Genetic Programming Toolbox for Multigene Symbolic Regression. In Ao S. I., Castillo O., Douglas C., Feng D. D., Lee J. (eds.) Proceedings of the International MultiConference of Engineers and Computer Scientists, Vol. I, IMECS 2010, 17-19 March 2010, Hong Kong.

[39] Shields A. (1936). Application of Similarity Principles and Turbulence Research to Bedload Movement. Transl. into English by Ott W. P. and Van Uchelen J. C. Pasadena, California: California Institute of Technology; 1936.

[40] van Rijn L. C. (1982). Computation of Bed-Load and Suspended Load. Report S487-II, Delft Hydraulics Laboratory, Delft, The Netherlands.

[41] van Rijn L. C. (1984). Sediment Transport, Part II: Suspended Load Transport. Journal of Hydraulic Engineering, ASCE; 110(11) 1613-1641.

[42] Vanoni V. A. (1978). Predicting Sediment Discharge in Alluvial Channels. In: Water Supply and Management. Oxford: Pergamon Press. p399-417.

[43] Vanoni V. A. (2006). ASCE Manuals and Reports on Engineering No. 54, Sedimentation Engineering. Virginia, U.S.A.: ASCE.

[44] Vanoni V. A., Brooks N. H. (1957). Laboratory Studies of the Roughness and Suspended Load of Alluvial Streams. Sedimentation Laboratory Report No. E68. Pasadena, California: California Institute of Technology.

[45] Velikanov M. A. (1955). Dynamics of Channel Flow – v.2. In: Sediments and the Channel, 3rd Edition. Moscow: State Publishing House for Tech.– Theoretical Lit.; p107-120 [in Russian].

[46] Witten I. H., Frank E., Hall M. A. (2011). Data Mining: Practical Machine Learning Tools and Techniques, 3rd Edition. Burlington, MA: Morgan Kaufmann.

[47] Yalin M. S. (1977). Mechanics of Sediment Transport. Oxford: Pergamon Press.

[48] Yang C. T. (1972). Unit Stream Power and Sediment Transport. Journal of the Hydraulics Division, ASCE; 98(HY10) 1805-1836.

[49] Yang C. T. (1973). Incipient Motion and Sediment Transport. Journal of the Hydraulics Division, ASCE; 99(HY10) 1679-1704.

[50] Yang C. T. (1977). The Movement of Sediment in Rivers. Surveys in Geophysics; 3(1) 39-68.

[51] Yang C. T. (2003). Sediment Transport: Theory and Practice. Original edition McGraw-Hill; 1996. Reprint edition by Krieger Publication Company.

[52] Yang C. T., Marsooli R., Aalami M. T. (2009). Evaluation of Total Load Sediment Transport Formulas Using ANN. International Journal of Sediment Research; 24(3) 274-286.

[53] Zakaria N. A., Azamathulla H. Md, Chang C. K., Ab Ghani A. (2010). Gene Expression Programming for Total Bed Material Load Estimation – A Case Study. Science of the Total Environment; 408(21) 5078-5085.

[54] Zanke U. (1977). Berechnung der Sinkgeschwindigkeiten von Sedimenten. Mitt. des Franzius – Instituts für Wasserbau, Heft 46, Seite 243. Technical University Hannover.

Chapter 11

Modelling Cohesive Sediment Dynamics in the Marine Environment

Katerina Kombiadou and Yannis N. Krestenitis

Additional information is available at the end of the chapter

1. Introduction

The inflow of fine sediments in the marine environment affects various processes and has significant socioeconomic and environmental impacts, the most straight-forward of which is related to sedimentation in deltaic and coastal areas. However, fine sediments have the ability to affect solar radiation in the water column and to absorb contaminants on their surface [1,2]. Therefore, fine sedimentary plumes also affect primary productivity and the distribution of pollutants in the column and the seabed. The distinction between fine ($<63\mu m$) and coarse-grained ($>63\mu m$) sediments is rather general and varies with the type of the sedimentary matter, but it is a fact that the dominance of inter-particle cohesion forces over gravitational forces varies inversely with the floc diameter. Thus, the effect of cohesion is much more pronounced on the behaviour of clays (diameter $<2\mu m$) than of silts (2-$63\mu m$); in fact, the development of cohesion in silty clays is mainly due to the clay fraction in the sediment [1]. Adhesion diversifies the behaviour of cohesive and sandy sediments in the aquatic domain, adding further complexities to modelling efforts for fine sedimentary plumes. Their movement is the result of the combined action of various physical processes and forces (Figure 1).

The processes that control sediment transport in coastal microtidal areas can be divided in the ones that take place in the water column (pelagic processes):

- advection

- dispersion

- coagulation – deflocculation

- settling

- effects of seawater stratification

and the ones that take place at the boundary of the seabed (benthic processes):

- deposition

- erosion

- self-weight consolidation – resuspension

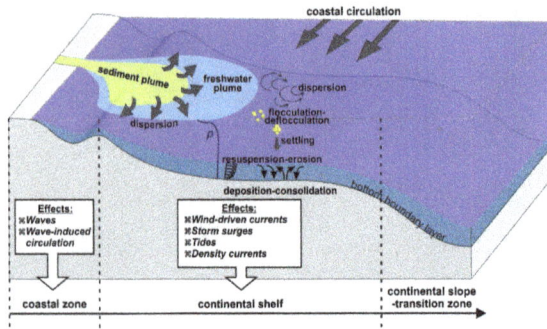

Figure 1. The main processes that control sediment transport in coastal, deltaic areas.

These processes are analyzed in the following sections in terms of physical interpretation and mathematical parameterization, along with the three basic modelling approaches used for the description of sediments in the aquatic domain. Finally, results from a particle-tracking model applied in a micro-tidal shelf sea are presented and discussed.

2. Sediment transport models

Models describing transport and dispersion of sediments in the aquatic domain can be divided in Eulerian (e.g. [2]), that describe the evolution of suspended matter concentration using finite differences and the mass conservation principle, Lagrangian (e.g. [3]) in which the advection and dispersion of a specific amount of sediment mass is being traced with computational time and in mixed Euler-Lagrange models (e.g. [4]), in which transport is expressed by particles and dispersion through finite differences. The differences between these modelling approaches and their advantages and disadvantages are given further down.

2.1. Eulerian models

The evolution of sediment concentration in an Eulerian Model (hereafter EM), assuming mass conservation and incompressible flow, is written:

$$\frac{\partial c}{\partial t}+\frac{\partial cu}{\partial x}+\frac{\partial cv}{\partial y}+\frac{\partial c(w+w_s)}{\partial z}=\frac{\partial}{\partial x}\left(K_H\frac{\partial c}{\partial x}\right)+\frac{\partial}{\partial y}\left(K_H\frac{\partial c}{\partial y}\right)+\frac{\partial}{\partial z}\left(K_V\frac{\partial c}{\partial z}\right)+S-L \tag{1}$$

In the equation, c is the mass concentration of suspended sediments, u, v and w are the fluid velocities along the x, y and z directions, w_s the sediment settling velocity, K_H and K_V the turbulent dispersion coefficients in the horizontal and vertical direction and S and L sediment source and loss terms, where applicable. The last three terms of the left-hand side express matter advection, while the first three terms of the right-hand side express matter dispersion. Regarding the source and loss terms, these can be mass exchanges in the bottom boundary, as well as sediment inflow (for example due to riverine discharges). Considering the free surface as the reference for the vertical axis (z=0), the bottom boundary condition reads:

$$K_H\frac{\partial c}{\partial z}\bigg|_{z=-H}=R_{er}-R_{dep} \tag{2}$$

where R_{er} και R_{dep} are the erosion or resuspension and deposition rates, respectively.

2.2. Lagrangian models

Largangian (or particle-tracking) Models (hereafter: LM) are stochastic models that describe suspended matter in the column not as mass concentration in each of the cells of the computational grid, but as passive particles of specific mass that are being advected and dispersed by the flow. Thus the movement of a particle along direction i (i=1, 2 and 3 for directions x, y and z, respectively) is given by formulas of the form:

$$\frac{dx_i}{dt}=\bar{u}_i+u_i'\ ,\ u_i'=\xi\sqrt{\frac{6K_i}{dt}},\ \ \xi\in[-1,\ +1] \tag{3}$$

In equation 3, the first term expresses the advective part of the particle movement, whose range is defined deterministically by the velocity field, and the second the dispersive part that is defined stochastically using a random number ξ that ranges from -1 to +1; the amplitude of the stochastic part of the motion is determined by the corresponding mass dispersion coefficient K_i. It is noted that for the case of the vertical transport (i=3), the deterministic particle velocity is the summation of the vertical flow velocity and the settling velocity of the particle:

$$\bar{u}_3=w+w_s \tag{4}$$

In a tracer model particles in the domain are traced with simulation time and bear information (like the mass or the characteristic diameter of the particle, its origin etc) that follows the particle throughout its movement, remaining constant or varying with time. Each particle accounts for a specified mass of sediment and, thus, a computational particle represents a cumulus of actual sedimentary particles that are considered to have the same properties and fate.

2.3. Mixed Euler-Lagrange models

Euler-Lagrange Models (hereafter: ELM) are derived from the classic Eulerian advection-dispersion equation, which, applying a simplified temporal discretization is written [4-6]:

$$\frac{\partial c}{\partial t} + u\nabla c = \nabla K \nabla c \Rightarrow \frac{c^n - c^{n-1}}{\Delta t} + \left[u\nabla c\right]^{n-1} = \left[\nabla K \nabla c\right]^n \tag{5}$$

Considering an arbitrary, auxiliary value, c^t, the former equation can be divided into two parts. One that describes advection:

$$\frac{c^t - c^{n-1}}{\Delta t} + \left[u\nabla c\right]^{n-1} = 0 \tag{6}$$

and one expressing dispersion:

$$\frac{c^n - c^t}{\Delta t} = \left[\nabla K \nabla c\right]^n \tag{7}$$

The advection equation states that concentration remains constant along lines of equal concentration that are defined by:

$$dx_i\big|_{i=1,2,3} = dt \cdot u_i \tag{8}$$

Thus, these lines are being 'traced' backwards in time for each note of the computational grid and the values for the temporal point $n-1$ are calculated for the corresponding ones of step n. The c^t concentrations for step n are determined using spatial interpolation. Finally, the dispersion equation is solved using finite differences or elements, through which the values of the concentration are updated for the next temporal step.

2.4. Common features and differences of the 3 models

One of the main disadvantages of the EM is that the advection-dispersion equation is solved in all of the grid points of the domain, which, combined with the necessity to adhere to the Courant-Friedrichs-Lewy (CFL) condition, increases the computational load. Contrastingly, a LM performs the necessary calculations only in the locations of particles while an ELM simulates the processes in a decoupled manner. It should be noted that, due to the stochastic nature of dispersion, simulating the same conditions with a random walk model can produce different results regarding the suspended matter in the domain. This effect is lower, the more the particles taken into account by the simulation. However, the number of particles in the domain is a parameter that should be taken into consideration, since increasing the number of particles increases the accuracy of the simulation but also increases computational load. Therefore, it is important to optimize the number of particles selected so that the simulation is accurate and also realizable within reasonable computational time. Even though ELMs present numerical advantages, their most significant disadvantage is that they do not inherently conserve mass, either locally or globally [6] due to tracking errors and non-conservative flow fields; these errors can be eliminated using very fine grids. Using concentrations, in both EM and ELM, provides erroneous results in the location of the sources and does not account for dispersion phenomena in a particular point, unlike LM [5] that describe the processes at the level of the cohesive floc. Furthermore, the resolution of a model that uses concentration to describe suspended mass in the domain is, unavoidably, defined by the discretization of the grid. Contrastingly, a particle-tracking model can describe the particulate matter distribution in scales much smaller than that of the spatial step. The transition from particle distribution to concentrations is easily performed, when and where this is required, from the mass of each particle and the number of particles in each grid cell. Finally, one of the most important and interesting advantages in using LM is the ability to assign information to particles (like origin, kind, state, values of characteristic parameters etc) in the form of indicators and to trace their evolution with time. Namely, knowing the 'history' of each sedimentary parcel allows the visualization the trajectory of each particle and of the changes it has undergone, as well as the segregation of particles according to common properties and the investigation of transport patterns related to sedimentary origin.

3. Physical processes for cohesive sediments in the marine environment

3.1. Pelagic processes

The pelagic processes are the ones that take place in the water column and can generally be described as advection and dispersion. However, flocculation and the corresponding changes in the particles' settling rate and stratification affect advective and dispersive matter transport in the column; these processes are described following.

3.1.1. Turbulent dispersion

It is important to clarify the distinction between diffusion and dispersion of suspensions. Mass diffusion can be molecular in the case of laminar flow, which is the case for example in the stochastic Brownian motion of particles, or turbulent, in the case that the flow in the field presents turbulent eddies. For the movement of particles to be considered diffusive, the existence of a concentration gradient in suspended matter is required. In the opposite case, when the concentration in the field is homogenous, dispersion theoretically has no effect since the lateral expansion of the plume takes place in the same rate in both directions. The mass diffusion coefficient ranges from 10^{-9}m^2/s to 10^2m^2/s for molecular and turbulent diffusion, respectively. Contrastingly, the case of mass dispersion requires mixing of the flow field due to velocity gradients and the corresponding coefficient ranges from 10^{-3}m^2/s to 10^2m^2/s. In coastal and marine systems the seawater velocities are variable with depth and along the direction of the flow and, therefore, turbulent dispersion is the main driving mechanism. The scale of turbulent eddies is important to the evolution of a suspended sedimentary plume. In cases that the eddy is much larger than the plume, the movement is essentially advective and the shape of the plume remains unchanged, whereas for intermediate eddies the plume is deformed and its boundaries are stretched; when turbulent

scales are much smaller than the plume itself, the main effect is dispersive and leads to eve-
ning of the shape of the plume [7].

Richardson [8] studied the evolution of sedimentary plumes from a point, non-constant
source considering the segregation between the particles of the plume. Defining standard
deviation of particles from their average position as the characteristic length-scale l, he
reached a correlation with the dispersion coefficient of the form:

$$K = 0.2 \cdot l^{4/3} \tag{9}$$

where l is the length scale of the system. This equation is known as the 'four-thirds power
law'. One of the most widely applied relationships for the calculation of the turbulent dispersion coefficient from the hydrodynamic conditions is the Smagorinski formula [9,10]:

$$K_H = C \cdot dx \cdot dy \sqrt{\left(\frac{\partial u}{\partial x}\right)^2 + \frac{1}{2}\left(\frac{\partial v}{\partial x} + \frac{\partial u}{\partial y}\right)^2 + \left(\frac{\partial v}{\partial y}\right)^2} \tag{10}$$

$$K_V = \frac{dz^2}{dx \cdot dy} K_H \tag{11}$$

C, in equation 10, is a constant that depends on the horizontal discretization step and varies
from ~0.01 for steps of a few meters to 0.2 for steps of the order of kilometres, while the subscripts H and V (eq. 10-11) denote horizontal and vertical directions, respectively.

The dispersion coefficients can also be estimated from the corresponding values of turbulent
viscosity of seawater (v_{ti}) using relationships of the form [11]:

$$K_i = \beta_i F_{mi} v_{ti} \tag{12}$$

The introduction of coefficients β_i and F_{mi} is due to the separation of seawater and sediment
mixing and to the reduction of vertical mixing, respectively. They are usually taken equal to
one, apart from the case of stratified environments, where the coefficient F_{mi} is estimated as a
damping function of the vertical transport; the effect of stratification is analyzed further down.

3.1.2. Flocculation - Settling

A basic distinction of fine-grained from sandy sediments is the development of cohesion
that enables the formation of bonds between particles and eventually the formulation of co-
agulants of larger diameter than their primary constituents. These bonds may be due to attractive van der Waals or electrochemical forces. Van der Waals forces are particularly
strong but weaken significantly with distance, which means that they are effective only in
the case that the particles are at very low distances. Suspended oblate argillaceous particles

are usually negatively charged in their surface and bear positive charge in the edges; the overall electrostatic charge in the surface is negative and therefore repulsive forces exist between particles. Free ions and cations in the marine environment form a cloud in the surface of clay particles, known as (Guoy) double layer [12], that balances the electrostatic charge of the particles and allows them to come closer together, collide and form bonds.

The flocculation process is mainly governed by three mechanisms [13-16]:

1. Random Brownian movement causes collisions between particles and their attachment to a larger coagulant.

2. Turbulence in flows causes particles to collide and aggregate, on one hand, while turbulent shear can induce the breakup of an aggregate, when the shear overcomes the shear strength of the bond between the primary particles.

3. Larger particles settling through a suspension can drift smaller particles, which attach to them, thus forming larger aggregates (differential settling).

It should be noted that the contribution of Brownian motion to flocculation is significant only for small particle diameters (~1-2μm) and results to aggregates of smaller strength, compared to the other two mechanisms [13]. As the diameter of the particle increases, shear becomes the main criterion for the evolution of the process [17].

Other parameters that affect the formation of clayey aggregates, favouring the collision of particles are [18]:

1. Salinity; it has been established that transition to a higher salinity environment leads to increase of the diameter of cohesive suspended matter. The critical salinity value for coagulation, S_0, related to the clay minerals contained the clay fraction are 0.6, 1 and 2.4 for kaolinite, illite and montmorillonite, respectively [16].

2. The existence of cations increases the attractive forces between particles and, therefore, the aggregate's strength.

3. The increase of temperature.

4. The existence of organic matter and detritus, bacteria, microphytes and macrophytes. It is essentially a form of biostabilization of the aggregate's structure by extracellular secretions of the microorganisms.

5. The increase of suspended matter concentration, which increases the frequency of collisions between particles and, therefore, the possibility of adhesion to a larger coagulate.

Dyer [19] established the theoretical background for the evolution of aggregate diameter under the effects of flow shear and suspended concentrations. Concentration increases the diameter of the floc, due to higher availability of matter, while shear stress increases flocculation at low values, favouring inter-particle collisions, and causes floc-breakup and corresponding decrease of the particle diameter for values higher than the strength of the aggregate. Mathematical expressions that can describe this behaviour are exponential formulas with Munk-Anderson type damping functions, of the form [20]:

$$D_{ag} = a_1 \frac{1 + b_1 G}{1 + b_2 G^{d_1}} c^{d_2} + a_2 \tag{13}$$

The constant parameters of equation 13 depend on the concentration of the suspension, c. For high concentrations (c=3.7g/l) these constants have been experimentally estimated at a_1=0.08, a_2=0.86, b_1=29, b_2=125, d_1=4.6 and d_2=-1.1, while for lower concentrations (c=0.8g/l) the same constants read a_1=0.06, a_2=0.88, b_1=33, b_2=138, d_1=3.8 and d_2=-3.5 [21]. The turbulent dissipation parameter G expresses the effects of shear to the diameter evolution and depends on the energy dissipation rate ε and kinematic viscosity v: $G = \sqrt{\varepsilon / v}$

A method widely applied for the parameterization of flocculation is based on the assignment of fractalic dimensions to the cohesive flocs, considering that every aggregate consists of self-similar fractalic entities. Accepting that the linear dimension of the characteristic particle is unitary, it follows that the particle volume is proportional to the number of primary particles. Thus, if n_f is its fractalic dimension, N the number of primary particles that form the aggregate and L its linear dimension, it follows that [15]:

$$n_f = \lim_{L \to \infty} \frac{\ln[N(L)]}{\ln[L]} \tag{14}$$

Thus, the number of particles in an aggregate is proportional to the diameter of the primary particle, D_p:

$$N \sim D_p^{n_f} \tag{15}$$

The fractalic dimension for cohesive particles ranges from 1.4 for very fragile flocs, like marine snow, to 2.6-2.8 for dense, consolidated beds. It is generally accepted that the fractalic dimension increases from marine snow (1.4-1.8) to estuarine sediments (1.7-2.2) and to oceanic sediments (1.94-2.14) [22,23]. An average fractalic dimension for estuarine and coastal areas is n_f=2 [15].

Applying the fractal theory to the evolution of cohesive aggregate diameter it follows that there is an analogy between the ratio of aggregate and primary particle dimensions and the corresponding volume concentration, ϕ [24]:

$$\frac{D_{ag}}{D_p} = \left(\frac{\phi}{\phi_p} \right)^{\frac{1}{3 - n_f}} \tag{16}$$

The relationship between volume concentration and aggregate mass is written [15]:

$$\phi = \frac{\rho_s - \rho_w}{\rho_t - \rho_w} \frac{c}{\rho_s} = f_s N D_{ag}^{\ 3} \tag{17}$$

ρ_s, ρ_w and ρ_t are solid phase, water and aggregate densities and f_s is a shape constant which, for the case of spherical particles equals to $\pi/6$. The increase in aggregate diameter also involves decrease in its density, ρ_t, compared to the primary particles, ρ_p, due to increase in interstitial water [12-14,16,17]. Considering the density difference of solid and liquid phase, the density of the aggregate can be calculated by [24]:

$$\rho_t = \rho_w + (\rho_s - \rho_w) \left(\frac{D_p}{D_{ag}} \right)^{3-n_f} \tag{18}$$

Based on the fractal theory, the evolution of the aggregate diameter in turbulent flow is [15]:

$$\frac{dD_{ag}}{dt} = k_A c G D_{ag}^{\ 4-n_f} - k_B G^{q+1} (D_{ag} - D_p)^p D_{ag}^{\ 2q-1} \tag{19}$$

The first term of the right hand-side of equation 19 expresses the effect of coagulation and the second the effect of floc break-up. The corresponding dimensional flocculation and deflocculation parameters (k_A [m^2/kg] and k_B [sec$^{1/2}$/m^2]) are defined as:

$$k_A = k_A' \frac{D_p^{\ n_f-3}}{n_f f_s \rho_s} \qquad k_B = k_B' \frac{D_p^{\ -p}}{n_f} \left(\frac{\mu}{F_y} \right)^q \tag{20}$$

In equations 18-20, c is the suspended sediment concentration, D_p the diameter of the primary flocs, q and p are constants ($q \sim 0.5$ and $p=3-n_f$), μ is the dynamic viscosity of water, F_y is the strength of the sediment ($F_y = O\{10^{-10}\}$) and k_A' and k_B' are non-dimensional flocculation and deflocculation coefficients ($k_A' = O\{10^{-1}\}$ and $k_B' = O\{10^{-5}\}$) [15].

The corresponding changes to the settling velocity of the aggregate can be calculated by the Stokes velocity, taking into account shape factors α and β ($\alpha=\beta=1$ for perfectly spherical particles):

$$w_s = \frac{\alpha}{\beta} \frac{(\rho_t - \rho_w) g}{18\mu} D_{ag}^{\ 2} \tag{21}$$

The shape factors can be simplified considering an irregularity parameter for the flocs, a_{irr}, ($a_{irr} \leq 1$) equal to the ratio of minimum to maximum aggregate dimensions. Following [25] and after some manipulations:

$$\begin{cases} \alpha = a_{irr} \\ \beta = 1 - \lambda_1 \left(1 - a_{irr}\right) \end{cases} \qquad (22)$$

where λ_1 is a coefficient related to the orientation of the particle during settling. It equals 0.2 or 0.4 when the particle settles with its maximum axis parallel or perpendicular to depth. Considering parallel orientation as the most frequent one, the settling velocity reads:

$$w_s = \frac{a_{irr}}{0.8 + 0.2 a_{irr}} \frac{(\rho_t - \rho_w)g}{18\mu} D_{ag}^{\ 2} \qquad (23)$$

The relation between settling velocity, flow shear stress and suspended sediment concentration (c in mg/l) for microflocs (<160μm) and macroflocs (>160μm), defined after experimental measurements from estuarine areas is [26-29]:

$$w_{s_{macro}} = \begin{cases} 0.644 + 0.000471c + 9.36\tau - 13.1\tau^2, & 0.04Pa \le \tau \le 0.65Pa \\ 3.960 + 0.000346c - 4.38\tau + 1.33\tau^2, & 0.65Pa < \tau \le 1.45Pa \\ 1.18 + 0.000302c - 0.49\tau + 0.057\tau^2, & 1.45Pa < \tau \le 5.00Pa \end{cases} \qquad (24)$$

$$w_{s_{micro}} = \begin{cases} 0.244 + 3.25\tau - 3.71\tau^2, & 0.04Pa \le \tau \le 0.52Pa \\ 0.65\tau^{0.541}, & 0.52Pa < \tau \le 10.0Pa \end{cases} \qquad (25)$$

The predominance of macroflocs versus microflocs can be defined from their concentration quotient, which was found to depend on the overall suspended sediment concentration through a relationship of the form [26]:

$$\frac{c_{macro}}{c_{micro}} = 0.815 + 0.00318c - 0.14 \cdot 10^{-6} c^2 \qquad (26)$$

The settling velocity of particles can be distinguished in three 'phases', depending on the suspended matter concentration [14]. For low concentrations, C_1=0.1-0.3g/lt, free (Stokes') settling dominates. For concentrations higher than the limit for free settling and lower than C_2=2-20g/lt settling is highly affected by flocculation and involves increase of the settling velocity due to increase of the particle diameter. The settling velocity can be thus expressed as an exponential function of concentration:

$$w_s = k_1 c^{n_1} \qquad , c_1 \le c \le c_2 \qquad (27)$$

where k_1 (~4/3) and n_1 O{10⁻¹} are experimentally determined constants. Taking into account the effect of shear stress, the settling velocity can be determined by the equation [30]:

$$w_s = w_{s,r} \frac{1 + aG}{1 + bG^2} \tag{28}$$

In equation 28, $w_{s,r}$ is the settling velocity of a single particle in still liquid and a and b ($O\{1\}$) are empirical coefficients related to the sediments.

For concentrations higher than C_2 the settling velocities reduce with increasing suspended matter, phenomenon known as hindered settling. Hindered settling essentially expresses the effect of neighbouring particles to the settling velocity of a specific particle, which is important especially for high concentration suspensions. This hindering effect is caused by eddy separation around settling particles, causing reduction to the settling rates of particles within this buoyant flow, but also due to inter-particle collisions and electrochemical interactions. The hindered settling velocity related to the volumetric concentration is:

$$w_s = w_{s,r} \left(1 - \phi\right)^n \tag{29}$$

n is a parameter that varies inversely with the particle Reynolds number within the range $2.5 < n < 5.5$. A corresponding expression [15] reads:

$$w_s = w_{s,r} \frac{\left(1 - \phi_*\right)^m \left(1 - \phi_p\right)}{1 + 2.5\phi}, \qquad \phi_* = \min\{1, \phi\} \tag{30}$$

where ϕ_p is the volumetric concentration of primary particles and m is a constant for the inclusion of possible non-linear effects. The expression that relates the mass concentration in the field to the hindered settling velocity is similar [14]:

$$w_s = w_{so} \left[1 - k_2 \left(c - C_2\right)\right]^{n_2} \tag{31}$$

For highly concentrated suspensions (~80-100g/lt) the settling velocity is practically zero.

A general formulation that can describe the evolution in conditions of flocculation and hindered settling is [14]:

$$w_s = \frac{a \cdot c^n}{\left(c^2 + b^2\right)^m} \tag{32}$$

The coefficients are determined experimentally; from experimental studies in various estuarine and riverine areas [31] the range of their values is: a=0.01-0.23, b=1.3-25.0, n=0.4-2.8

and $m=1.0$-2.8. The maximum settling velocity in this case (for concentration equal to C_2) is [14]:

$$w_{so} = a \cdot b^{n-2m} \frac{\left(\dfrac{2m}{n} - 1\right)^{m-2n}}{\sqrt{\dfrac{2m}{n}}}, \quad C_2 = \frac{b}{\sqrt{\dfrac{2m}{n} - 1}} \tag{33}$$

3.1.3. Effects of stratification

One of the most widely applied methods for the determination of the stability of the water column is the use of the gradient Richardson number [32]:

$$Ri = \frac{N^2}{\left(\dfrac{\partial U}{\partial z}\right)^2} > \frac{1}{4}, \quad N = \sqrt{-\frac{g}{\rho_w}\frac{\partial \rho_w}{\partial z}} \tag{34}$$

N is the Brunt-Väisälä frequency. In flows with Ri values lower that $\frac{1}{4}$ Kelvin-Helmholtz instabilities can occur. Based on this criterion, the effect of stratification to mass dispersion, based on the mixing-length theory, is often expressed using Munk-Anderson type velocity and mass dispersion damping functions (F_m) [7]:

$$F_m = (1 + \beta Ri)^\alpha \qquad \{\alpha \approx -0.5, \beta \approx 10\} \tag{35}$$

The flux Richardson number, which is the ratio of the buoyant forces to the turbulent kinetic energy production [7], is also frequently used; the mass dispersion coefficient K_m is [33]:

$$K_m = \frac{R_f}{1 - R_f}\frac{\bar{\varepsilon}}{N^2} \xrightarrow{R_f \geq R_{f,crit}=0.15} K_m \leq 0.2 \frac{\bar{\varepsilon}}{N^2} \tag{36}$$

The overbar in equation 36 indicates time averaging. Regarding the mixing of freshwater plume, expressed as entrainment velocity to the underlying layer (W_e), it can be estimated as [34]:

$$W_e = \frac{e_1 U_o}{Ri} \tag{37}$$

where U_o is the velocity of the underlying layer and e_1 ($O\{10^{-3}\}$) an empirical coefficient.

Another indicator for the strength of the stratification is the Peclet number (Pe) that expresses the predominance of settling over dispersion (h: width of the mixing zone):

$$Pe = \frac{w_s h}{K_m} \ll 1 \tag{38}$$

It has recently been established that, even in the case that density increases with depth, the hydrostatic stability of the system is not guaranteed [35], since inhomogeneities in the density field can lead to transporting water masses that can induce a process of matter removal from suspension. The prognosis of such instabilities cannot be described using a Richardson number [33]. Double diffusion is the development of instable density gradients due to differential molecular diffusion of the properties of the fluid, which are capable of changing its density. Given that the thermal diffusion coefficient ($K_T \approx 10^{-3} m^2/s$) is two orders of magnitude larger than the salinity coefficient ($K_S \approx 10^{-5} m^2/s$), it follows that temperature diffuses faster; this can lead to areas in the flow with higher density than the underlying fluid, thus causing mixing of the water column. Double diffusive instabilities can form as salt fingers, in the case that salinity is the destabilizing parameter (T and S decreasing with depth), or as diffusive layers, when temperature is the destabilizing factor (T and S increasing with depth). Salt fingering instability takes the form of tightly packed blobs of sinking salty and rising fresher water masses near the thermocline, which quickly develops to transport away from the interface. Diffusive layering convection involves a buoyant thermal diffusion flux through the pycnocline that is higher (in terms of density) than the corresponding salinity flux, thus leading to a downward density flux that causes transport from the stratified zone to the homogenous layer. The susceptibility of the column to double diffusive phenomena can be estimated using the stability ratio (R_p), which is the ratio of the stabilizing parameter gradient to the destabilizing one, expressed in terms of density:

$$R_p = \begin{cases} \alpha \dfrac{\partial T}{\partial z} \Big/ \beta \dfrac{\partial S}{\partial z} \to \text{salt fingering} \\[2mm] \beta \dfrac{\partial S}{\partial z} \Big/ \alpha \dfrac{\partial T}{\partial z} \to \text{diffusive layering} \end{cases}, \quad \begin{cases} \alpha = -\dfrac{1}{\rho_w}\dfrac{\partial \rho_w}{\partial T} \\[2mm] \beta = \dfrac{1}{\rho_w}\dfrac{\partial \rho_w}{\partial S} \end{cases} \tag{39}$$

In equation 39, α is the thermal expansion coefficient and β is the haline contraction coefficient. The gravitational stability of the column increases with the stability ratio; it has been established that double diffusive instabilities develop within the range $1 < R_p < 10$ [36,37], while areas where $R_p \geq 10$ are highly stable, with low settling rates of fine sedimentary plumes (~cm/h) [37]. Thus, $R_p = 10$ can be used as a threshold value for the inhibition of double diffusive instabilities and, therefore, for the indication of a highly stable column.

These effects can be taken into account in sediment transport modelling, using damping functions to the vertical advection (F_w) and dispersion (F_{Kv}) of matter [38]:

$$F_w = \begin{cases} 1 & , \quad R_p < 10 \\ (1 + \; R_p) & , \quad R_p \geq 10 \end{cases} \tag{40}$$

$$F_{Kv} = \begin{cases} 1 & , \quad R_p < 10 \\ \exp(- \; \cdot \; R_p) & , \quad R_p \geq 10 \end{cases} \tag{41}$$

3.2. Benthic processes

Modelling the transport of cohesive sediments in coastal and deltaic areas requires the description of the processes of mass exchange between the column and the seabed through the processes of erosion and deposition. In practice, formulating this module is one of the most challenging tasks in modelling fine sediment transport, mainly due to the lack of general and widely applicable formulae between erosion resistance and bed shear stress. However, even if the main questions regarding mass exchange quantification were answered, the fact that the characteristic parameters vary within the sediment with time and space (horizontally and vertically) still remains [39].

There are three main distinctions regarding fine sediment transport in the boundary of the bed, compared to coarse-grained sediments [40]:

1. Cohesive matter is transported exclusively in suspension, while coarser matter is also transported in semi-contact with the bed, as bed load.

2. Cohesive sediments are not transported as dispersed particles. Flocculation increases settling rates and is responsible their deposition.

3. Cohesive sediment beds undergo self-weight settling and consolidation. In cases of rapid sedimentation the deposited sediments are light and present minimal resistance to shear stress. Cohesive beds can be homogenous or vertically stratified regarding density and strength.

Various modules have been proposed for the exchange processes between seabed and water column, which are mainly based on the transition of the matter to different states. Four characteristic states for a sediment-water mixture are defined in [1]:

1. Horizontally advected mobile suspension.

2. Concentrated benthic suspension (CBS), horizontally static but with vertical mobility.

3. Consolidating soft deposit (fluid mud).

4. Stable, consolidated part of the seabed.

The static suspension is formed by the deposition of the transported suspension, especially under low hydrodynamic conditions, and presents minimal mechanical strength. As deposi-

tion progresses, a CBS is formed; CBSs are suspensions with strong interactions with the turbulent flow field through buoyant effects, which however continue to behave as Newtonian
fluids [41]. In time, consolidation and the related physicochemical changes lead to strength
development and, eventually, to the formation of a fluid mud layer with lower interstitial
water, higher shear strength and more stable structure. It must be noted that the transition
from one phase to another is gradual and that a change in the flow conditions may lead to
phenomena comparable to ones that an increase in concentration would produce. Whether
these phenomena take place or not depends on the temporal scales of the processes; for example the time scale of deposition and flocculation define if fluid mud is formed under low
flow conditions and the temporal scale of consolidation determines if re-entrainment or erosion phenomena partake during accelerated flow. Regarding the concentration/density (dry
and bulk, ϱ_d and ϱ_b) ranges of the aforementioned suspensions and deposits, the following
values are cited [42]:

- Dilute suspension (low concentration C=0.1-10kg/m³)

- CBS (high concentration C=10-100kg/m³)

- Fluid mud (C=ϱ_d=100-250kg/m³, ϱ_b=1050-1150kg/m³)

- Partially consolidated bed (ϱ_d=250-400kg/m³, ϱ_b=1150-1250kg/m³)

- Fully consolidated bed (ϱ_d=400-550kg/m³, ϱ_b=1250-1350kg/m³)

The most straight-forward treatment of the mass-exchange in the boundary of the bed is derived from the balance between settling, deposition and erosion, through conditions of the
form [39]:

$$\varepsilon_s \frac{\partial c}{\partial z} - w_s c = S_E - S_D \tag{42}$$

In the former relationship, the right hand-side is the balance between turbulent dispersion
and settling and the right is the balance of erosion and deposition rates.

3.2.1. Bottom shear stress

The shear stress velocity in the bed can be related to the depth-averaged flow velocity (U)
through a Chezy coefficient, Ch [11]:

$$u_* = U \sqrt{g}/Ch, \qquad Ch = 18 \log(12h/k_S) \tag{43}$$

where h is the flow depth and k_s an equivalent roughness height. The direct dependence if
the shear stress to the flow velocity at a small distance from the bed can also be used [7]:

$$\tau_b = \rho \cdot C_d \cdot u(1m)^2 \tag{44}$$

In equation 44, $u(1m)$ is the velocity at a distance of 1m from the bed and C_d the bottom friction coefficient ($O\{10^{-3}\}$). This coefficient can be estimated assuming logarithmic velocity profile through equations of the form [2]:

$$C_d = \max\left\{\left[\frac{1}{\kappa}\ln\frac{(H+z_b)}{z_0}\right]^{-2}, 0.0025\right\}$$ (45)

z_0 is the roughness length of the bed ($O\{10^{-3}m\}$). High density gradients near the bed result to reduction of the apparent roughness with corresponding increase in transport for the same bottom shear stress. Thus, the actual erosion rates for the same flow conditions are lower, compared to the ones calculated not taking into account these buoyancy effects in the bottom boundary conditions. A significant modelling improvement is achieved using damping factors in the bottom boundary conditions; the calculation of the bottom shear can be performed through [43]:

$$u_* = F_t\kappa z\frac{\partial U}{\partial z}, \quad F_t = (1+100Ri)^{-\frac{1}{3}}$$ (46)

where F_t is the velocity damping factor, κ the von Karman constant, $\partial U/\partial z$ the velocity gradient at the marginal grid of the bed and Ri the gradient Richardson number. The damping function essentially expresses the ratio of the actual turbulence to the turbulence in the absence of suspended matter. Based on this approach, the gradient at the bed is calculated explicitly from the velocity profile and not using the log-law. Therefore, the 'corrected' velocity is:

$$U = \frac{u_*}{\kappa z}\ln\left(\frac{z}{\alpha z_0}\right), \quad \frac{1}{F_t} = 1 - \frac{z}{\alpha}\frac{\partial\alpha}{\partial z}$$ (47)

where α is a bottom roughness correction coefficient. The reduction in bottom roughness corresponds to drag reduction, observed both in nature and in the laboratory. Series of numerical experiments [43] suggest that α can be estimated through exponential functions of the form:

$$\alpha = \exp\left[-(1+\beta w_s/u_*)\left(1 - \exp\left(-bRi^m\right)\right)\right], \quad \{\beta = 7.7, b = 1.25, m = 0.6\}$$ (48)

In coastal areas, where the activity of both waves and currents is significant, their combined effect to the range of the shear stresses can be addressed through equations of the form [2]:

$$\tau_b = \frac{1}{2}f_{cw}\rho_w|u_c + u_w|^2$$ (49)

In the former relationship u_c and u_w are the current and wave amplitude velocities near the bed and f_{cw} the wave-current friction factor.

3.2.2. Deposition

The sediment deposition rate (S) is typically defined using the Krone formula [44]:

$$S = \begin{cases} 0 & , & |\tau_b| \ge \tau_{cr,dep} \\ C_b \cdot W_s \left(\dfrac{|\tau_b|}{\tau_{cr,dep}} - 1 \right), & |\tau_b| < \tau_{cr,dep} \end{cases} \tag{50}$$

The particulate matter that can be transported by turbulent flow may be expressed by the equilibrium concentration C_e [15,41]:

$$C_e = K_s \frac{\rho_b}{g} \frac{\rho_s - \rho_w}{\rho_s} \frac{u_*^3}{hW_s} \tag{51}$$

where K_s is a first-order proportionality coefficient ($K_s{\sim}0.7$) and W_s is the characteristic (mean) settling velocity. It can be assumed that, in the case that the system is in balance, this velocity equals the entrainment velocity from a CBS, with the sediment suspension to be in equilibrium. At the fluid mud-water interface little or no turbulence is produced. Therefore, as turbulent mixing reduces, C_e is further reduced. This results to a cumulative effect and the complete breakdown of the vertical suspended concentrations' profile. For cohesive sediments the equilibrium concentration can also be considered as saturation concentration.

The critical shear stress for deposition can be related to the strength of the cohesive floc, F_c ($O\{10^{-8}N\}$), through relationships of the form [24]:

$$\tau_{cr,dep} = \frac{1}{5\pi} \frac{F_c}{D^2} \approx 0.06 \frac{F_c}{D^2} \tag{52}$$

Critical shear velocity for deposition can also be expressed with respect to the settling velocity of the particles [44]:

$$u_{*cr,dep} = \begin{cases} 0.008 & , W_s \le 5 \cdot 10^{-5} \\ 0.023 & , W_s \ge 3 \cdot 10^{-4} \\ 0.094 + 0.02\log_{10}(W_s) & , 5 \cdot 10^{-5} < W_s < 3 \cdot 10^{-4} \end{cases} \tag{53}$$

3.2.3. Erosion

The sediment that enters suspension during erosion of the seabed is typically quantified using the erosion rate, which is the eroded mass per unit surface and unit time $[ML^{-2}T^{-1}]$. For homogenous, consolidated beds, the well known Partheniades formula applies [46]:

$$\varepsilon = \varepsilon_o \left(\frac{\tau_b}{\tau_{cr,er}} - 1 \right) \tag{54}$$

The critical erosion shear stress ($\tau_{cr,er}$) and the erosion rate (ε_o) constant are considered to be constant with time and within the sediment. However, resistance to erosion depends on various parameters, among which sediment composition, porosity and degree of consolidation of the deposit [47]. The seabed may be soft, partially consolidated, with high water content (>100%), or may be a denser, more stable deposit. The way in which erosion takes place also varies with the range of the applied shear stress and the stress history of the deposit. Three erosion processes have been identified [14]:

1.   floc-to-floc erosion, or surface erosion of the bed,

2.   high concentration fluid mud entrainment and

3.   mass erosion of the bed.

A soft bed that usually consists of freshly deposited clayey matter that is under consolidation presents non-uniform characteristics and is therefore vertically stratified. The erosion rate in this case is [14]:

$$\varepsilon = \varepsilon_f \cdot \exp\left[\alpha \left(\tau_b - \tau_{cr,er}(z) \right)^{\beta} \right] \tag{55}$$

In equation 55, ε_f is the floc erosion constant $[kg/m^2/s]$ (equals the erosion rate in the case that the shear stress equals the threshold value), $\tau_{cr,er}(z)$ is the erosion resistance profile with depth z [Pa] and β ($\beta \sim 0.5$) and α $[Pa^{-\beta}]$ are constants. The values of ε_f (10^{-4}-10^{-7} $gcm^{-2}min^{-1}$) and α (5-20mN$^{-1/2}$) depend on the sediment-water mixture.

It is clear that one of the most significant parameters in modelling erosion of cohesive bed is the determination of the critical shear threshold. Various experimentally determined formulas have been proposed, among which many relate the erosion threshold to the density of the bed through relationships of the form:

$$\tau_{cr,er} = \xi \cdot \rho_d^{\varsigma} \tag{56}$$

There is a high variability in the values of the constants involved. We indicatively mention that [48] estimated them at ξ=5.42 10^{-6} and ζ=2.28, while [49] defined ξ=1.2 10^{-3} and ζ=1.2. It is noted that, alternatively to the use of the sediment dry density, bulk density is also used in self-similar expressions. From insitu measurements in a macrotidal mudflat [50] a positive correlation of the critical erosion shear stress was found with the sediment water content (W), the bulk density (ρ_b) and the mass loss on ignition (LOI):

$$\tau_{cr,er} = \begin{cases} -0.0012W + 0.51 \\ 0.0013\rho_b - 1.4 \\ -0.041LOI + 0.65 \end{cases} \tag{57}$$

It follows that the erosion rate constant and the critical shear threshold are decisive parameters for the quantification of eroding masses. These parameters are usually experimentally determined; however the values proposed in literature present high variability. The main reason for this variability is the dependence of the erosion process on various factors, like:

• Density: the resistance to erosion increases with the density of the bed.

• Floc diameter – fine fraction content: the effect of the percentage of fines in the sediment to the critical erosion stress is strongly non-linear.

• Clay composition: the lithological composition of clay affects the behaviour of the sediment under shear.

• Stress history: the consolidation degree of the sediment is a significant parameter since clayey sediments change their mechanical behaviour in cases of preloading and overconsolidation.

• Organic content: experimental measurements have shown that the erosion threshold is significantly lower for the case of organic-rich beds.

• Biostabilisation: bacteria and benthic diatoms increase the cohesion of the bed through secretion of extracellular polymeric substances, while the existence of macrophytes in the surface of the bed prohibits erosion reducing the flow velocities through their canopy and at the same time stabilizing the sediment through their root system.

• Bioturbation: the existence of micro-fauna in the surface layer of the sediment reduces the resistance to erosion through loosening of the sediment structure during grazing and the production of faecal pellets.

• Measurement method (laboratory or insitu): the measurement method itself introduces uncertainties regarding the accuracy and the general applicability of the values, given that there is high variability for similar areas using different instrumentation. The differences are even higher between insitu and laboratory experiments, with the former to provide significantly higher erodiblity parameters than the latter.

3.2.4. Self-weight consolidation – Resuspension

A fresh mud deposit has the form of a sediment-water mixture with sediment concentrations of the order of a few tens of grams per liter. This mixture evolves in three stages [12,42]:

1. Deposition of aggregates during the first hours forming fluid mud

2. Depletion of the interstitial water in a period of one to two days

3. Depletion of the intra-particle pore water (very slow consolidation)

Typical consolidation time-scales for a deposit with bulk density of 1150-1250kg/m³ are of the order of 1 week, 1 month, 3 months and 0.5-1year for a layer thickness of 0.25m, 0.5m, 1m and 2m, respectively [42].

The resistance of the deposit to resuspension increases with consolidation time, process that can be parameterised considering a simplified vertical structure of the bed. In its simplest form, only two layers are taken into account, a surface, non-consolidated, low-strength layer and a deeper, fully consolidated one; different densities and erodibility parameters are assigned to each layer. This approach can be extended to a polystromatic representation of the vertical composition of the bed and of the consolidation process.

One of the most widely applied parameterizations for the process of consolidation is the Gibson equation [51]:

$$\frac{\partial \phi}{\partial t} - \frac{\rho_s - \rho_w}{\rho_w} \frac{\partial}{\partial z}\left(k\phi^2\right) - \frac{1}{\rho_w g}\frac{\partial}{\partial z}\left(k\phi\frac{\partial \sigma'}{\partial z}\right) = 0 \tag{58}$$

Equation 58 is written in Eulerian form and using the solid volume concentration [51], while k and σ' are the permeability and the effective stress, respectively. Under the reasoning that the critical failure stress for the sediment comes as a result of failure of the bonds between and within the aggregates, a Mohr-Coulomb type criterion applies:

$$\tau_{cr,er} = c' + \tan\varphi\sigma' \tag{59}$$

The first term of equation 59 expresses the effect of cohesion (c'), which is the shear strength in zero effective stress, and φ is the friction angle.

Exponential functions that relate the compactibility of the sediment through parameters like the void ratio, e, to effective stress are also used [52]:

$$e = \left(e_0 - e_\infty\right)\exp\left(-\lambda\sigma'\right) + e_\infty \tag{60}$$

Another empirical relationship [53] between the critical erosion shear velocity and the yield strength of muddy deposits reads:

$$\begin{cases} u_{cr,er} = 0.013\tau_y^{1/4}, & \tau_y = b\rho_d^{3} < 3\,N/m^2 \\ u_{cr,er} = 0.009\tau_y^{1/2}, & \tau_y = b\rho_d^{6} > 3\,N/m^2 \end{cases} \tag{61}$$

The first of the equations refers to fluid mud and the second to mud that has exceeded the plasticity limit. Accepting that the critical shear stress threshold for resuspension ($\tau_{cr,res}$) of deposited sediments ranges between the corresponding values for deposition ($\tau_{cr,dep}$) and erosion ($\tau_{cr,er}$), at deposition (t_d=0) and after full consolidation (t_d=t_{fc}), respectively, its value can be estimated through exponential equations of the form [38]:

$$\tau_{cr,res} = \tau_{cr,dep} + \left(\tau_{cr,er} - \tau_{cr,dep}\right)\cdot\left[1 - e^{-n\cdot t_d}\right], \quad n \approx 7/t_{fc}\,[T^{-1}] \tag{62}$$

3.3. The fine sediment transport model

Based on the preceding analysis, the formulation of a sediment transport module involves the mathematical description of all the aforementioned pelagic and benthic processes and the interactions between them. A typical outline of such a module is presented in figure 2, showing the transition of the sedimentary plume to various conditions, the governing processes at each stage and the interactions between the processes. These considerations were taken into account in formulating the Fine Sediment Transport Model (FSTM); the model is outlined in the present work and characteristic results from simulations in Thermaikos Gulf (NW Aegean Sea) with point (rivers, mechanical erosion) and distributed (atmosphere) sources are presented and discussed.

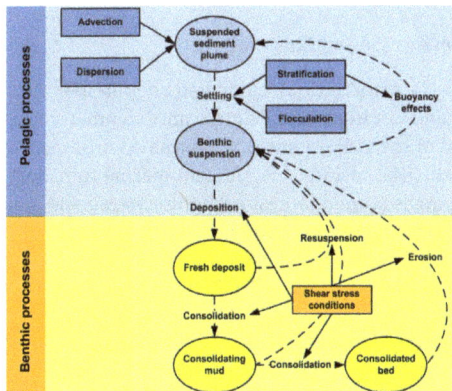

Figure 2. Schematic representation of a cohesive sediment transport module. Dashed vectors denote transition from one state (ovals) to another and solid vectors indicate effect of a process (rectangles) to another.

FSTM [38] was formulated using the particle-tracking method (LM), thus describing sediments in the marine environment as particles of specific mass and characteristic properties (assigned to each particle in the form of indicators) that are being passively advected and dispersed by the currents. FSTM accounts for all the pelagic and benthic processes that were described in the previous section. More specifically, the position of the particle in the next temporal step is calculated from eq. [3-4], using the dispersion coefficients for the amplitude of the stochastic particle displacement defined in eq. [10-11]. The evolution of the cohesive particle diameter is defined using the fractal method (eq. 19-20), while the changes in the bulk density of the aggregate are calculated from equation 18. These parameters are used to determine the settling velocity of the particle through the modified Stokes law (eq. 23). The effects of stratification to the vertical propagation of the sedimentary plume are taken into account using the double diffusion theory and assigning damping functions (eq. 40-41) to the vertical deterministic and stochastic displacements. The constants γ, δ and κ were estimated at 0.18, -0.9 and 0.03 [38]. Regarding near-bed processes, the benthic shear stress is defined explicitly, from the horizontal velocity profiles (eq. 46), and the critical shear velocity for deposition of a particle is determined from its settling velocity, using equation 53. After the deposition of a particle onto the bed, the critical shear stress for resuspension of the particle is considered to increase with depositional time (eq. 62). After full consolidation ($t_d=t_{fc}$) the particle is considered to be part of the seabed, which is assumed homogenous and fully consolidated (eq. 54). Regarding the boundary conditions of FSTM, radiation is applied at the open boundaries of the domain (particles are allowed to cross the boundary and are excluded from following computational cycles) and reflection is applied at the surface (particles are not allowed to escape to the atmosphere), at the bed (in cases that a particle does not deposit it is reflected to its previous vertical position) and at 'dry' grid cells (along one or both horizontal directions). The sources of particles may be the topmost grid cell in the vertical sense, in cases of river-borne or aeolian transported particles, or the bottom cell (for physically or mechanically eroded matter).

4. Characteristic applications

4.1. The application domain: Thermaikos Gulf

Thermaikos Gulf (Figure 3), a micro-tidal, elongated shelf in the NW Aegean Sea, is an area of high socio-economic and environmental significance. Numerous anthropogenic activities take place in the vicinity of the gulf, forcing the marine environment with residues from agricultural, urban and industrial residues. Significant marine transport loads also exist, since the port of Thessaloniki is the second largest harbor in Greece and a gateway to the Balkans. Due to its trophic state (mesotrophic, compared to the oligotrophic open Aegean Sea) and its mild-sloped bottom relief, Thermaikos is also one of the key marine sites for aquaculture and fishing (trawling and otherwise) of the Hellenic region. Sources of freshwater and fine sediments by rivers and smaller streams are distributed along the greater part of its northern and western coastlines; four are the main rivers discharging in the area, Axios, Pinios, Aliakmon and Loudias (Figure 3), of which the former two are the largest. It should be noted that the deltaic area of Axios, Aliakmon and Loudias is protected by the Ramsar convention (Wetlands of International Importance) and is included in the EU network of protected areas Natura 2000 (GR1220010).

FSTM was applied in the domain of Thermaikos, covering the aquatic area that extends southwards to longitude 39.6°N and eastwards to latitude 23.5°E (Figure 3). The grid is horizontally curvilinear ($dx=dy=1/60°$) and vertically Cartesian ($dz=2m$). The necessary input hydrodynamic and physical parameters (seawater velocities, salinity and temperature fields) were derived by the North Aegean Sea (NAS) model [54]. Time-series of sediment fluxes from the 4 main rivers, used in the simulations, were recorded during the METRO-MED project [55]. Results from three applications are presented further down, for river-borne sediments, aeolian transported dust and mechanically (trawling) eroded matter. It is noted that the hydrodynamics for the former case are after results of NAS with climatological forcing, thus expressing the typical circulation features in the domain.

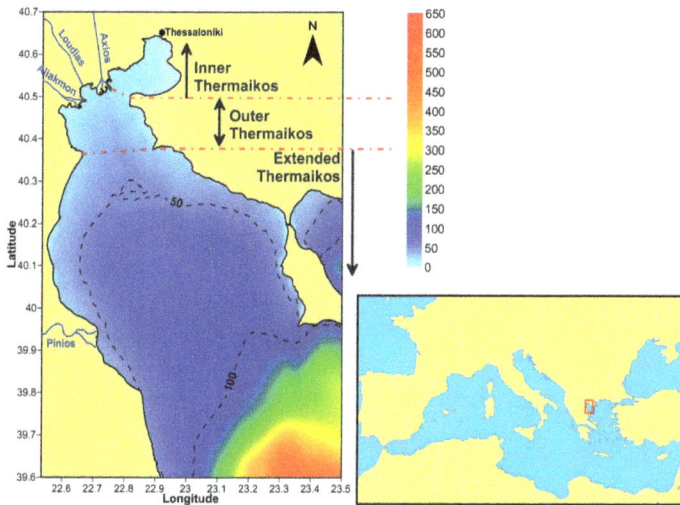

Figure 3. Bathymetry and geomorphological parts Thermaikos Gulf and its location in the Mediterranean Sea

4.2. Point source (river-borne sediment) simulation

4.2.1. Transport patterns and particulate matter distribution

One of the major advantages of formulating FSTM using the EL method is the ability to track the history of each particle and to segregate sediment in the domain according to common properties. An example of visualizing such information is given in figure 4; the figure shows the evolution of suspended and surface (top 2m) particles, 30, 120, 210 and 300 days from the beginning of the simulation (corresponding to late January, April, July and October, respectively). The chromatic encoding denotes the riverine origin of the particles. These representations give information regarding the seasonal variability and the transport patterns of sediments by each of the 4 rivers.

It can be noted that, during winter (Figure 4b), Suspended Particulate Matter (SPM) enter the enclosed inner (northmost) part of the gulf, matter that mainly originates from Axios. At the same time, stratification is strong in the northern river-dominated area, leading to high presence of SPM in the surface layer (Figure 4a). The northward expansion of sediment from Axios is probably due to anticyclonic movement of the surficial waters in the gulf during the wintry months [56]. This movement is also responsible for the northward expansion of the Pinios plume, ascertainment that is in accordance with [57]. Matter from Aliakmon is mainly transported towards the south along the western coastline, together with part of the sediment supply of Axios. During the vernal period (Figure 4d) a distinct cyclonic transport pattern is detected in the outer part that forces sediments to move southwards along the west coast, while SPM from Axios are still present in the inner gulf. The same transport pattern is observed for surficial sediments (Figure 4c) with a strong effect of freshwater plumes along the western coastline, effect that keeps sediments in the surface layer at high distances from the outflows. The dispersion of particles is high near the northerly river-outflow system, forcing particles (mainly from Axios) to move eastwards in addition to the prevailing cyclonic movement. The presence of SPM during summer is low (Figure 4f) due to reduced river-borne sediment fluxes; suspended particles appear mainly in the northern part of the outer gulf, with low SPM in the surface layer (Figure 4e). During autumn (Figure 4h), the increase of sediment supplies causes related increase to SPM and a predominant cyclonic transport pattern exists in the northern part of the gulf; a small part of the Axios sediment supply, again, spreads eastwards and towards the inner gulf. Stratification is strong not only for the northern part, but also near the outfalls of Pinios (Figure 4g), with a significant spreading of the surficial plume, which is the most profound of the whole year.

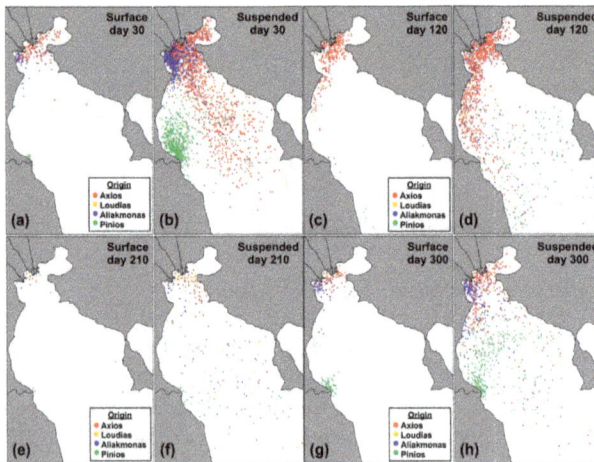

Figure 4. Variation of surficial and suspended particles at the 30th (mid-winter) (a, b), 120th (mid-spring) (c, d), 210th (mid-summer) (e, f) and 300th day of the simulation (mid-autumn) (g, h). The chromatic encoding denotes the different riverine origin of the particles.

4.2.2. The contribution of the rivers to the sedimentation of the gulf

The existence of transport and sedimentation patterns related to the riverine origin of SPM and the geomoprhological parts of the gulf becomes evident. The following table (Table 1) presents the percentages of deposition in each region for each of the four rivers, to facilitate the investigation of the spatial distribution of the sediments with respect to its riverine origin.

Part of Thermaikos	River			
	Axios	Loudias	Aliakmon	Pinios
Inner part	28.2%	19.8%	5.8%	0.0%
Outer part	43.5%	44.7%	61.5%	2.4%
Extended part	28.3%	35.5%	32.7%	97.6%

Table 1. Sedimentation percentages of the rivers in each part of the gulf

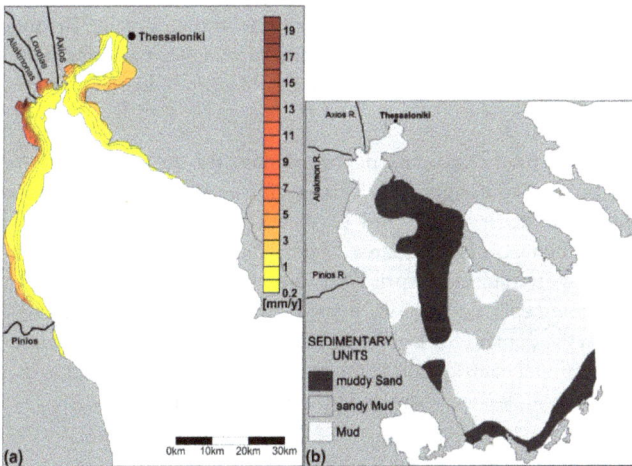

Figure 5. Simulated sedimentation rate [mm/y] (a) and sediment types after [58] (b).

Sediments from Aliakmon present a constant, throughout the year, transport pattern, with the prevalence of advection against dispersion, and predominant movement along the low-depth areas of the western coast. The majority of the sediment supply of Aliakmon deposits in the outer gulf, while a very small part of the supply reaches the enclosed inner Thermaikos. Contrastingly, Axios appears as the main sediment supplier for the inner gulf, along with a much smaller contribution from the Loudias. Matter from Axios presents a great dispersal in the outer part of the gulf, accumulating sediments along the eastern and western coasts; nearly half of the sediment fluxes of Axios settle in the outer gulf (Table 1) while the remaining half equally supplies the coastal areas of the inner and extended Thermaikos. Similar transport and sedimentation patterns are exhibited by Loudias, mainly due to the proximity of the two river outflows. Particles from Pinios supply the coastal zone along the western coastline, primarily northwards and secondarily southwards from the outflow, remaining almost exclusively in the extended part of the domain. The northward deflection of the Pinios plume is due to the strong bottom slope near the mouth of the river [57] and to the effect of the Sporades eddy, which frequently turns anti-cyclonic [56].

The simulated sedimentation rates (Figure 5a) are in good correlation with corresponding values defined after core-sampling in the area [58]. Furthermore, the areas of high river-borne sediment accumulation coincide with the locations in the gulf where the sediment ranges from mud to sandy mud (Figure 5b), validating the modelling approach. In the coastal zone of the eastern outer Thermaikos, where the benthic material grades from sandy mud to muddy sand, the simulated depositing trends are reducing towards the south.

4.3. Distributed source (aeolian transported dust) simulation

Results of the SKIRON/Eta forecasting system and the Eta/NCEP regional atmospheric model [58] were used for the determination of dust inflow in the domain. The model provides prognoses for total dust deposition over the Mediterranean Sea, parameter that was used after two-linear interpolation to transfer data from the domain of the Mediterranean to the domain of Thermaikos.

The time-series of the total dust mass deposited onto the aquatic surface for the period 13/04-13/06/2005 is depicted in figure 5 b. This specific period was selected on account of an intense dust storm that occurred on April 17, 2005. During this incident thick bands of Saharan dust covered the whole Hellenic area, as depicted in the satellite image of figure 5 a. It can be noted that the dust storm of April 17th is predicted by the atmospheric model (Figure 5b) and that three more episodes of lower intensity followed in May (7, 12 and 21). The total mass introduced to the domain during the period in question is $0.72 \ 10^{-6}$t, value that exceeds the corresponding annual contribution of the rivers of the area, estimated at $0.65 \ 10^{-6}$t/y [55]. It is noted that the simulation was extended by 1 month (up to 13/07).

4.3.1. Dust distribution

Investigating the temporal evolution of total suspended and deposited dust masses (Figure 6a) it can be noted that the peak-values of the suspended dust masses, as expected, follow the related peaks of dust inflow (Figure 5b), while deposition increases with time. Maximum total suspended dust masses are recorded in mid-May, while almost all of the matter has settled on the bed at the end of the simulation. The average suspended concentrations at the surface (Figure 6b) follow the morphology of the input time-series, which is expected since the surface is the source of the particles, with values ranging from 8 to 16mg/l at peak 'loading'. The depth-averaged concentrations (Figure 6b) show a small temporal hysteresis in peak values compared to the inflow time-series, of the order of 1-2 days, and range between 0.05 and 1.5mg/l. The maximum suspended concentrations in the field are of the order of

21mg/l and 50mg/l for the water column and the surface layer. The average suspension times of dust in the domain were estimated at 17 days. At the same time, the residence time of dust in deposition increases almost exponentially with time, indicating that deposition prevails over resuspension; near the end of the simulation, the majority of particles settled mainly in the low-depth areas along the coastlines.

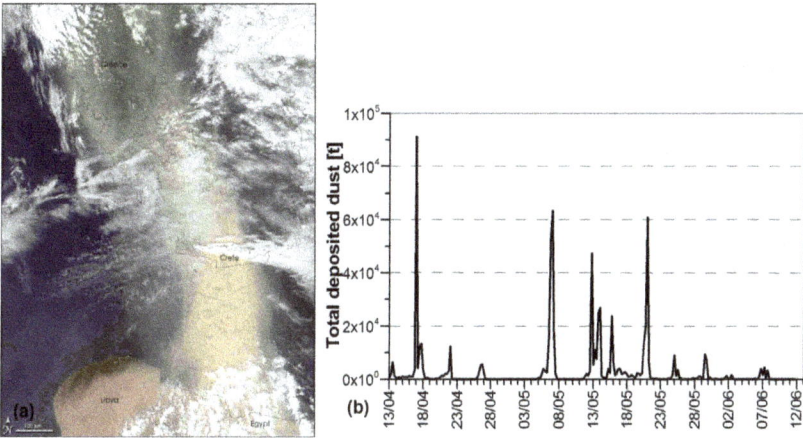

Figure 6. MODIS satellite imagery of the dust storm on April 17, 2005 (a) (the image is a property of NASA) and temporal evolution of the total deposited mass in the domain during the period 13/04-13/06/2005 (b).

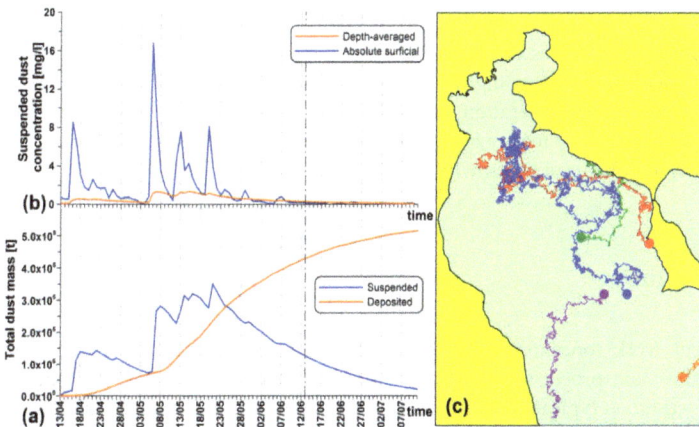

Figure 7. Temporal evolution of total suspended and deposited mass (a) and of suspended depth-averaged and absolute concentrations in the surface layer (b) and horizontal trajectories of randomly selected dust particles (c): the location of entrance in the field is denoted as a dot.

4.3.2. Dust particle trajectories

As in any simulation using a LM, it is possible to track the movement and changes in characteristics of sediments with time using FSTM. Such an example is given in figure 6c as horizontal trajectories of randomly selected dust particles; the location of entrance of the particle in the surface layer of the domain is noted with a dot. Some of these particles escaped the domain through the open boundaries (orange and mauve curves), while others settled on the bed after performing particularly arbitrary movements and presenting very different suspension times. Emphasising on the particle with the maximum residence time in the field (blue curve in figure 6c), it can be noted that it entered the field on 19/04(18:00) and initially moved cyclonically for 4 days, period during which its characteristics remained stable. Following, the particle entered an anti-cyclonic motion for 5 days, at the end of which 28/04(17:30) it started to move cyclonically; this motion lasted for 5 days, during which the particle performed a full circle. Afterward, the particle transported westwards for 3.3 days, period after which its movement became particularly arbitrary. In the vertical sense (data not shown) the particle retained an almost stable settling rate, with an average value of 50mm/hr. The particle finally settled onto the bed 38.5 days after its entrance in the field.

4.4. Mechanically eroded matter

FSTM was also applied for the investigation of mechanically eroded (trawled) masses in Thermaikos, application whose main findings are described briefly in the present chapter; the reader is referred to [60] for further information. The mechanical erosion rates were estimated at $430gr/m^2$, value directly comparable to similar coastal areas, while the sedimentary matter mobilized due to the activity of the benthic trawlers during one trawling season was found significantly higher than the respective contribution of the rivers in the gulf. The sediments, after mobilization by the trawls, continued to move at low distances from the bed, forming Benthic Nefeloid Layers (BNLs), finding that supports the claim that part of the BNLs found in Thermaikos originates from the trawling activity in the area. The redistribution of sediments due to trawling was found generally low, with the matter to redeposit at a small distance from its generation, typically lower than 5km and with average suspension times of 1 to 5 days. The mass export rates from the gulf were also found to be small, especially for the initial part of the trawling period.

5. Conclusions

The present work concentrates on the physical interpretation and modelling approaches used for the main processes that drive sediment transport on coastal and shelf seas. The analysis focused on typical modelling methods and on the physical processes that take place in the water column and near the bed, including the corresponding mass-exchanges. More specifically, advection-dispersion, flocculation and settling mechanisms were analyzed as the processes that take place in the column, along with the effects of stratification to the vertical movement of fine sediments. Regarding near-bed processes, the aspects addressed include shear stress conditions, particle deposition and gradual consolidation and erosion of the bed. In order to accurately describe the fate of cohesive sediments in the marine environment, a numerical tool must generally include all of these processes. However, there are cases in which some of the processes have minimal or no effect; for example, hindered settling does not partake for low concentration environments (<20mg/l) and can therefore be ignored, while stratification becomes significant only in the presence of strong density gradients, typical in estuarine areas. Likewise, wind-generated wave impacts on bottom shear become significant in low-depth (<10m), coastal areas and need to be accounted for only in high resolution regional scale models.

The FSTM model, formulated based on the particle-tracking method and the considerations analyzed in the chapter, was briefly described and indicative results from applications in Thermaikos Gulf, a microtidal shelf sea of the Northern Aegean Sea, regarding point and distributed sedimentary sources were presented and discussed; transport and sedimentation patterns in the gulf were identified for the river-borne sediments of the area. The prevailing cyclonic circulation of the area appears as the main mechanism controlling sedimentation pathways in the domain and forcing higher sediment accumulation along the shallow areas of the western coast. Sediment from Aliakmon is the most characteristic example of this behaviour, with consistent, throughout the year, cyclonic movement and deposition; the greater part of the Aliakmon sediment influx ($\sim$⅔) accumulates in the outer gulf, while a significant amount ($\sim$⅓) reaches the southern part of the shelf. The geomorphology of Thermaikos also plays a significant part in the phenomena; the location of the Axios delta to the centre of the northern coastline exposes its sediments to stronger currents of alternating directions and, thus, its sediments present much higher dispersion in the field, compared to the other rivers of the area. This behaviour is enhanced by seasonal stratification, which forces sediment to remain suspended longer at periods of high freshwater supplies from the rivers. Nearly half of the sediment from Axios settles in the outer part of the gulf, along the shallow areas of both the western and eastern coasts. The remaining part of its outflow co-equally replenishes the sediment of the extended and inner parts. In fact, the mathematic investigation showed that Axios is the main sediment supplier of the enclosed inner Thermaikos. The steep bottom relief near the outflows of Pinios, combined with the hydrodynamics of the area, forces its particles to settle along the western coast close to the delta, with a higher northward spreading. The mass export towards the deep Sporades basin was estimated by the simulation at approximately 10% of the total riverine sediment supply, which corresponds to around 85 10^3t/yr; this high sediment accumulation in the shelf is consistent with the characterisation of Thermaikos as a sediment trap.

Results from the three-month simulation of aeolian transported dust during the dust storm reported in mid-April of 2005, showed concentrations of suspended dust between 8 and 16mg/l at maximal loading and corresponding depth-averaged values ranging from 0.05 to 1.5mg/l. The mean suspension times of dust in the water column were around 17 days, while dust deposition prevailed over resuspension. These high concentrations and suspension times reveal the significance of the process of aeolian transport to the suspended masses in the gulf that can exceed the contribution of river-borne matter considerably in cases of dust storms.

Author details

Katerina Kombiadou* and Yannis N. Krestenitis

Laboratory of Maritime Engineering and Maritime Works, Civil Engineering Department, Aristotle University of Thessaloniki, 54124, Thessaloniki

References

[1] Mehta, A. J., Hayter, E. J., Parker, W. R., Krone, R. B., & Teeter, A. M. (1989). Cohesive sediment transport I: Process Description. *Journal of Hydraulic Engineering*, 115(8), 1076-1093.

[2] Wang, X. H., & Pinardi, N. (2002). Modeling the dynamics of sediment transport and resuspension in the northern Adriatic Sea. *Journal of Geophysical Res*, 107, 3225.

[3] Rodean, H. C. (1996). Stochastic Lagrangian Models of Turbulent Diffusion. *American Meteorological Society*, 84.

[4] Barros, A. P., & Baptista, A. M. (1990). An Eulerian-Lagrangian Model for Sediment Transport. In: Spaulding M., editor. Estuaries, in Estuarine and Coastal Modeling. *ASCE*, 102-112.

[5] Suh, S-W. (2006). A hybrid approach to particle tracking and Eulerian-Lagrangian models in the simulation of coastal dispersion. *Environmental Modelling and Software*, 21(2), 234-242.

[6] Oliveira, A., & Baptista, A. M. (1998). On the role of tracking on Eulerian-Lagrangian solutions of the transport equation. *Advances in Water Resources*, 21, 539-554.

[7] Lewis, R. (1997). Dispersion in estuaries and coastal waters. *John Wiley & Sons Ltd. England*, 312.

[8] Richardson, L. F. (1926). Atmospheric diffusion shown on a distance-neighbour graph. *Proceedings of the Royal Society*, A110, 709-737.

[9] O'Brien, J. J. (1986). The diffusive problem. In: O' Brien JJ, editor. Advanced physical oceanographic numerical modelling. *ASI series. D. Reidel Publ. Comp.*, 127-144.

[10] Mellor, G. L. (1996). Introduction to physical oceanography Princeton University, New Jersey ., 260.

[11] O'Connor, B. A. (1993). Three-dimensional sediment-transport models. In: Abbott MB, Price WA, editors. *Estuarial and harbour engineer's reference book*, Chapman and Hall Pubs. London, 199-214.

[12] Partheniades, E. (2009). Cohesive sediments in open channels. *Properties, transport and applications*, Elsevier, 358.

[13] Krone, R. B. (1972). A field study of flocculation as a factor in estuarial shoaling processes. *Technical Bulletin 19. U.S. army Corps of Engineers*, 32-62.

[14] Mehta, A. J. (1993). Hydraulic behaviour of fine sediment. In: Abbott MB, Price WA, editors. *Coastal, Estuarial and Harbour Engineer's Reference Book*, Chapman and Hall. London, 577-584.

[15] Winterwerp, J. C. (1999). On the dynamics of high-concentrated mud suspensions. Judels Brinkman & Ammerlaan. Delft. , 172.

[16] McAnally, W. H., & Mehta, A. J. (2001). Collisional aggregation of fine estuarial sediment. In: McAnally WH, Mehta AJ, editors. *Coastal and Estuarine Sediment. Proc. in Marine Science 3*, Elsevier, 19-39.

[17] McAnally, W. H., & Mehta, A. J. (2002). Significance of aggregation of fine sediment particles in their deposition. *Estuarine Coastal and Shelf Science*, 54, 643-653.

[18] Eisma, D. (1986). Flocculation and de-flocculation of suspended matter in estuaries. *Netherlands Journal of Sea Research*, 20(2-3), 183-199.

[19] Dyer, K. R. (1989). Sediment processes in estuaries: future research requirements. *Journal of Geophysical Research*, 94(C10), 14327-14339.

[20] Spearman, J. R., & Roberts, W. (2002). Comparison of flocculation models for applied sediment transport modelling. In: Winterwerp JC, Kranenburg C editors. *Fine Sediment Dynamics in the Marine Environment. Proc. in Marine Science 5*, Elsevier, 277-293.

[21] Bale, A. J., Widdows, J., Harris, C. B., & Stephens, J. A. (2006). Measurements of the critical erosion threshold of surface sediments along the Tamar Estuary using a miniannular flume. *Continental Shelf Research*, 26, 1206-1216.

[22] Chen, S., & Eisma, D. (1995). Fractal geometry of in situ flocs in the estuarine and coastal environments. *Netherlands Journal of Sea Research*, 32(2), 173-182.

[23] Logan, B. E., & Wilkinson, D. B. (1990). Fractal geometry of marine snow and other biological aggregates. *Limnology and Oceanography*, 35(1), 130-136.

[24] Kranenburg, C. (1999). Effects of floc strength on viscosity and deposition of cohesive sediment suspensions. *Continental Shelf Research*, 19(13), 1665-1680.

[25] Sternberg, R. W., Berhane, I., & Ogston, A. S. (1999). Measurement of size and settling velocity of suspended aggregates on the northern California continental shelf. *Marine Geology*, 154, 43-53.

[26] Manning, A. J. (2004). Observations of the properties of flocculated cohesive sediment in three Western European Estuaries. *Journal of Coastal Research*, SI41, 70-81.

[27] Manning, A. J., & Dyer, K. R. (2002). The use of optics for the in-situ determination of flocculated mud characteristics. *Journal of Optics A: Pure and Applied Optics*, 4, S71-S81.

[28] Baugh, J. V., & Manning, A. J. (2007). An assessment of a new settling velocity parameterisation for cohesive sediment transport modelling. *Continental Shelf Research*, 27(13), 1835-1855.

[29] Manning, A. J., & Dyer, K. R. (2007). Mass settling flux of fine sediments in Northern European estuaries: measurements and predictions. *Marine Geology*, 245, 107-122.

[30] Van Leussen, W. (1994). Estuarine macroflocs and their role in fine grained sediment transport. *PhD thesis. University Utrecht*, 488.

[31] McAnally, W. H., & Mehta, A. J. (2007). Fine grained sediment transport. In: Garcia MH, editor. Sedimentation engineering: processes, management, modeling, and practice. *ASCE manuals and reports on engineering practice* [110], ASCE, 253-306.

[32] Turner, J. S. (1973). Buoyancy Effects in Fluids. *Cambridge University Press*, 91-126.

[33] Gargett, A. E. (2003). Differential diffusion: an oceanographic primer. *Progress in Oceanography*, 56, 559-570.

[34] Burrows, R., & Ali, K. H. M. (2001). Entrainment studies towards the preservation (containment) of 'freshwaters' in the marine environment. *Journal of Hydraulic Research*, 39(6), 591-599.

[35] Kunze, E. (2003). A review of oceanic salt-fingering theory. *Progress in Oceanography*, 56, 399-417.

[36] Kelley, D. E., Fernando, H. J. S., Gargett, A. E., Tanny, J., & Özsoy, E. (2003). The diffusive regime of double-diffusive convection. *Progress in Oceanography*, 56, 461-481.

[37] Schladow, S. G., Thomas, E., & Koseff, J. R. (1992). The dynamics of intrusions into a thermohaline stratification. *Journal of Fluid Mechanics*, 236, 127-165.

[38] Kombiadou, K., & Krestenitis, Y. N. (2012). Fine sediment transport model for river influenced microtidal shelf seas with application to the Thermaikos Gulf (NW Aegean Sea). *Continental Shelf Research*, 36, 41-62.

[39] Dearnaley, M. P., Roberts, W., Jones, S., Leureur, K. C., Lintern, D. G., Merckelbach, L. M., Sills, C. G., Toorman, E. A., & Winterwerp, J. C. (2002). Measurement and modelling of the properties of cohesive sediment deposits. In: Winterwerp JC, Kranenburg C, editors. Fine Sediment Dynamics in the Marine Environment. *Proc. in Marine Science 5. Elsevier*, 57-89.

[40] Teeter, A. M., Johnson, B. H., Berger, C., Stelling, G., Scheffner, N. W., Garcia, M. H., & Parchure, T. M. (2001). Hydrodynamic and sediment transport modelling with emphasis on shallow-water vegetated areas (lakes, reservoirs, estuaries and lagoons). *Hydrobiologia*, 444, 1-23.

[41] Winterwerp, J. C., Bruens, A. W., Gratiot, N., Kranenburg, C., Mory, M., & Toorman, E. A. (2002). Dynamics of concentrated benthic suspension layers. In: Winterwerp JC, Kranenburg C, editors. Fine Sediment Dynamics in the Marine Environment. *Proc. in Marine Science 5*, Elsevier, 41-55.

[42] Van Rijn, L. C. (1998). Mud Coasts. *Principles of coastal morphology*, Aqua Pubs. Amsterdam. Netherlands, 6.1-6.51.

[43] Toorman, E. A., Bruens, A. W., Kranenburg, C., & Winterwerp, J. C. (2002). Interaction of Suspended Cohesive Sediment and Turbulence. In: Winterwerp JC, Kranenburg C, editors. Fine Sediment Dynamics in the Marine Environment. *Proc. in Marine Science 5. Elsevier.*, 7-23.

[44] Krone, R. B. (1962). Flume studies of the transport of sediment in estuarial shoaling processes. Final Report. Hydraulic Engineering Laboratory and Sanitary Engineering Research Laboratory. University of California. Berkeley. USA. , 110.

[45] Pohlmann, T., & Puls, W. (1994). Currents and transport in water. In: Sündermann J, editor. *Circulation and contaminant fluxes in the North Sea. Springer Verlag. Berlin*, 345-402.

[46] Partheniades, E. (1965). Erosion and deposition of cohesive soils. *Journal of the Hydraulics Division*, 91(HY2), 105-138.

[47] Parchure, T. M., & Mehta, A. J. (1985). Erosion of Soft Cohesive Sediment Deposits. *Journal of Hydraulic Engineering*, 111(10), 1308-1326.

[48] Thorn, M. F. G., & Parsons, J. G. (1980). Erosion of cohesive sediments in estuaries:. *An engineering guide, 3rd International Symposium on Dredging Technology. Paper F1. BHRA Fluid Engineering*, 349-358.

[49] Ockender, M. C., & Delo, E. A. (1988). Consolidation and erosion of estuarine mud and sand mixtures-An experimental study. HR Wallingford. Report No SR 149.

[50] Mitchener, H. J., & O'Brien, D. J. (2001). Seasonal variability of sediment erodibility and properties on a microtidal mudflat, Peterstone Wentlooge, Severn Estuary, UK. In: McAnally WH, Mehta AJ, editors. *Coastal and Estuarine Fine Sediment Processes. Proc. in Marine Science 3. Elsevier*, 301-322.

[51] Merckelbach, L. M., Kranenburg, C., & Winterwerp, J. C. (2002). Strength modeling of consolidating mud beds. In: Winterwerp JC, Kranenburg C, editors. *Fine Sediment Dynamics in the Marine Environment. Proc. in Marine Science 5. Elsevier*, 359-373.

[52] Hawlader, B. C., Muhunthan, B., & Imai, G. (2008). State-dependent constitutive model and numerical solution of self-weight consolidation. *Géotechnique*, 58(2), 133-141.

[53] Migniot, C. (1968). A study of the physical of different very fine sediments and their behaviour under hydrodynamic action. *La Houille Blance*, 7, 591-620.

[54] Kourafalou, V. H., & Tsiaras, K. (2007). A nested circulation model for the North Aegean Sea. *Ocean Science*, 3, 1-16.

[55] Karamanos, H., & Polyzonis, E. (2000). Thermaikos Gulf: the adjacent land area and its potentiality for the freshwater, sediment and pollutant supply. *In: Final Scientific Report of the EU-MAST-III ELOISE Project METRO-MED. Part B. CT960049*, 55-67.

[56] Kourafalou, V. H., & Barbopoulos, K. A. (2003). High resolution simulations on the North Aegean Sea seasonal circulation. *Annales Geophysicae*, 21(1), 251-265.

[57] Kontoyannis, H., Kourafalou, V. H., & Papadopoulos, V. (2003). Seasonal characteristics of the hydrology and circulation in the northwest Aegean (eastern Mediterranean): Observations and modelling. *Journal of Geophysical Research*, 108(C9), 3302.

[58] Karageorgis, A., & Anagnostou, Ch. (2001). Particulate matter spatial-temporal distribution and associated surface sediment properties: Thermaikos Gulf and Sporades Basin, NW Aegean Sea. *Continental Shelf Research*, 21, 2141-2153.

[59] Nickovic, S., Kallos, G., Papadopoulos, A., & Kakaliagou, O. (2001). A model for the prediction of desert dust cycle in the atmosp here. *Journal of Geophysical Research*, 106, 18113-18129.

[60] Kombiadou, K., & Krestenitis, Y. N. (2011). Simulating the fate of mechanically eroded masses in the Thermaikos Gulf. *Continental Shelf Research*, 31, 817-831.

Chapter 12

Modelling of Sediment Transport in Shallow Waters by Stochastic and Partial Differential Equations

Charles Wilson Mahera and Anton Mtega Narsis

Additional information is available at the end of the chapter

1. Introduction

Sediment transport in coastal waters take place in near-shore environments due to the motions of waves and currents resulting in the formation of characteristic coastal landforms such as beaches, barrier islands, and capes, etc. Though sediment transport modelling has been carried for over three decades[26], the study remains a challenging topic of research since a unified description of the process is still to be achieved[25]. A brief review of sediment transport modelling and the widely used classical as well as most recent methods and their limitations in applications are given by [22, 34]. It is observed that sediment transport studies (e.g. the underlying physics and methods used) and modes of sediment transport are yet to be fully studied. Both experimental approaches and mathematical modelling coupled with advanced numerical solutions are needed for better understanding of the how fundamental sediment transport processes is significant for environmental researchers to provide practical and scientifically sound solutions to hydraulic engineering problems. According to [34], the choice of a method for solving a specific problem depends on the nature and complexity of the problem itself, the capabilities of the chosen model to simulate the problem adequately, data availability for model calibration and verification and overall available time and budget for solving the problem. [34] also found that discrepancies between hydrodynamic/sediment transport model predictions and measurements can be attributed to different causes. They include over simplification of the problem by using an inappropriate model 1D versus 2D or 2D versus 3D, the use of inappropriate input data, lack of appropriate data for model calibration, unfamiliarity with the limitations of the hydrodynamic/sediment transport equations used in developing the model, and computational errors in source codes because of approximations in the numerical schemes used in solving the governing equations (boundary condition problems/truncation errors because of discretization.

Modelling sediment transport basically involves interaction of hydrological, hydrodynamic and sedimentology processes. The interactions cause variability of the process parameters which are partly deterministic (having a known structure in space/time, e.g. the yearly storm/calm weather period) and partly stochastic (e.g. sediment pick-up or turbulent viscosity, and also weather variations), both leading to slow spatial/temporal variations in the sand wave-dynamics[31]. Therefore, stochastic characteristics of the governing parameters (such as, suspended sediment inflow concentration) and parameters describing the beginning and rate of erosion and sedimentation respectively are considered as stochastic variables. The observations of the stochastic nature of bedload transport have given impetus to a probabilistic formulation of bedload transport equations.

In stochastic modelling spatio-temporal behavior of phenomena is modelled with random components. The random walk simulation model enables the observation of the phenomena in scales much smaller than the grid size, as well as the tracing of the movement of individual particles, thereby describing the natural processes more accurately. Concentrations of particles are easily calculated from the spatial positions of the particles and, more importantly, when and where required. Furthermore, errors due to numerical diffusion observed in methods such as Finite differences or Finite elements, are avoided and there is considerable reduction in computational time since the calculating load is restricted to the domain parts where the majority of the parcels are gathered.

In a random walk model the displacement of an arbitrary particle, at each time step consists of an advective, deterministic component and an independent, stochastic component. In a simplified one-dimensional transport model the Brownian motion of a particle can be described by the Langevin equation[32].

In order to investigate the fate of suspended sediment in coastal and estuarine waters as well as the evolution of sea or river beds, sediment dynamics need to be represented at a scale relevant to the numerical discretized solution, and significant effort is devoted to parameterize sediment processes. Sediment diffusivities, settling velocities, and cohesive processes such as flocculation all have an impact on suspended sediment throughout the water column. The approaches implemented in these coastal models may present distinct strengths and shortcomings with regard to some important issues for coastal zones, both numerical and physical. While these detailed limitations need to be considered as part of model assessment, more general issues also hinder present state of the art models. In particular, sediment transport is inherently highly empirical, which is further compounded by issues arising from turbulence closure schemes.

In this chapter we focus on deterministic (i.e., process based) coastal ocean models, which are being increasingly used to study coastal sediment dynamics and coastal morphological evolution[23, 27–29]. These models usually treat the short term (hours to days) to medium term (days to months) evolution. Historically, they were first based on depth averaged equations (two dimensional horizontal (2DH) models) and were applied both to riverine see [33] and coastal[24] environments.

We also study the methods based on probability concept appeared to be superior for predicting local transport of bedload. Although several deterministic methods show comparable performance for predicting total sectional transport rate, their performances are significantly reduced for predicting lateral variation of local transportation rates [22]. Development of a stochastic theory based model that can explicitly present the random term of sediment concentrations could be achieved. With the development of better numerical techniques, the stochastic differential equations can be solved using Itô's integration technique without the need to rely on analytical solutions under simplified conditions.

It is known that bedform changes have a significant effect on the flow dynamics contributed by imbalance between sediments in and out from those areas. The imbalance can easily be disturbed by the external factors, such as extreme storm events, mean sea-level rise, changes in tidal regime, human interferences and so on. Therefore, a better prediction of these bedforms is required to be able to understand their sensitivity to external conditions. In this chapter we developed and describe a particle-based approach to simulate entrainment,transport, and settling of non-cohesive sediments in shallow waters. Sediment distributions are modelled as a set of particles that are tracked on an individual basis by solving Lagrangian transport equations that account for the drift part by the mean flow, settling, and random horizontal motions.

The rest of this chapter is organised as follows. The brief description of an Eulerian transport model is done in Section 3. The particle model for sediment transport is discussed in Section 4. The interpretation of the partial differential equation called Fokker-Planck equation into the well-known Eulerian transport model for sediment transport is described in Section 5. The description and discussion of the two dimensional channel for a test case of sediment transport is carried out in Section 7.

In our work we do carry out the estimation of the change in bedforms. Nevertheless, we do not yet recompute the flow velocities when a change in the shallow water depth occurs.

2. Shallow water flow equations

In order for particle models to describe transport problems in shallow waters, the inputs such as water flow velocities $[U(x, y, t), V(x, y, t)]^T$, water levels ξ, water depths $H(x, y, t)$ and so forth are required. In our application, the inputs are often computed by the hydrodynamic model, which can solve the depth-averaged shallow water equations or 3 dimensional shallow water [15]. The generated results in this case are written into a matlab format that can be loaded and read in the particle model for simulation of sediment transport. The inputs are assumed to satisfy the shallow water equations. The momentum equations are represented by the following equations:

$$\frac{\partial U}{\partial t} + U\frac{\partial U}{\partial x} + V\frac{\partial U}{\partial y} + g\frac{\partial \xi}{\partial x} - fV + g\frac{U(U^2 + V^2)^{\frac{1}{2}}}{(C_z)^2 H} = 0 \tag{1}$$

$$\frac{\partial V}{\partial t} + U\frac{\partial V}{\partial x} + V\frac{\partial V}{\partial y} + g\frac{\partial \xi}{\partial y} + fU + g\frac{V(U^2 + V^2)^{\frac{1}{2}}}{(C_z)^2 H} = 0. \tag{2}$$

The velocity is uniform over the vertical, therefore, for that reason, the rise and fall of free surface is given by equations of conservation of mass called the continuity equation:

$$\frac{\partial H}{\partial t} + \frac{\partial(UH)}{\partial x} + \frac{\partial(VH)}{\partial y} = 0, \tag{3}$$

where

$H = h + \zeta$ is the total depth;

ζ is the water-level with respect to a reference;

h depth of the water with respect to a reference;

C_z bottom friction coefficient (Chezy coefficient);

g acceleration of gravity;

f Coriolis parameter.

The shallow water equations are entirely described by equations (1)-(3), provided the closed and open boundary conditions and initial fields are given [9].

3. Eulerian sediment transport model

In this section we briefly introduce the Eulerian model for sediment transport. We consider noncohesive type of sediment particle. The dynamics of the suspended particles can be described by well-known Eulerian transport model with the source and sink terms included. The following Eulerian sediment transport model is similar to that in [20], for example:

$$\frac{\partial(HC)}{\partial t} + \frac{\partial(HUC)}{\partial x} + \frac{\partial(HVC)}{\partial y} - \frac{\partial}{\partial x}(D\frac{\partial HC}{\partial x}) - \frac{\partial}{\partial y}(D\frac{\partial HC}{\partial y})$$
$$= -\gamma HC + \mathcal{E}(U,V) \cdot \lambda_s. \qquad (4)$$

Where $C(x,y,t)$ is depth averaged concentration, γ is the deposition coefficient, $\mathcal{E}(U,V) = (U^2 + V^2)(m^2 s^{-2})$ is a function of flow velocities and the term $\lambda_s \cdot \mathcal{E}(U,V)$ models erosion of sediment particles. The particle pick up function is parameterized as $\lambda_s \cdot \mathcal{E}(U,V)$, where, λ_s is the erosion coefficient, it can be related to sediment properties (grain size, grain shape). This parameterisation is motivated by the analysis of field observations reported in [14] and reference therein. Typically, $\lambda_s \approx 3 \times 10^{-2}(kgsm^{-4})$ for fine sand. In this article $\lambda_s = 0.0001(kgm^{-4}s)$ is within the range reported in literature (see e.g., [14, 16]). Note that the term γHC models the deposition of sediment and γ is the deposition coefficient, it is reported that $\gamma \approx 4 \times 10^{-3}s^{-1}$ [14] for fine sand.

3.1. Determination of bedlevel changes by Eulerian transport model

In addition to suspension and deposition processes, the following equation is used to determine the depth changes and therefore the change of bed-level in each grid cell i, j:

$$\frac{\partial h}{\partial t} = \frac{1}{(1 - po)\rho_s}(d_e - s_e). \qquad (5)$$

Where $s_e = \gamma HC$ stands for deposition and $d_e = \lambda_s \cdot \mathcal{E}(U,V)$ stands for erosion (term responsible for suspending particles). Sea bed porosity is represented by po, ρ_s is the density of sediment particles.

In this section we have constructed a simplified transport model which is derived from the Eulerian transport model (4). This simple model is then made consistent with the simplified Lagrangian particle model. In this way it becomes easy to compare the bedlevel changes to see if they are similar. We simplified Equation (4) by assuming that the deposition and erosion processes balance:

$$\gamma CH = (U^2 + V^2) \cdot \lambda_s. \tag{6}$$

Quite often, transport in water is defined as the product of the concentration of sediment particles C and a velocity U or V as well the depth of water in that grid cell. Thus, transport along x and y directions is respectively given by;

$$q_x = UCH \text{ and } q_y = VCH,$$

where in vector form $\bar{q} = [q_x, q_y]^T$, using equation (6), it follows that

$$\bar{q} = [(U^2 + V^2)U \cdot f_d, \ (U^2 + V^2)V \cdot f_d]^T, \tag{7}$$

where $f_d = \frac{\lambda_s}{\gamma}$ stands for the drag force. This depends on the properties of a particle for example its size or its area. Thus, in order to determine how much mass exits or comes into a given location, it is important to consider the divergence. The divergence determines the average rate of how much mass comes into the cell(change of mass per second per area):

$$\frac{\partial m}{\partial t} = -\text{div}(\bar{q}),$$
$$\tag{8}$$

where $\bar{q} = \frac{1}{T} \int_0^T q dt$, div stands for divergence. Consequently, Equation (8) represents the rate of how much mass stays behind or leaves the cell by assuming the absence of destruction or creation of a matter. Since we want to determine the effects of sediment transport on the sea bedforms, the equation for the bed level is represented by;

$$\frac{\partial h}{\partial t} = \frac{1}{(1 - po)\rho_s} \frac{\partial m}{\partial t}. \tag{9}$$

In the present application the determination of the bed level change using finite difference scheme is estimated by the following equations:

$$\frac{\partial h}{\partial t} = -\frac{1}{\rho_s(1 - po)} \cdot \text{div}(\bar{q}).$$

For cases where flow is in one direction for example when $v = 0$, transport is given by the following equation:

$$\frac{\partial h}{\partial t} = -\frac{f_d}{\rho_s(1 - po)} \cdot \text{div}(\overline{U^3}).$$

With the aid of Equation (8)–(9), accordingly the determination of bed level changes is now done by using the following equation:

$$\Delta h \approx \frac{-f_d T}{\rho_s(1 - po)} \cdot \left(\frac{\partial \mathbf{u_m}}{\partial x} + \frac{\partial \mathbf{v_m}}{\partial y}\right). \tag{10}$$

where $\mathbf{u_m} = \frac{1}{T}\int_0^T(U^2 + V^2)Udt$ and $\mathbf{v_m} = \frac{1}{T}\int_0^T(U^2 + V^2)Vdt$. Next, let us now discuss the Lagrangian particle model in the following section.

4. A particle model for sediment transport in shallow waters

A particle model is a description of a transport process by means of random walk models. Random walk model is defined as the stochastic differential equation that describes the movement of a particle that subsequently undergoes a displacement, which consists of the drift part and a stochastic(diffusive) part [4, 13].

4.1. Integration of particle movement

In this section, the following 2-dimensional stochastic differential equations is developed:

$$dX(t) \overset{\text{Itô}}{=} \left[U + \frac{D}{H}\left(\frac{\partial H}{\partial x}\right) + \frac{\partial D}{\partial x}\right]dt + \sqrt{2D}dW_1(t),$$

$$dY(t) \overset{\text{Itô}}{=} \left[V + \frac{D}{H}\left(\frac{\partial H}{\partial y}\right) + \frac{\partial D)}{\partial y}\right]dt + \sqrt{2D}dW_2(t), \tag{11}$$

where the Brownian process $W_1(t)$ and $W_2(t)$ are Gaussian [3], and $D(x, y, t)$ is the horizontal dispersion coefficient for sediment transport. Typically, $D = \mathcal{O}(10 - 100)m^2/s$ [14]. Note that $U(x, y)$ and $V(x, y)$ are the flow velocities along the x and y direction respectively given in m/s, $H(x, y)$ is the averaged total depth plus relative water levels due to waves, $dW_1(t)$ and $dW_2(t)$ are independent increments of Brownian motions with mean $(0,0)^T$ and covariance $E[dW_1(t)dW_2(t)^T] = Idt$ where I is an identity matrix ([3, 10]). The simulation of sediment transport is initiated with zero number of particles.

4.2. Deposition of sediment particles

We associate with each sediment particle a binary state which at any time t is given by

$$S_t = \begin{cases} 1 & \text{particle is in suspension} \\ 0 & \text{otherwise (particle is on the sea bed).} \end{cases}$$

Given a particle in suspension, we are interested in the transition of state 1 to state 0. In continuous form, this transition can be modelled by the following equation

$$\frac{dP(S_t = 1)}{dt} = -\gamma \cdot P(S_t = 1), \qquad \text{where initially} \qquad P(S_0 = 1) = 1 \qquad (12)$$

where $\gamma(x, y, t)$ is the deposition coefficient, in this chapter $\gamma = \gamma(x, y, t)$ is constant, $P(S_t = 1)$ is the probability that the state of the particle at time t is 1. When the particle is in the flow, its evolution is described by the following transition probability equation in discrete form:

$$\begin{aligned} P(S_{t+\Delta t} = 1 \mid S_t = 1) &= P(S_0 = 1) \cdot [1 - \gamma \cdot \Delta t] \\ &= [1 - \gamma \Delta t] \end{aligned} \qquad (13)$$

Assuming the system state (e.g. flow field and turbulence patterns) to be constant during the period of the time step, it follows that the probability that a particle will be sedimented is given by

$$P(S_{t+\Delta t} = 0 \mid S_t = 1) = 1 - [P(S_{t+\Delta t} = 1 \mid S_t = 1)]. \qquad (14)$$

4.3. Suspension of sediment particles

Mass represents concentration of a group of particles at a certain location. A source term is included in our particle model such that the expected number of suspended particles (*enp*) in grid cell i, j at time t is given by

$$enp_{(i,j,t)} = \frac{\Delta x \cdot \Delta y \cdot \Delta t \cdot (U^2 + V^2) \cdot \lambda_s}{\mathcal{M}_p}, \qquad (15)$$

where $\mathcal{M}_p$ is the mass of each particle, Δx and Δy are the width of the grid cells along x and y directions respectively, Δt is the time step size, and λ_s is the erosion coefficient. For each grid cell i, j we use the expected number of particles from Equation (15) to determine the actual number of particles to be suspended. This is done by drawing a number from a Poisson distribution function.

5. The relationship between the Kolmogorov Forward Partial differential and The Advectiion Diffusion transport model

In this section the relationship between the Kolmogorov Forward Partial differential corresponding to the 2-dimensional Stochastic differential equations and the well known two dimensional advection diffusion equations for sediment transport is discussed in detail. Since we are interested in the particle being in suspension, we assume that the particle at position (x, y) at time t has expectation of their mass $\langle \cdot \rangle$ defined by;

$$\langle m(x, y, t) \rangle = f(x, y, t) \cdot P(S_t = 1). \qquad (16)$$

This is known as mass density of particles per unit area of the grid box. The Kolmogorov Forward Partial differential is known as the Fokker-Planck equation(FPE)[3]. In this section we incorporate the processes such as suspension and sedimentation states of the particles in the model for transport of sediments in shallow water. We let D be the diffusion coefficient, $P(S_t = 1)$, is the probability that particle is in suspension, $(S_t = 1)$ denotes a state that a particle is in suspension and $(S_t = 0)$ denotes the state that the particle is deposited on the sea bed or bed of the shallow water. The stochastic process (X_t, Y_t) is a Markov process. The probability density function of the particle position $f(x, y, t)$ based on the two dimensional Stochastic Differential equations (SDEs) (11) evolves according to the following Fokker-Planck equation [9].

$$
\begin{aligned}
\frac{\partial f(x,y,t)}{\partial t} = &-\frac{\partial}{\partial x}\left(\left[U+(\frac{\partial H}{\partial x}D)/H+\frac{\partial D}{\partial x}\right]\cdot f(x,y,t)\right) \\
&-\frac{\partial}{\partial y}\left(\left[V+(\frac{\partial H}{\partial y}D)/H+\frac{\partial D}{\partial y}\right]\cdot f(x,y,t)\right) \\
&+\frac{1}{2}\frac{\partial^2}{\partial x^2}(f(x,y,t)\cdot 2D)+\frac{1}{2}\frac{\partial^2}{\partial y^2}(f(x,y,t)\cdot 2D). \quad (17)
\end{aligned}
$$

The resulting sediment transport model (11) is an extension of two dimensional particle model for pollutant dispersion in the shallow waters developed by [9]. The extension in the present model includes the erosion and deposition terms. It is possible to derive the Fokker-Planck equation that describes the probabilistic transport of the sediments from one location to another. The derivation of the Fokker-Planck equation is e done as follows, we first differentiate equation (16) with respect to time t, to obtain

$$
\frac{\partial}{\partial t}\langle m(x,y,t)\rangle = P(S_t = 1)\frac{\partial}{\partial t}f(x,y,t) + f(x,y,t)\frac{\partial}{\partial t}P(S_t = 1),
$$

next with the aid of Equation (12), it follows that,

$$
\frac{\partial}{\partial t}\langle m(x,y,t)\rangle = P(S_t = 1)\frac{\partial}{\partial t}f(x,y,t) - \gamma f(x,y,t)\cdot P(S_t = 1). \quad (18)
$$

Therefore, we also add the erosion term to Equation (18) and come up with

$$
\frac{\partial}{\partial t}\langle m(x,y,t)\rangle = P(S_t = 1)\frac{\partial}{\partial t}f(x,y,t) - \gamma f(x,y,t)\cdot P(S_t = 1) + \lambda_s\cdot\mathcal{E}(U,V) \quad (19)
$$

next we multiply on both sides of Equation (17) by $P(S_t = 1)$ to obtain

$$
\begin{aligned}
P(S_t = 1)\frac{\partial}{\partial t}f(x,y,t) = &-\frac{\partial}{\partial x}\left(\left[U+\frac{D}{H}\frac{\partial H}{\partial x}+\frac{\partial D}{\partial x}\right]f(x,y,t)\cdot P(S_t = 1)\right) \\
&-\frac{\partial}{\partial y}\left(\left[V+\frac{D}{H}\frac{\partial H}{\partial y}+\frac{\partial D}{\partial y}\right]f(x,y,t)\cdot P(S_t = 1)\right) \\
&+\frac{1}{2}\frac{\partial^2}{\partial x^2}(2D\cdot f(x,y,t)P(S_t = 1))+\frac{1}{2}\frac{\partial^2}{\partial y^2}(2D\cdot f(x,y,t)\cdot P(S_t = 1)). \quad (20)
\end{aligned}
$$

The substitution of Equation (20) into equation (19) gives the following Fokker-Planck equation which now have extra terms that model the deposition and erosion of sediments.

$$
\begin{aligned}
\frac{\partial \langle m(x,y,t) \rangle}{\partial t} = {} & -\frac{\partial}{\partial x} \left(\left[U + \frac{D}{H} \frac{\partial H}{\partial x} + \frac{\partial D}{\partial x} \right] \cdot \langle m(x,y,t) \rangle \right) \\
& - \frac{\partial}{\partial y} \left(\left[V + \frac{D}{H} \frac{\partial H}{\partial y} + \frac{\partial D}{\partial y} \right] \cdot \langle m(x,y,t) \rangle \right) \\
& + \frac{1}{2} \frac{\partial^2}{\partial x^2} \left(2D \cdot \langle m(x,y,t) \rangle \right) + \frac{1}{2} \frac{\partial^2}{\partial y^2} \left(2D \cdot \langle m(x,y,t) \rangle \right) \\
& - \gamma \cdot \langle m(x,y,t) \rangle + \lambda_s \cdot \mathcal{E}(U,V). \quad (21)
\end{aligned}
$$

The last two terms in the transport equation (21) represents the process of sediment deposition and erosion of sediments respectively. An average mass $\langle m(x,y,t) \rangle$ per unit area (kg/m^2) of a particle is related to a particle depth averaged concentration $C(x,y,t)$ in mass per unit volume (kg/m^3) ([9, 18]).

The concentration of materials is given in kg/m^3, therefore, the expected mass of a particle at position (x,y) can be related by the concentration in that grid cell(location) as follows:

$$
\langle m(x,y,t) \rangle = H(x,y,t) \cdot C(x,y,t). \quad (22)
$$

Substitution of equation (22) into the Fokker-Planck equation (21) leads to the two dimensional Advection diffusion partial differential equation commonly known as Eulerian sediment transport model (4). Consequently, the transport equation (4) which was discussed in Section 3, is consistent with the particle model for sediment transport (11)-(26).

Therefore, after having constructed the particle model for sediment transport, it is now necessary to develop the equations that cater for the bed level changes using the particle model.

In the next section we shall briefly discuss the numerical approximation of our particle model.

6. Numerical approximation of the particle model

Euler scheme is used in the numerical implementation of the particle model. The scheme is convergent in the weak sense with accuracy of order $\mathcal{O}(\Delta t)$ and it is half order accurate in the strong sense. Higher order schemes for stochastic differential equations are described in [11]. The discretisations of the hydrodynamic flow models is widely discussed by [15], for example. The particle model (11)-(26) is discretised and uses the following Euler scheme to approximate the numerical solutions. We discretise the two dimensional stochastic differential equations for integrating the movement of the particle in similar way to that as in [9] with the modifications by the inclusion of the sedimentation and deposition parts:

$$
\bar{X}(t_{k+1}) = \bar{X}(t_k) + \left[U + (\frac{\partial H}{\partial x} D)/H + \frac{\partial D}{\partial x} \right] \Delta t_k + \sqrt{2D} \Delta W_1(t_k) \quad (23)
$$

$$\bar{Y}(t_{k+1}) = \bar{Y}(t_k) + \left[V + (\frac{\partial H}{\partial y}D)/H + \frac{\partial D}{\partial y}\right]\Delta t_k + \sqrt{2D}\Delta W_2(t_k) \tag{24}$$

$$P_{k+1}(S_t = 1) = (1 - \gamma(x, y, t)\Delta_k)P_k(S_t = 1). \tag{25}$$

Where $\bar{X}(t_{k+1})$ and $\bar{Y}(t_{k+1})$ are the numerical approximations of $X(t)$ and $Y(t)$ respectively, while $\bar{X}(t_0) = X(t_0) = x_0$ and $\bar{Y}(t_0) = Y(t_0) = y_0$ are initial locations of a particle. In addition to Eqns. (23)-(25), we also use Eqns. (15)-(26) to make the simulation of sediment transport complete. There are several schemes that can be used for simulation process for instance, Euler, Heun, Milstein scheme and Runge kutta methods. Much detailed work on numerical methods for stochastic models can be found in (e,g., [11]).

Numerical schemes such as the Euler scheme often show very poor convergence behaviour. This implies that, in order to get accurate results, small time steps are needed thus requiring much computation. Another problem with the Euler (or any other numerical scheme) is its undesirable behaviour in the vicinity of boundaries; a time step that is too large may result in particles unintentionally crossing boundaries. Therefore, the treatment of boundary condition for particle Models is often done in the following section as follows.

6.1. Boundaries

One problem with numerical integration of particle positions arises in the vicinity of boundaries. Given the current location, $(X(t), Y(t))$, we may find that the new location, $(X(t + \Delta t), Y(t + \Delta t))$, is on the other side of a boundary, i.e. the particle has crossed a boundary. Depending on the type of boundary this may be physically impossible. We consider two types of boundaries. The first type, closed boundaries, represents boundaries intrinsic to the domain such as banks, sea bed, and coastal lines. The second type of boundaries are open boundaries, which arise from the modeller's decision to artificially limit the domain because particles are not expected to reach any further or simply because no domain information is available at those locations. It is clear that it is undesirable to have particles cross the first type of boundary, whereas for the second type it is quite natural. Based on this classification, we apply the following rules to particles crossing borders during integration:

- In case an open boundary is crossed by a particle, the particle remains in the sea but is now outside the scope of the model and is therefore removed.

- In case a closed boundary is crossed by a particle during the drift step of integration, the step taken is cancelled and the time step halved until the boundary is no longer crossed. However, because of the halving, say n times, the integration time is reduced to $2^{-n}\Delta t$, leaving a remaining $(1 - 2^{-n})\Delta t$ integration time, which, at a constant step size, requires at least another $2^n - 1$ steps in order to complete the full time-step Δt. Note that at each of these steps it may be needed to further reduce the step size. This further reduction applies only to the current time step, leaving the step size of following sub-steps unaffected. This method effectively models shear along the coastline.

- If a closed boundary is crossed during the diffusive part of integration, the step size halving procedure described above is maintained with the modification that in addition to the position, the white noise process is also restored to its state prior to the invalidated integration step. The process of halving the time step and continuing integration with the reduced step size is repeated until the full Δt time step has been integrated without crossing a boundary.

In addition, this chapter have also considered the pick up of sediment particles at the inflows this will be discussed in the next section.

6.2. Particle flux at open boundaries

We now consider particle flux at open boundaries. This flux is the difference between particles flowing into and out of the domain. The number of particles flowing out should not be controlled as it is a natural consequence of the movement of a particle. As soon as a particle crosses an open boundary it is considered gone and further integration is no longer possible as no data outside the domain is given. For this very reason, however, we do need to explicitly model the particles flowing in. We determine the expected number of particles entering the domain as follows:

$$
enp_{(i,j,t)} = \begin{cases} \frac{\Delta y \cdot \Delta t \cdot V \cdot (U^2 + V^2) \cdot \lambda_s}{\gamma \mathcal{M}_p} & \text{inflow parallel to y-axis} \\[3mm] \frac{\Delta x \cdot \Delta t \cdot U \cdot (U^2 + V^2) \cdot \lambda_s}{\gamma \mathcal{M}_p} & \text{inflow parallel to x-axis,} \end{cases} \tag{26}
$$

where, γ is the deposition coefficient. The actual number of particles added at the domain boundary at each iteration is obtained by drawing a value from the Poisson distribution parameterised by the above expectation value.

To determine the actual number of particles to be suspended in a grid cell i, j we draw a number from a Poisson distribution function with mean $enp_{(i,j,t)}$ determined by Equation (15). We assume that particles are infinitely many on the sea bed. However, the particle that is suspended is not the same as the one that is deposited.

Before implementing the particle model (11)-(26), we first required to show the consistency between the Fokker-Planck equation and its the Eulerian transport model. This will be described in the next Section.

7. Determination of bedlevel changes using particle models

Comparing with a simplified form of Equation (4), where in this case we assume for the local change in mass $\frac{\partial m}{\partial t} \approx d_e - s_e$. We also assume that the deposition and erosion processes balance (see Section 3.1). The approximation of the change in mass with respect to time, in each grid cell i, j in the this particle model is determined by the following equations:

$$
\frac{\partial m}{\partial t} = \frac{\Delta \mathcal{N}_p}{\Delta t} \frac{1}{\Delta x \Delta y} \mathcal{M}_p. \tag{27}
$$

Using equation (5) and the fact that $\frac{\partial m}{\partial t} \approx d_e - s_e$, the equation for the bed level change can be derived as follows,

$$\frac{\partial h}{\partial t} = \frac{\Delta \mathcal{N}_p}{\Delta t} \frac{1}{\rho_s (1 - po) \Delta x \Delta y} \mathcal{M}_p.$$

Where, $\mathcal{M}_p$ is the mass of each particle, ρ_s and po denote respectively the density of an individual grain particle and the bed porosity. From [14], we find $\rho_s = 2650 kg/m^3$ and $po \approx 0.5$. $\Delta \mathcal{N}_p$ is the difference between the number of deposited and suspended particles at each iteration in each grid cell i, j. Hence the cumulated (integrated) change of the level of the sea Δh in m for all time steps is determined by the following equation:

$$\Delta h = \int_0^T \frac{\partial h}{\partial t} dt$$

$$\Delta h \approx \sum \frac{\Delta \mathcal{N}_p}{\rho_s (1 - po) \Delta x \Delta y} \mathcal{M}_p. \tag{28}$$

More information about the effect of parameters on the sea bed level changes can be found in [16], for example.

7.1. Primary input of the model

The initial field is defined as

$$depth = h_0 + h_1 \cdot \exp(-((x - 0.0).^2)/(2 \cdot (wd)^2)),$$

where the initial amplitude of the disturbance is $h_1 = 10.0$, width of disturbance $(wd) = 2000m$, the tidal period $T = 720$ minutes. Constant for sediment transport rate $K = 0.16 kgs^3/m^6$, porosity $p0 = 0.5$, $\rho = 2600[kg/m^3]$, density of sediment, horizontal domain length of x goes from $-10000m$ to $10000m$. The horizontal domain length of y goes from $2m$ to $4500m$.

Diffusion constant $D = 10m^2/s$, starting time $Tstart = 0s$, number of seconds in a year $Tyear = 365 \cdot 24 \cdot 60 \cdot 60s/year$. Final time s $Tstop = 100 \cdot Tyear$, $M = 50$ is the number of grid points across the channel, $dt = 1.0 \cdot year$ is the time-step s. Tidal mean discharge per unit width m^2/s is given by $q_0 = h_0 \cdot u_0$, tidal amplitude discharge per unit width m^2/s is given by $q_1 = h_0 \cdot u_1$. We then determine the changes in the bedforms as described in Section 3.1. The rate of the changes in the level of the floor of the sea is described by the divergence of transport in each grid cell. For comparison purposes, the divergence of transport is we computed by using the original data from the hydrodynamic model as described in this section.

Note that the two dimensional channel in a Matlab code is used whereby in the routine, we first compute the flow fields U, V as well as its depths in each grid cell. Then compute average load transport and finally compute the divergence using a finite difference methods in a Matlab code. Where the divergence of the U-vector field along x and the V-vector field along y is evaluated using central differences wherever possible and forward or backward differences on the boundaries. Note that the average change of the bedforms in each grid cell are estimated by using Equation (10) in Section 3.1. The results obtained are eventually compared with those obtained when the estimations of the changes of bedforms are carried out using Equation (28) with the aid of the two dimensional SDEs or sometimes in this chapter we call it as the Lagrangian particle model. This stochastic model is discussed in Section 7. Note that both results due to Equation (10) and Equation (28) are similar interms of shapes, (see Fig. 2).

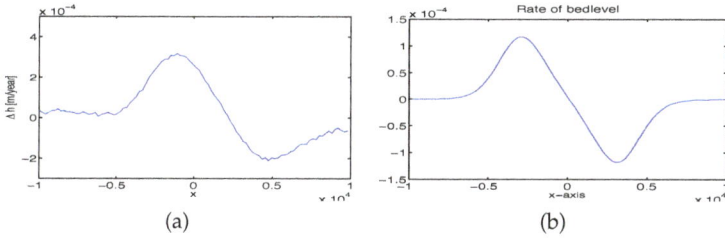

(a)   (b)

Figure 1. Change of bed level in $m/year$ for a two dimensional channel (a) is due to the particle model, while (b) the result is computed by using the Eulerian approach.

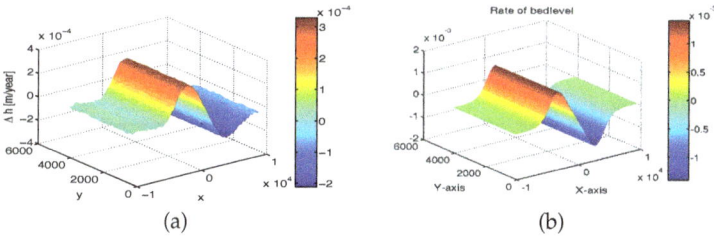

(a)   (b)

Figure 2. Change of bed level in $m/year$ for a two dimensional channel (a) is due to the particle model, while (b) the result is computed by using the Eulerian approach.

The Figures 1 and 2 represent the results obtained by solving the same problem by using two approaches. Fig. 1(a) and Fig. 2(a) are due to a simplified particle model with very small effect of the diffusion. On the other hand, in Figures 1(b) and Fig. 2(b) are due to a simplified Eulerian model, no diffusion is considered. We should expected to get deposition at the retardation of the flow and erosion at the acceleration of the flow. Resulting in the net migration of the channels in the direction of $(\overline{U^3})$ as in [19]. The positive sign on the colorbar in the figures imply that deposition is taking place while the negative sign implying the occurrence of erosion of sediments.

Some results, in Fig. 3(a,b), represent the local depth change in two selected cells. Part (a) shows a steady deposition in the grid cell at the location $(x, y) = (5km, 2km)$ whereas another grid cell in the location $(x, y) = (-2km, 0.8km)$, part (b), there is also a steady deposition. The diffusion coefficient in the test case for the particle model is $0.00001m^2/s$, $\gamma = 0.00013s^{-1}$.

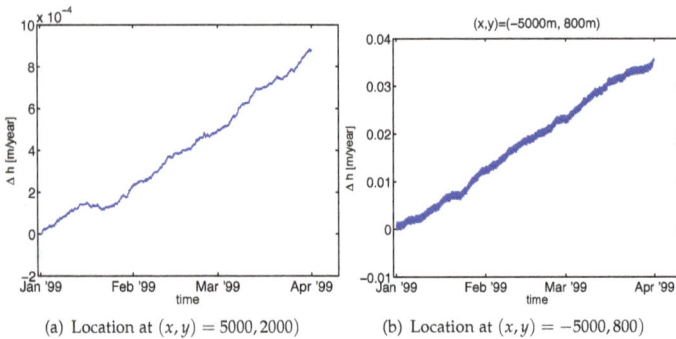

(a) Location at $(x, y) = 5000, 2000$

(b) Location at $(x, y) = -5000, 800$

Figure 3. Cumulative local changes in depth by using data of 90 days for two selected grid cells in the ideal two dimensional channel domain light particles

The parameters for the test case experiments using ideal domain with the PM are found in (Table 1.)

Constant	Unit	Value	Constant	Unit	Value
porosity	-	0.50	γ	s^{-1}	0.00013
grid offset	m	$(-10000, 2)$	λ_s	$kg{\cdot}s{\cdot}m^{-4}$	0.001
grid size	-	22×101	D	$m^2 \cdot s^{-1}$	0.000010
cell size	m	200×200	δ	-	0
sand density	kgm^{-3}	2650	$\mathcal{M}_p$	kg	3000
initial location	m	$(1800, 400)$	$f = \frac{\lambda_s}{\gamma}$	$kg \cdot s^2 \cdot m^{-4}$	$\mathcal{O}(10^{-4})$

Table 1. Parameters used by particle model for the sediment transport in the test case

8. Transport of heavy particles in shallow water

As mentioned earlier, sediment transport is a complex process determined by various properties of the sediment materials. Let us in this section consider that we deal with heavy particles. Heavy particles unlike the lighter ones, tend to attain the equilibrium much faster than the tidal cycle. The description of the derivation of the equations for the heavy particles have followed similar lines using the same equations as those for the lighter particles.

$$\frac{\partial(CH)}{\partial t} + \frac{\partial(HUC)}{\partial x} + \frac{\partial(HVC)}{\partial y} = S_e - S_d \tag{29}$$

where in this case $S_e = \lambda_s U^k$ is erosion term, $-S_d = \gamma CH$, is the deposition term

where λ_s, γ, k are constants. In addition, an equation for the change of the bedlevel is presented:

$$\frac{\partial h}{\partial t} = \frac{1}{\rho_s(1 - po)} \cdot (S_e - S_d). \tag{30}$$

However, the difficulty with the heavy (coarser) material is that the parameter γ becomes larger (settling velocity) and this makes the equation (29) 'stiff'. The maximum allowed time step thus becomes very small and since we want to make long simulation this is not very convenient. Therefore, the solution is to make a first order approximation and assume that the two source terms S_e and S_d are much bigger than the right side terms of Equation (29). Therefore, we can write equation (29):

$$\delta \left(\frac{\partial (CH)}{\partial t} + \frac{\partial (HUC)}{\partial x} + \frac{\partial (HVC)}{\partial y} \right) = S_e - S_d, \tag{31}$$

now we substitute a Taylor series expansion of the depth averaged concentration CH

$$CH = Q = Q_0 + \delta Q_1 + \mathcal{O}(\delta^2),$$

into Equation (31) to get the following equations such that 0-order is given by :

$$S_e - S_d = 0, \tag{32}$$

while the 1st order is:

$$\frac{\partial (Q_0)}{\partial t} + \frac{\partial (UQ_0)}{\partial x} + \frac{\partial (VQ_0)}{\partial y}) = -(\gamma Q_1). \tag{33}$$

In other words, in Equation (33), we are looking for a source term (γQ_1) that balances the advective transport of the 0 -order solution. Although this is easy in finite difference approach, however, in the particle model, Equation (33) can be approached as follows. The best we can do so far is to solve:

$$\frac{\partial Q}{\partial t} + \frac{\partial (UQ)}{\partial x} + \frac{\partial (VQ)}{\partial y} = 0, \tag{34}$$

with (32) as an initial condition for every time step separately and then set

$$\gamma Q_1 \approx \frac{\partial Q}{\partial t}. \tag{35}$$

In other words, Equation (33) should be more or less balanced. If we now omit the source term and measures the rate of changes, these should be approximately equal. Therefore for the heavy particles, between the beginning and end of the integration time loop should do the following

(i) First we remove all deposited particles.

ii) Followed by generating particles according to (see Equation 32)

ii) Store net change in number of particles (concentration).

v) Next we do one time step integration of the particle using Equation (34).

v) Compute differences over the previous step using (iii).

For now we have assumed that

$$\left|\frac{\partial Q_0}{\partial t}\right| \ll |\gamma Q_1|.$$

Note that

$$\text{transport vector } q_x = UHC = \frac{\lambda_s}{\gamma}U(U^2 + V^2) \tag{36}$$

$$\text{transport vector } q_y = VHC = \frac{\lambda_s}{\gamma}V(U^2 + V^2), \tag{37}$$

since $\lambda_s U^2 - \gamma CH = 0$, the equation for the rate change of the bedlevel due to the transport coarse material is given by

$$\frac{\partial h}{\partial t} = -\left(\frac{\partial q_x}{\partial x} + \frac{\partial q_y}{\partial y}\right) \cdot \frac{1}{\rho_s(1 - po)}, \tag{38}$$

where $f_d = \frac{\lambda_s}{\gamma}$ is the pick up function which depends into the characteristics of the sediment/sand materials. For example in case of a mixture of larger sand of volume=l^3 will have a mass=$\rho_s l^3$. While a sand of double size whose volume is $8l^3$ will have mass=$\rho_s 8l^3$. Therefore, the two particle require different value of the drag force particle f_d. But in this chapter we assume that all particles have the same mass, l=length.

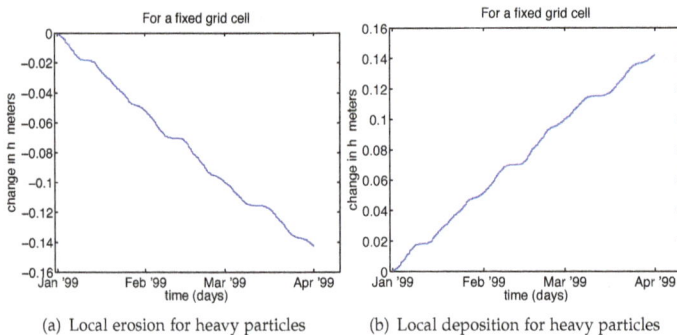

(a) Local erosion for heavy particles   (b) Local deposition for heavy particles

Figure 4. Cumulative local changes in depth by using data of 90 days for two selected grid cells in the ideal two dimensional channel domain for heavy particles

9. Discussion and conclusions

In this chapter we have developed a two dimensional stochastic differential equations. This type of Mathematical model is known as Particle model in some cases. In particular we have developed a two dimensional particle model for sediment transport in which we have added two more equations to model erosion and deposition processes. The description of Transport of materials in shallow waters or in atmosphere can be described by Eulerian approach. This is a deterministic approach in which Partial Differential Equations are used. Numerical methods are usually implemented to approximate the solutions of the PDEs. In this case, one can be faced with computation problems if the dimensional of the model is high. An in most cases if you modify the model to include in more processes it may not be possible to get a closed solution. An alternative approach called the Lagrangian approach can be also used to describe the transport of sediments[35]. This is a probabilistic approach that uses the transition probabilities to derive the state of the transported material. The transition probabilities or density functions is the solution of the Fokker -Planck equation which is also a type of PDEs.

The crucial part is to show that FPE is consistent with ADEs, once that is done then one can derive the underlying SDEs. The derived SDEs can be used as a particle model for the simulation of the transport of materials in the shallow or atmosphere.

In this chapter we have derived the Fokker-Planck equation and included the deposition and erosion terms based on the developed particle model. Moreover, we have also shown that by interpreting the Eulerian sediment transport as the Fokker Planck equation with the additional terms, it becomes possible to derived the underlying particle model that is consistent with the Eulerian sediment transport model. Furthermore, the results of sediment transport due to particle model has been compared to that obtained by computing the transport using Eulerian model in their simplest form. We have got some results for the changes in the bed level for the data of 90 days for an idealized two dimension domain. We have also used our model to test the prediction of bed-level changes by using the approach of parallel computations of bedforms using the real data of the Dutch North sea[5]. Therefore, at least for now we can say that we have solved the set of mathematical equations called particle model for sediment transport. These equations have given us reasonable results for the sea bedlevel changes. Nevertheless, the determination of the morphological changes is a complicated process that depends on several factors such as waves, the size of sand, mass and density[36]. The particle model in this work has been simplified, what we can say is that the results are reasonable. But more factors will have to be taken into account. For instance, in the particle model we have considered that all particles have equal mass while in reality each particle has different mass. For better predictions of the complex behaviour sediment trasnsport that vary with time and space a feedback among **water motion, sediment transport and bottom changes**, we recommend the coupling of both the hydrodynamic and the transport models. Moreover, make sure the grid mesh are of the same form for the hydrodynamic and particle model. Sometimes you may also be required to change the number of particles into concentration. That can be done by using a function called point spread function(PSF).

Author details

Charles Wilson Mahera and Anton Mtega Narsis

1 University of Dar-es-salaam, College of Natural and Applied Sciences, Department of Mathematics, Dar-es-salaam, Tanzania
Department of General Studies, Dar es Salaam Institute of Technology, Dar es Salaam, Tanzania

References

[1] P. S. Addison, Bo Qu, A. Nisbet, and G. Pender.(1997). A non-Fickian particle-tracking diffusion model based on fractional Brownian motion. *International Journal for numerical methods in fluids*, Vol, 25, No. 0, pp. 1373–1384.

[2] B.Øksendal (2005). *Stochastic Differential Equations. An introduction with applications*, 6th Ed, Springer, New York,.

[3] Arnold. L. (1974). *Stochastic differential equations: Theory and applications*, Wiley, London.

[4] R. W. Barber & N. P. Volakos (2005). Modelling depth-integrated pollution dispersion in the gulf of thermaikos using a lagrangian particle technique, *Proceedings of Inter. Confer. on Water Resources Management*, pp. 173–184, ISBN, Portugal, April and 2005, WIT Trans. on Ecol. and the Env., Vol 80, Algarve

[5] W. M. Charles, E.van den Berg, H. X. Lin, A. W. Heemink, and M. Verlaan (2008). Parallel and distributed simulation of sediment dynamics in shallow water using particle decomposition approach. *International Journal of Parallel and Distributed Computing*,Vol 68 717Ú728.

[6] H. B. Fischer, E.J. List, R.C.Y. Koh, J. Imberger, & N.H. Brooks (1979). *Mixing in Inland and Coastal Waters*, 6th Ed,Academic Press, New York.

[7] C. W. Gardiner, (2004). *Handbook of Stochastic Methods for physics, Chemistry and the Natural Sciences*, Springer-Verlag, Berlin.

[8] J. Fredsøe and R. Deigaard, (1992). *Mechanics of Coastal Sediment transport*, World Scientific, Singapore.

[9] A. W. Heemink (1990). Stochastic modeling of dispersion in shallow water *Stochastic Hydrology and hydraulics*, Vol. 4, :161-174

[10] A. H. Jazwinski, (1970). *Stochastic Processes and Filtering Theory*, Academic Press, New York.

[11] P. E. Kloeden and E. Platen, (1999). *Numerical solutions of Stochastic Differential equations. Application of Mathematics,Stochastic Modelling and applied probability*,Springer-Verlag, Berlin Heidelberg.

[12] J. Fredsøe and R. Deigaard, (1992). *Mechanics of Coastal Sediment transport*, World Scientific, Singapore.

[13] A. Matheja (2000). Sediment transport using the random walk method on top of cobra *Proc. of the 4th of Inter. Confer. on Hydroinformatics, Iowa City, USA*, 1–7

[14] H. M. Schuttelaars and H. E. de Swart (1996). An idealized long-term morhpodynamic model of a tidal embayment *Eur. J.Mech., B/Fluids* Vol. 15(1), :55–80

[15] G. S. Stelling (1983). Communications on construction of computational methods for shallow water flow problems *PhD Dissertation*, Delft University of Technology, The Netherlands

[16] K. de Jong (1998). Tidally Averaged Transport Models *PhD Dissertation*, Delft University of Technology, The Netherlands.

[17] A. F. B. Tompson and L. W. Grelhar(1990). Numerical simulation of the solute transport in three-dimensional,randomly heterogeneous porous media *Water Resources Research*, Vol. 26(10) :2541–2562.

[18] H. F. P. van den Boogaard, M. J. J . Hoogkamer, and A. W. Heemink (1993). Parameter identification in particle models *Stochastic Hydrology and hydraulics*, Vol. 7(2), :109-130.

[19] J. van den Kreeke, S. E. Hoogewoning, and M.Verlaan,H. F. P. van den Boogaard, M. J. J . Hoogkamer, and A. W. Heemink (1993). Analytical model for the morphodynamics of a trench in the presence of tidal currents *Continental Shelf Research*, Vol. 22(11-12), :1811-1820.

[20] L. C. van Rijn (1993). Principles of sediment transport in rivers, estuaries and

[21] Amoudry, L. O., and A. J. Souza, A. J. (2011) .Deterministic coastal morphological and sediment transport modelling: A review and discussion", Rev. Geophys., 49, RG2002, doi:10.1029/2010RG000341.

[22] Bhuiyan, F. 2011 Stochastic and Deterministic Methods of Computing Graded Bedload Transport, Sediment Transport - Flow and Morphological Processes", Prof. Faruk Bhuiyan (Ed.), ISBN: 978-953-307-374-3.

[23] Brown, J. M., & Davies, A. G. (2009) Methods for medium term prediction of the net sediment transport by waves and currents in complex coastal regions", Cont. Shelf Res., 29, pp.1502-1514.

[24] de Vriend, H. J. (1987)2DH mathematical modelling of morphological evolutions in shallow water, *Coastal Eng.*, Vol 11, pp. 1-27.

[25] Dewals, B, Rulot, F., Erpicum, S., Archambeau, P. & Pirotton, M. (2011) *Advanced Topics in Sediment Transport Modelling: Non-alluvial Beds and Hyperconcentrated Flows, Sediment Transport*, Dr. Silvia Susana Ginsberg (Ed.), ISBN: 978-953-307-189-3.

[26] Frey, P. & Church, M. (2009) How River Beds Move, *Science*, Vol 325(5947), pp. 1509 - 1510.

[27] Harris, C. K., Sherwood, C. R., Signell, R. P., Bever, A. J., & Warner, J. C. (2008) Sediment dispersal in the northwestern Adriatic Sea, *J. Geophys. Res.*, 113, C11S03, doi:10.1029/2006JC003868.

[28] Hu, K., Ding, P., Wang, Z., and Yang, S. (2009) A 2D/3D hydrodynamic and sediment transport model for the Yangtze Estuary, China, *J. Mar. Syst.*, Vol 77, pp. 114-136.

[29] Lumborg, U. (2005) Modelling the deposition, erosion, and flux of cohesive sediment through Øresund, *J. Mar. Syst.*, Vol 56, pp. 179-193.

[30] Man, C. (2007) *Stochastic Modelling of Suspended Sediment Transport in Regular and Extreme Flow Environments*, PhD Dissertation, State University of New York at Buffalo, p. 182.

[31] Németh, A. A. & Hulscher, S. J. M. H., (2003) Finite amplitude sand waves in shallow seas, *IAHR-symposium on River, Coastal and Estuarine Morphodynamics*, Barcelona, Spanje, editors Sánchez-Arcilla, S. en Bateman, A., pp. 435-444.

[32] Rodean, H.C. (1996) Stochastic Lagrangian Models of Turbulence Diffusion, *American, Meteorological Society*, 84pp, Boston.

[33] Struiksma, N., Olesen, K. W., Flokstra, C., & de Vriend, H. J. (1985) Bed deformation in curved alluvial channels, *J. Hydraul. Res.*, Vol 23(1), pp. 57-79.

[34] Papanicolauo, A. N., Elhakeem, M., Krallis, G., Prakash, S., & Edinger, J., (2008) Sediment Transport Modeling Review-Current and Future Developments, *Journalof Hydraulic Engineering, ASCE, 14pgs*

[35] Hunter, J. R,Crais, P.D and Phillips H. E.(1993) On the use of random walk models with spatially variable diffusivity.*Computation Physics*, 106:366–376, 1993.

[36] H. M. Schuttelaars and H. E. de Swart (2000) Multiple morphodynamic equilibria in tidal embayments.*J. Geophysical Research*, 105(c10):14,105–14,118.

Chapter 13

Numerical Modeling of Flow and Sediment Transport in Lake Pontchartrain due to Flood Release from Bonnet Carré Spillway

Xiaobo Chao, Yafei Jia and A. K. M. Azad Hossain

Additional information is available at the end of the chapter

1. Introduction

Lake Pontchartrain is a brackish estuary located in southeastern Louisiana, United States. It is the second-largest saltwater lake in U.S. The lake covers an area of 1630 square km with a mean depth of 4.0 meters. It is an oval-shaped quasi-enclosed water body with the main east-west axis spanning 66 km, while the shorter north–south axis is about 40 km. It is connected to the Gulf of Mexico via Rigolets strait, to Lake Borgne via Chef Menteur Pass, and to Lake Maurepas via Pass Manchac. These lakes form one of the largest estuaries in the Gulf Coast region. It receives fresh water from a few rivers located on the north and northwest of the lake. The estuary drains the Pontchartrain Basin, an area of over 12,000 km^2 situated on the eastern side of the Mississippi River delta plain.

Lake Pontchartrain has served the surrounding communities for more than two centuries. The coastal zone of the Lake and its basin has offered opportunities for fishing, swimming, boating, crabbing and other recreational activities. The Lake Basin is Louisiana's premier urban estuary and nearly one-third of the state population live within this area. Over the past decades, rapid growth and development within the basin have resulted in significant environmental degradation and loss of critical habitat in and around the Lake. Human activities associated with pollutant discharge and surface drainage have greatly affected the lake water quality (Penland et al., 2002).

In order to protect the city of New Orleans from the Mississippi River flooding, the Bonnet Carré Spillway (BCS) was constructed from 1929 to 1936 to divert flood water from the river into Lake Pontchartrain and then into the Gulf of Mexico. However, a BCS opening event may cause many environmental problems in the lake. To evaluate the environmental impacts of the flood water on lake ecosystems, it is important to understand the hydrodynamics as well as sediment transport in the lake.

Lake Pontchartrain is a large shallow lake and the water column is well mixed. In general, the water movements within the lake are affected by wind and tide. During the BCS opening for flood release, the flow discharge over the spillway produces significant effects on the lake hydrodynamics.

Numerical models generally are cost-effective tools for predicting the flow circulation and pollutant transport in a lake environment. In recent years, numerical models have been applied to simulate the flow and pollutant distribution in Lake Pontchartrain. Hamilton et al. (1982) developed a 2D-vertical integrated model to simulate the flow circulation in Lake Pontchartrain. The model is an explicit numerical difference scheme based on leap-frog integration algorithm, and a coarse uniform mesh with a spacing of 1 km was selected for numerical simulation. Signell (1997) applied the coastal and ocean model (ECOM), developed by Hydroqual to simulate the tide and wind driven circulation processes in the lake. It was found that water levels in Mississippi Sound influence the circulation patterns in the eastern part of the lake, while the wind force dominates the flow pattern of the western part. McCorquodale et al. (2005) applied a 3D coastal ocean model, ECOMSED, to simulate the flow fields and mass transport in Lake Pontchartrain. Dortch et al. (2008) applied the CH3D-WES hydrodynamic model, developed by the US Army Corps of Engineers, to simulate the lake flow field and post-Hurricane Katrina water quality due to the large amount of contaminated floodwater being pumped into the Lake. McCorquodale et al. (2009) developed a 1D model for simulating the long term tidal flow, salinity and nutrient distributions in the Lake. Most of those researches focus on hydrodynamics and pollutant transport in the Lake.

In this study, the flow fields and sediment transport in Lake Pontchartrain during a flood release from BCS was simulated using the computational model CCHE2D developed at the National Center for Computational Hydroscience and Engineering (NCCHE), the University of Mississippi (Jia and Wang 1999, Jia et al. 2002). This model can be used to simulate free surface flows and sediment transport, and the capabilities were later extended to simulate the water quality, pollutant transport and contaminated sediment (Chao et al. 2006, Zhu et al. 2008). CCHE2D is an integrated numerical package for 2D-depth averaged simulation and analysis of flows, non-uniform sediment transport, morphologic processes, water quality and pollutant transport. There are several turbulence closure schemes available within the model for different purposes, including the parabolic eddy viscosity, mixing length, k–ε and nonlinear k–ε models. A friendly Graphic User Interface (GUI) is available to help users to setup parameters, run the simulation and visualize the computational results. In addition to general data format, CCHE2D has capabilities to produce the simulation results in ArcGIS and Google Earth data formats (Hossain et al., 2011). Those capabilities greatly improve the model's applications.

The simulated flow and sediment distribution during the BCS opening were compared with satellite imagery and field measured data provided by the United States Geological Survey (USGS) and the United States Army Corp of Engineers (USACE). Good agreements were obtained from the numerical model. This model provides a useful tool for lake water quality management.

2. Bonnet Carré Spillway (BCS) opening for flood release

In response to the high flood stage of the Mississippi River and to protect the city of New Orleans, the Bonnet Carré Spillway (BCS) was built to divert Mississippi River flood waters to the Gulf of Mexico via Lake Pontchartrain (Fig. 1).

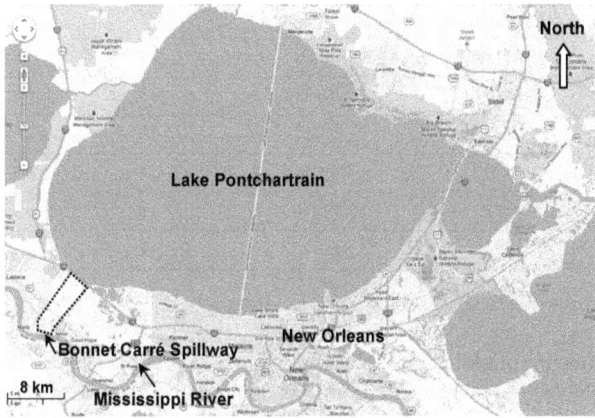

Figure 1. The location of Bonnet Carré Spillway (BCS)

The construction of the spillway was completed in 1931. It is located in St. Charles Parish, Louisiana - about 19 km west of New Orleans. The spillway consists of two basic components: a 2.4 km long control structure along the east bank of the Mississippi River and a 9.7 km floodway that transfers the diverted flood waters to the lake. The design capacity of the spillway is 7080 m³/s and will be opened when the Mississippi river levels in New Orleans approached the flood stage of 5.2 m. It was first operated in 1937 and nine times thereafter (1945, 1950, 1973, 1975, 1979, 1983, 1997, 2008 and 2011). The maximum flow discharges and days of opening for each event are listed in Table 1 (USACE 2011; GEC 1998).

During the BCS opening, a large amount of fresh water and sediment discharged from the Mississippi River into Lake Pontchartrain and then into the Gulf of Mexico. The flow discharge over the spillway produces significant effects on the lake hydrodynamics. It also changes the distributions of salinity, nutrients and suspended sediment (SS) in the lake dramatically. During a flood releasing event, the fresh water dominated the whole lake and the lake salinity reduced significantly. A lot of sediment deposited into the lake or was transported into the Gulf of Mexico. The contaminated sediment from Mississippi River could bring a lot of pollutants, such as nutrients, Al, Cu, Cr, Hg, Pb, Zn, etc., to the lake, and caused a lot of environmental problems. The algal bloom occurred in a large area of the lake after a flood release event. The blooms produced high levels of heptatoxins and caused decreases of dissolved oxygen in the lake (Dortch et al., 1998; Penland et al., 2002).

Year	Date opened	Date Closed	Days opened	Max. discharge m³/s
1937	Jan28	Mar 16	48	5975
1945	Mar 23	May 18	57	9005
1950	Feb 10	Mar 19	38	6315
1973	Apr 8	Jun 21	75	5522
1975	Apr 14	Apr 26	13	3115
1979	Apr 17	May 31	45	5409
1983	May 20	Jun 23	35	7589
1997	Mar 17	Apr 18	31	6881
2008	Apr 11	May 8	28	4531
2011	May 9	June 20	42	8892

Table 1. Information of Bonnet Carré Spillway opening for flood release

Due to a large amount of sediment discharged /deposited into the lake, the bed form of the lake changed. The BCS opening event produced significant changes in flow pattern, salinity and water temperature, which greatly affected the lake fish habitat, and caused negative impacts to oyster beds and fishery nursery grounds in the lake. In response to the dynamic changes in the salinity, temperature, water surface elevation, and bed form of the lake, it was observed that some species, particularly brown shrimp, shifted and moved. It may take a long time for the fisheries resources to recover from the flood release event.

To understand the impact of the BCS flood release event on the ecosystem of Lake Pontchartrain, the flow circulation and sediment transport are most important key tasks to be studied.

3. Model descriptions

To simulate the flow field and sediment transport in Lake Pontchartrain, a two-dimensional depth-averaged model, CCHE2D, was applied. CCHE2D is a 2D hydrodynamic and sediment transport model that can be used to simulate unsteady turbulent flows with irregular boundaries and free surfaces (Jia and Wang 1999, Jia et al. 2002). It is a finite element model utilizing a special method based on the collocation approach called the "efficient element method". This model is based on the 2D Reynolds-averaged Navier-Stokes equations. By applying the Boussinesq approximation, the turbulent stress can be simulated by the turbulent viscosity and time-averaged velocity. There are several turbulence closure schemes available within CCHE2D, including the parabolic eddy viscosity, mixing length, k–ε and nonlinear k–ε models. In this model, an upwinding scheme is adopted to eliminate oscillations due to advection, and a convective interpolation function is used for this purpose due to its simplicity for the implicit time marching scheme which was adopted in this model to solve the unsteady equations. The numerical scheme of this approach is the second order. The velocity correction method is applied to solve the pressure and enforce mass conservation. Provisional velocities are solved first without the pressure term, and the final solution of the velocity is obtained by correcting the provisional velocities with the pressure solution. The system of the algebraic equations is solved using the Strongly Implicit Procedure (SIP) method (Stone 1968).

3.1. Governing equations

The free surface elevation of the flow is calculated by the continuity equation:

$$\frac{\partial h}{\partial t} + \frac{\partial uh}{\partial x} + \frac{\partial vh}{\partial y} = 0 \tag{1}$$

The momentum equations for the depth-integrated two-dimensional model in the Cartesian coordinate system are:

$$\frac{\partial u}{\partial t} + u\frac{\partial u}{\partial x} + v\frac{\partial u}{\partial y} = -g\frac{\partial \eta}{\partial x} + \frac{1}{h}\left(\frac{\partial h\tau_{xx}}{\partial x} + \frac{\partial h\tau_{xy}}{\partial y}\right) + \frac{\tau_{sx} - \tau_{bx}}{\rho h} + f_{Cor}v \tag{2}$$

$$\frac{\partial v}{\partial t} + u\frac{\partial v}{\partial x} + v\frac{\partial v}{\partial y} = -g\frac{\partial \eta}{\partial x} + \frac{1}{h}\left(\frac{\partial h\tau_{yx}}{\partial x} + \frac{\partial h\tau_{yy}}{\partial y}\right) + \frac{\tau_{sy} - \tau_{by}}{\rho h} - f_{Cor}u \tag{3}$$

where u and v are the depth-integrated velocity components in x and y directions, respectively; t is the time; g is the gravitational acceleration; η is the water surface elevation; ρ is the density of water; h is the local water depth; f_{Cor} is the Coriolis parameter; τ_{xx}, τ_{xy}, τ_{yx} and τ_{yy} are depth integrated Reynolds stresses; and τ_{sx} and τ_{sy} are surface share stresses in x and y directions, respectively; and τ_{bx} and τ_{by} are shear stresses on the interface of flow and bed in x and y directions, respectively.

The turbulence Reynolds stresses in equations (2) and (3) are approximated according to the Bousinesq's assumption that they are related to the main rate of the strains of the depth-averaged flow field and an eddy viscosity coefficient v_t which is computed using the Smagorinsky scheme (Smagorinsky 1993):

$$v_t = \alpha \Delta x \Delta y \left[\left(\frac{\partial u}{\partial x} \right)^2 + \frac{1}{2} \left(\frac{\partial v}{\partial x} + \frac{\partial u}{\partial y} \right)^2 + \left(\frac{\partial v}{\partial y} \right)^2 \right]^{1/2}$$

(4)

The parameter α ranges from 0.01 to 0.5. In this study, it was taken as 0.1.

In CCHE2D model, three approaches are adopted to simulate non-uniform sediment transport. One is the bed load transport, which is to simulate the bed load only without considering the diffusion of suspended load. The second approach is the suspended load transport, which simulates suspended load and treats bed-material load as suspended load. The third approach is to simulate bed load and suspended load separately (Jia and Wang 1999, Jia et al. 2002, Wu 2008).

In this study, CCHE2D was used to simulate sediment transport in Lake Pontchartrain during the BCS opening for flood release. In this period, sediment transport in the lake is primarily dominated by suspended sediment. So the second sediment transport approach, suspended load, was used for this study, and the non-uniform suspended sediment (SS) transport equation can be written as:

$$\frac{\partial c_k}{\partial t} + u \frac{\partial c_k}{\partial x} + v \frac{\partial c_k}{\partial y} = \frac{\partial}{\partial x} \left(D_{cx} \frac{\partial c_k}{\partial x} \right) + \frac{\partial}{\partial y} \left(D_{cy} \frac{\partial c_k}{\partial y} \right) + S_{ck}$$

(5)

Where c_k is the depth-averaged concentration of the kth size class of SS; D_{cx} and D_{cy} are the mixing coefficients of SS in x and y directions, respectively; S_{ck} is the source term and can be calculated by:

$$S_{ck} = -\frac{\alpha_t \omega_{sk}}{h} (c_k - c_{t*k})$$

(6)

Where c_{t*k} is the equilibrium sediment concentration of the kth size class of suspended load; ω_{sk} is the settling velocity of the kth size class; α_t is the adaptation coefficient of suspended load, and it can be estimated using the formula proposed by Wu (2008).

Settling velocity is calculated using Zhang's formula (Zhang and Xie 1993):

$$w_{sk} = \sqrt{ \left(13.95 \frac{v}{d_k} \right) + 1.09 \left(\frac{\gamma_s}{\gamma} - 1 \right) g d_k } - 13.95 \frac{v}{d_k}$$

(7)

where v is the kinematic viscosity; d_k is the diameter of the kth size class of sediment; γ_s and γ are the densities of water and sediment; g is the gravity acceleration.

The equilibrium sediment concentration c_{i*k} can be calculated based on sediment transport capacities of fractional suspended load and bed load. Based on field and laboratory data, Wu et al (2000) proposed a formula to calculate the fractional suspended load transport capacity φ_{sk}:

$$\varphi_{sk} = 0.0000262 \left[\left(\frac{\tau}{\tau_{ck}} - 1 \right) \frac{U}{\omega_{sk}} \right]^{1.74} \tag{8}$$

where U is the depth-averaged velocity; τ is the shear stress; τ_{ck} is the critical shear stress and can be calculated by

$$\tau_{ck} = 0.03(\gamma_s - \gamma)d_k \left(\frac{p_{hk}}{p_{ek}} \right)^{0.6} \tag{9}$$

in which p_{hk} and p_{ek} are the hiding and exposure probabilities for the k-th size class of sediment, they can be defined as:

$$p_{hk} = \sum_{j=1}^{N} p_{bj} \frac{d_j}{d_k + d_j} \tag{10}$$

$$p_{ek} = \sum_{j=1}^{N} p_{bj} \frac{d_k}{d_k + d_j} \tag{11}$$

where N is the total number of particle size classes in the non-uniform sediment mixture; p_{bj} is the probability of particles d_j staying in front of particles d_k. A relationship between p_{hk} and p_{ek} is known as:

$$p_{hk} + p_{ek} = 1 \tag{12}$$

In Eq. (8) φ_{sk} can also be expressed by:

$$\varphi_{sk} = \frac{q_{s*k}}{p_{bk} \sqrt{\left(\frac{\gamma_s}{\gamma} - 1 \right) g d_k^3}} \tag{13}$$

in which p_{bk} is the bed material gradation; and q_{s*k} is the equilibrium transport rate of the k-th size class of suspended load per unit width. Based on Eqs. (8) and (13), the following equation can be obtained to calculate q_{s*k} :

$$q_{s*k} = \varphi_{sk} p_{bk} \sqrt{\left(\frac{\gamma_s}{\gamma} - 1\right) g d_k^3} = 0.0000262 \left[\left(\frac{\tau}{\tau_{ck}} - 1\right) \frac{U}{\omega_{sk}}\right]^{1.74} p_{bk} \sqrt{\left(\frac{\gamma_s}{\gamma} - 1\right) g d_k^3} \qquad (14)$$

Wu et al. (2000) also proposed a formula to calculate the fractional bed load transport capacity φ_{bk}:

$$\varphi_{bk} = 0.0053 \left[\left(\frac{n'}{n}\right)^{1.5} \frac{\tau}{\tau_{ck}} - 1\right]^{2.2} \qquad (15)$$

where n is the Manning's roughness coefficient; n' is the Manning's coefficient corresponding to the grain roughness, $n' = d_{50}^{1/6}/20$; the transport capacity φ_{bk} can also be expressed as:

$$\varphi_{bk} = \frac{q_{b*k}}{p_{bk} \sqrt{\left(\frac{\gamma_s}{\gamma} - 1\right) g d_k^3}} \qquad (16)$$

in which q_{b*k} is the equilibrium transport rate of the k-th size class of bed load per unit width. Based on Eqs. (15) and (16), the following equation can be obtained to calculate q_{b*k}:

$$q_{b*k} = \varphi_{bk} p_{bk} \sqrt{\left(\frac{\gamma_s}{\gamma} - 1\right) g d_k^3} = 0.0053 \left[\left(\frac{n'}{n}\right)^{1.5} \frac{\tau}{\tau_{ck}} - 1\right]^{2.2} p_{bk} \sqrt{\left(\frac{\gamma_s}{\gamma} - 1\right) g d_k^3} \qquad (17)$$

Based on Eqs.(14) and (17), the equilibrium sediment concentration c_{t*k} in Eq. (6) can be calculated by:

$$c_{t*k} = \frac{q_{s*k} + q_{b*k}}{Uh} = \frac{p_{bk} \sqrt{\left(\frac{\gamma_s}{\gamma} - 1\right) g d_k^3}}{Uh} \left\{0.0000262 \left[\left(\frac{\tau}{\tau_{ck}} - 1\right) \frac{U}{\omega_{sk}}\right]^{1.74} + 0.0053 \left[\left(\frac{n'}{n}\right)^{1.5} \frac{\tau}{\tau_{ck}} - 1\right]^{2.2}\right\} \qquad (18)$$

The wind shear stresses (τ_{sx} and τ_{sy}) at the free surface are expressed by

$$\tau_{sx} = \rho_a C_d U_{wind} \sqrt{U_{wind}^2 + V_{wind}^2} \qquad (19)$$

$$\tau_{sy} = \rho_a C_d V_{wind} \sqrt{U_{wind}^2 + V_{wind}^2} \qquad (20)$$

where ρ_a is the air density; U_{wind} and V_{wind} are the wind velocity components at 10 m elevation in x and y directions, respectively. Although the drag coefficient C_d may vary with wind speed (Koutitas and O'Connor 1980; Jin et al. 2000), for simplicity, many researchers assumed the drag coefficient was a constant on the order of 10^{-3} (Huang and Spaulding 1995; Rueda and Schladow 2003; Chao et al 2004; Kocyigit and Kocyigit 2004). In this study, C_d was taken as 1.5×10^{-3}.

In this study, the decoupled approach was used to simulate sediment transport. At one time step, the flow fields, including water elevation, velocity components, and eddy viscosity parameters were first obtained using the hydrodynamic model, and then the suspended sediment concentration was solved numerically using Eq. (5).

4. Model verification

4.1. Tide-induced flow

A comparison of model simulation to an analytical solution was performed for a tidally forced flow in a square basin with constant water depth and no bottom friction. It was assumed that right side of the basin is an open boundary, and the other three sides are the closed solid wall. The tidal flow was simulated by driving the free surface elevation at the open boundary of the basin. A standing cosine wave with the maximum amplitude of A_m at the open boundary was introduced. This case has been used by other researchers to verify their models (Huang and Spaulding 1995; Zhang and Gin 2000). The analytical solution for surface elevation (ξ) and velocity (U) were given by Ippen (1966):

$$\xi = A_m \cos(\omega t) \qquad (21)$$

$$U(x,t) = \frac{A_m \omega x}{H} \sin(\omega t) \qquad (22)$$

in which A_m is the tide amplitude; ω is the angular frequency, $\omega = 2\pi/T$; T is the tidal period; t is the time; x is the distance from the left closed basin boundary; U is the velocity in x direction; and H is the base water depth. In the numerical simulation, the following values were adopted: basin length is 12km, width is 12km, H=10m, T=12h, A_m=0.5m. Figs.2 and 3 show the comparison of the analytical solution and the numerical simulation results for water surface elevation and velocity. The numerical results are in good agreement with analytical solutions.

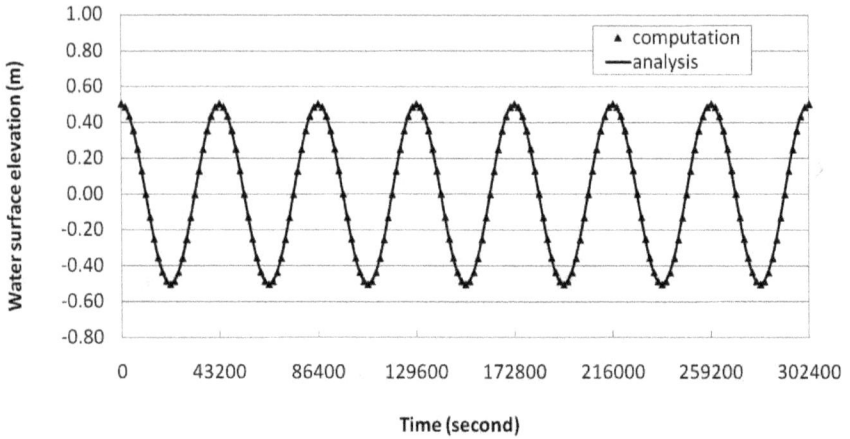

Figure 2. Comparison of analytical solution and numerical simulation for water surface elevation

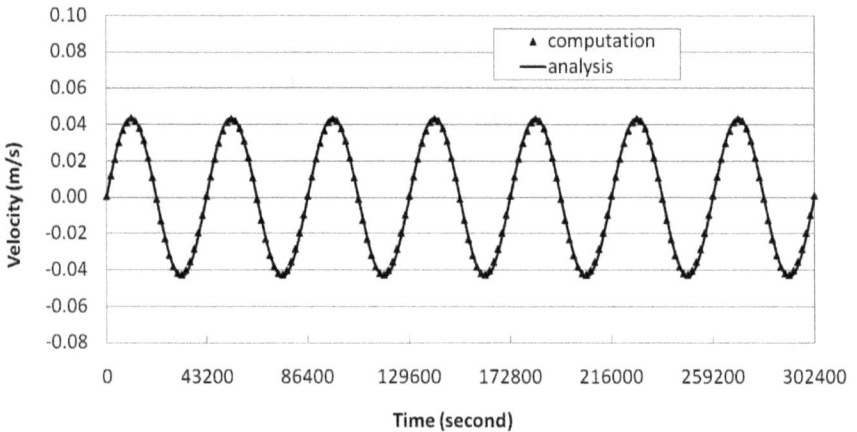

Figure 3. Comparison of analytical solution and numerical simulation for depth-averaged velocity

4.2. Verification of mass transport

To verify the transport simulation model, the numerical results were tested against an analytical solution for predicting salinity intrusion in a one-dimensional river flow with constant depth. It was assumed that the downstream end of the river is connected with the
ocean with salt water. At the end of the river reach ($x=0$), there is a point source with a constant salinity, S_0, from the ocean, and the salt water may intrude into the river due to dispersion (Fig.4). Under the steady-state condition, the salinity in the river can be expressed as:

$$U\frac{\partial S}{\partial x} = D_x \frac{\partial^2 S}{\partial x^2} \tag{23}$$

where U is the velocity (no tidal effect); S is the salinity in river; D_x is the dispersion coefficient; and x is the displacement from downstream seaward boundary (point O). An analytical solution given by Thomann and Mueller (1988) is:

$$S(x) = S_0 \exp(\frac{Ux}{D_x}) \tag{24}$$

in which S_0 is the salinity at downstream seaward boundary. In this test case, it was assumed that the water depth = 10 m, $U=0.03$m/s, $D_x=30$ m²/s, and S_0 = 30 ppt. Fig. 5
shows the salinity concentration distributions obtained by the numerical model and analytical solution. The maximum error between the numerical result and analytical solution
is less than 2%.

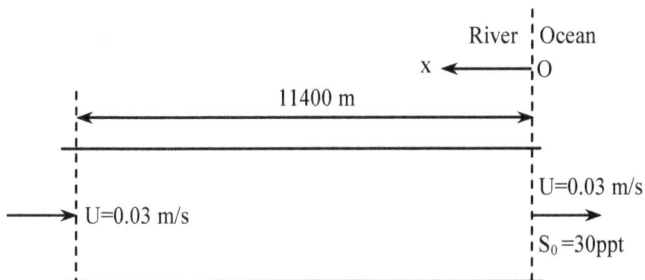

Figure 4. Test river for model verification

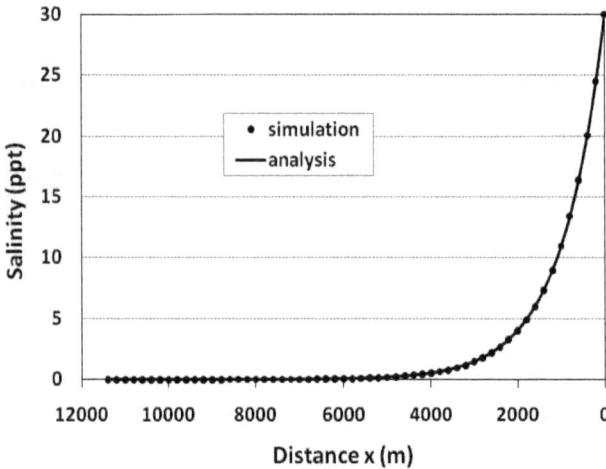

Figure 5. Salinity distribution along the river

5. Model application to lake pontchartrain

5.1. Study area

Fig. 6 shows the bathymetry and locations of field measurement stations of the study site — Lake Pontchartrain. The circulation in Lake Pontchartrain is an extremely complicated system. It is affected by tide, wind, fresh water input, etc. The lake has a diurnal tide with a mean range of 11 cm. Higher salinity waters from the Gulf of Mexico can enter the lake through three narrow tidal passes: the Rigolets, Chef Menteur, and a man-made Inner Harbor Navigation Canal (IHNC). Freshwater can discharge into the lake through the Tchefuncte and Tangipahoa Rivers, the adjacent Lake Maurepas, and from other watersheds surrounding the lake. The Bonnet Carré Spillway (BCS) is located at the southwest of the lake.

Based on the bathymetric data, the computational domain was descritized into an irregular structured mesh with 224×141 nodes using the NCCHE Mesh Generator (Zhang and Jia, 2009).

5.2. Boundary conditions

As shown in Fig.6, there are two inlet boundaries located at the northwest of the lake, and three tidal boundaries located at the south and east of the lake. The flow discharges at Tchefuncte and Tangipahoa Rivers obtained from USGS were set as two inlet boundary conditions. The hourly water surface elevation data at the Rigolets Pass obtained from USGS was set as a tidal boundary. Due to the lack of measured surface elevation data at Chef Menteur Pass, the Rigolets data was used (McCorquodale et al., 2005). After the BCS was opened for flood release, the flow discharge at BCS was set as inlet boundary conditions.

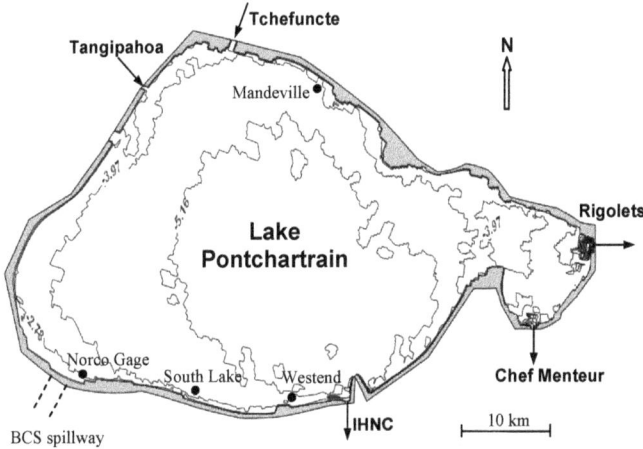

Figure 6. The bathymetry and field measurement stations in Lake Pontchartrain

The other tidal pass, IHNC, is a man-made canal which connects the Lake Pontchartrain and Mississippi River with a lock structure. It is also connected with both the Gulf Intracoastal Waterway and the Mississippi River Gulf Outlet (MRGO). The measured daily water surface elevation data is the only available data at IHNC. In general, the daily water surface elevation data can not represent the variations of tidal boundary. It would cause problems if the measured daily data was directly set as tidal boundary conditions at IHNC. To resolve this problem, the relationship between measured daily water surface elevations at Rigolets and IHNC tidal passes were established and adopted to convert the hourly data at Rigolest to the hourly data at IHNC. Since both IHNC and Rigolets tidal passes are connected with the Gulf of Mexico, the tide effects at these two places are assumed to be similar. By comparing the measured daily water surface elevations at Rigolets and IHNC, no obvious phase differences were observed (McCorquodale et al., 2005). Fig.7 shows the comparison of measured daily water surface elevations at Rigolets and IHNC tidal passes.

The measured results show that the surface elevations at the two locations have a close linear relation with the correlation coefficient r^2 being 0.92:

$$\eta_i = 1.0484\eta_r - 0.0055 \tag{25}$$

Where η_i and η_r are the daily surface elevations at IHNC and Regolets, respectively. It was assumed that the hourly water surface elevations at IHNC and Regolets have the similar relationships, and Eq. (25) was adopted to calculate the hourly water surface elevations at IHNC from the measured hourly surface elevation at Rigolets. The calculated hourly data was plotted together with the measured daily data at IHNC (Fig.8), a similarity distribution was observed. So the calculated hourly surface elevation was used as tidal boundary condition at IHNC.

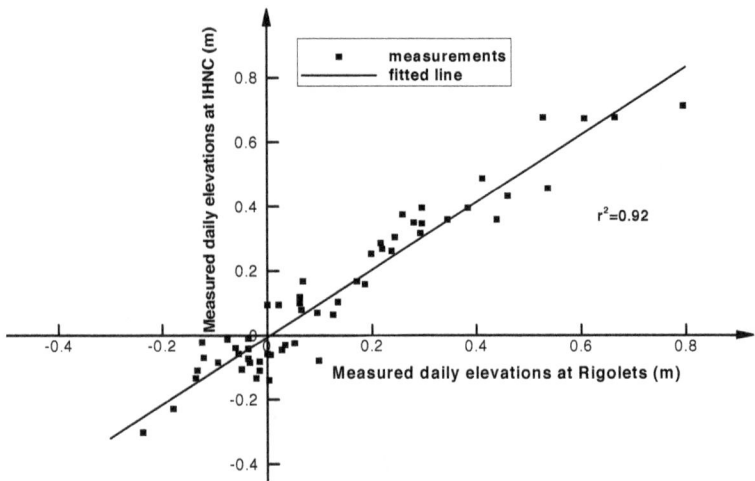

Figure 7. Comparison of measured daily water surface elevations at Regolets and IHNC

5.3. Model calibration

After obtaining the inlet boundaries, outlet boundaries, and wind speeds and directions, the developed model was applied to simulate the flow circulation and sediment transport in Lake Pontchartrain. Some field measured data sets were used for model calibration and validation.

A period from March 1 to 31, 1998, was selected for model calibration. For calibration runs, several parameters, such as drag coefficient C_d, Manning's roughness coefficient, and the parameter α in Smagorinsky scheme (Eq.4), were adjusted to obtain a reasonable reproduction of the field data. In this study, $C_d = 0.0015$, Manning's roughness coefficient = 0.025, and $\alpha = 0.1$. Simulated water surface elevations and depth-averaged velocities were compared with the field measured data. Fig. 9 shows the simulated and measured water surface elevations at the Mandeville. Fig. 10 and Fig.11 show the simulated and measured depth-averaged velocities in x and y directions at the South Lake Site, respectively.

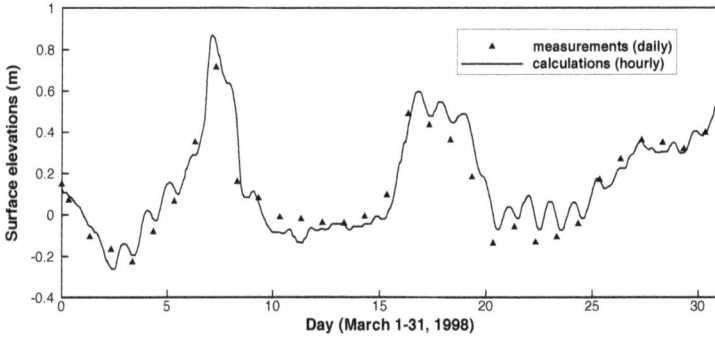

Figure 8. The calculated hourly water surface elevation at IHNC using Eq. (25)

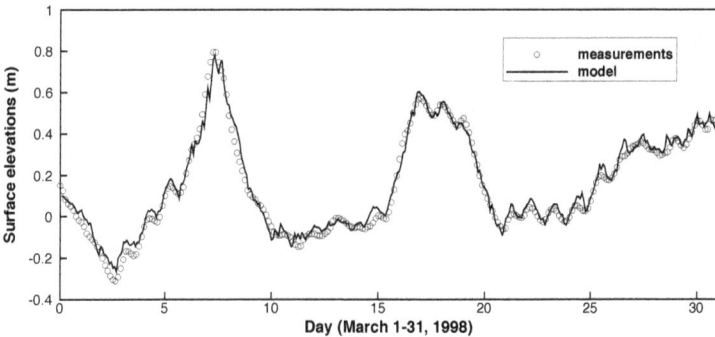

Figure 9. Simulated and measured water surface elevations at the Mandeville Station

A set of statistics error analysis, including root mean square error (RMSE), relative RMSE (RMSE/range of observed data) and correlation coefficient (r^2), were used to assess the performance of the model for the calibration case (Table 2). The RMSE between simulated and observed water surface elevations at Mandeville Station was 0.037m and the relative RMSE of water surface elevations at this station was 3.4%. The r^2 of simulated and observed water surface elevations at this station was 0.98. The measured velocity data set at South Lake station was used for model comparison. It can be observed that the RMSE of u-velocity and v-velocity were 0.044 m/s and 0.019 m/s. The relative RMSE for u-velocity and v-velocity were 15% and 12%; and r^2 of simulated and observed u-velocity and v-velocity were 0.52 and 0.41, respectively. In general, the flow fields produced by the numerical model are in agreement with field measurements.

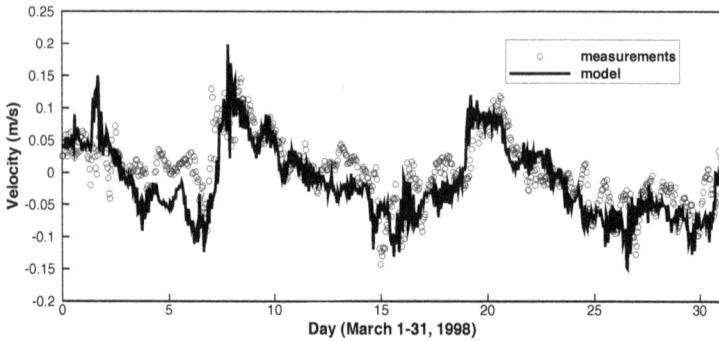

Figure 10. Simulated and measured depth-averaged velocities in west-east direction at the South Lake Site

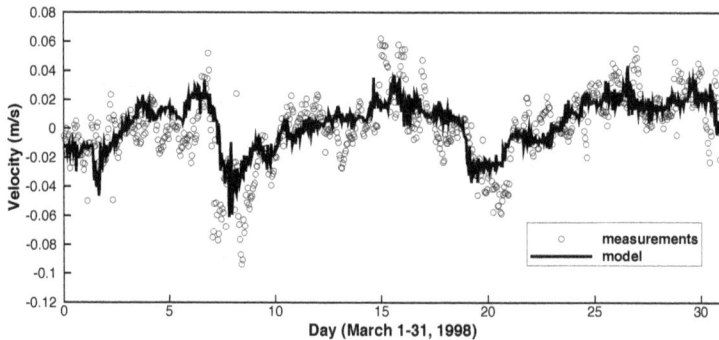

Figure 11. Simulated and measured depth-averaged velocities in south-north direction at the South Lake Site

Station	Variable	RMSE	Observed range	Relative RMSE	r^2
Mandeville	Water level	0.037 m	1.11 m	3.40%	0.98
South Lake	u-velocity	0.044 m/s	0.29 m/s	15%	0.52
South Lake	v-velocity	0.019 m/s	0.16 m/s	12%	0.41

Table 2. Statistics error analysis of the model calibration case

5.4. Modeling the suspended sediment during the BCS opening for 1997 flood release

After the Bonnet Carré Spillway (BCS) was built to divert Mississippi River flood waters to the Gulf of Mexico via Lake Pontchartrain, there were 10 times opening events occurred from 1937 to 2011. In this study, the 1997 flood release event was selected for model simulation.

In 1997, the BCS was opened for flood release from 3/17 to 4/18. The maximum flow discharge was about 6881 m³/s, and over 31 days of flood release. The average discharge was about 4358 m³/s. Fig. 12 shows the flow hydrograph at the spillway (Department of Natural Resources, 1997; McCorquodale et al. 2007). The total volume of sediment-laden water entering Lake Pontchartrain was approximately 1.18×10^{10} m³, or twice the volume of the lake (Turner et al., 1999). The total amount of sediment entering the lake was about 9.1 million tons, more than 10 times as much as the normal yearly sediment loads of the lake. The suspended sediment (SS) concentration at the spillway gate was about 240 mg/l (Manheim and Hayes, 2002).

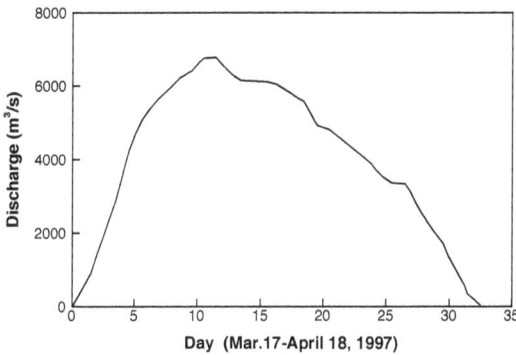

Figure 12. The flow hydrograph at the Bonnet Carré Spillway during the 1997 event

The calibrated CCHE2D model was applied to simulate the lake flow fields and sediment transport during the BCS opening in 1997. In this period, the flow discharge was very strong, and the "suspended load approach" was adopted for simulating sediment transport in the lake. The observed flow discharge was set as inlet boundary condition at BCS. The water surface elevations at Rigolets and Chef Menteur were set as tidal boundaries. The wind speeds and directions at the New Orleans International Airport were used for model simulation. The observed SS concentration was set as inlet sediment boundary condition at BCS. In general, the sediment in Lake Pontchartrain is cohesive sediment. However, during the BCS opening, sediment concentration in Lake Pontchartrain is dominated by the sediment coming from the Mississippi River. It was assumed that the effect of sediment cohesion on suspended sediment transport is not significant. Due to the lack of measured sediment data, the classes of non-uniform sediment size at BCS were estimated based on the observed sediment data in the lower Mississippi River (Thorne et al. 2008). Four size classes, including 0.005mm, 0.01mm, 0.02mm and 0.04mm were assumed to represent the non-uniform sizes of suspended sediment discharged into the lake from BCS. The fall velocity of each size class of sediment was calculated using the Eq. (7) proposed by Zhang and Xie (1993). During this period, the flow discharge over the spillway dominated the lake hydrodynamics and suspended sediment transport. The bottom shear stress due to water flow as well as wind driven flow were obtained using the hydrodynamic model. The critical shear stress was calculated using Eq. (9) proposed by Wu (2008). The equilibrium sediment concentration c_{i*k} was calculated using Eq. (18).

Fig. 13 shows the computed flow circulations in Lake Pontchartrain during the BCS opening. Due to the flood release, the entire lake water were moved eastward through Rigolets and Chef Menteur into the Gulf of Mexico, which was completely different from the flow patterns induced by tide and wind. Fig. 14 shows the comparisons of SS concentration obtained from the numerical simulation and remote sensing imageries (AVHRR data) provided by NOAA. The simulated SS concentrations are generally in good agreement with satellite imageries. The transport processes of SS in the lake were reproduced by the numerical model. The simulated results and satellite imageries revealed that a large amount of sediment discharged into the lake, moved eastward along the south shore and gradually expanded northward, eventually affecting the entire Lake after one month of diversion.

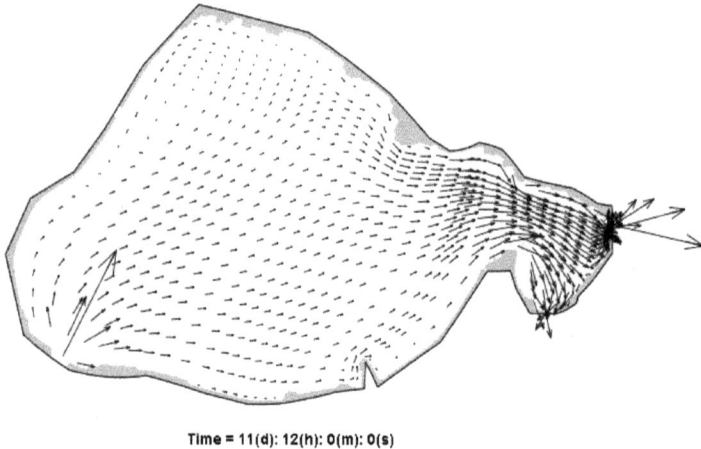

Time = 11(d): 12(h): 0(m): 0(s)

Figure 13. Flow circulations in Lake Pontchartrain during BCS opening in 1997

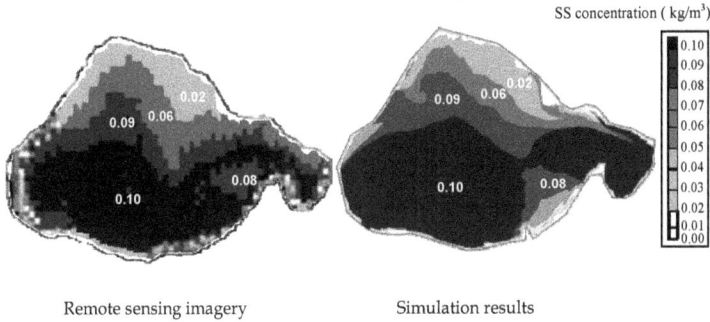

Remote sensing imagery   Simulation results

Figure 14. Comparisons of simulated depth-averaged SS concentration and remote sensing imageries (4/7/1997)

6. Discussion

In general, wind and tide are the major driving mechanisms of circulation in Lake Pontchartrain. When the tidal level changes, most of the water that enters or leaves the lake must come through the three narrow tidal passes at the east and south end of the lake. Since the tidal passes are very narrow, the tidal force may affect the flow fields near the tidal passes. When the wind blows over the lake, it may affect flow circulations of the whole lake. Fig. 15 shows the general flow pattern of the lake induced by tide and wind. It was completely different from the one when the BCS was opened for flood release and caused the entire lake water to be moved eastward into the Gulf of Mexico (Fig. 13). Due to the effects of tide and wind, the stronger currents occur along the shoreline where the water depth is shallow and near the narrow tidal passes, and weaker currents are in the center of the lake. These results are similar to results obtained by other researchers (Signell and List 1997, McCorquodale et al., 2005).

Under the normal condition, sediment in Lake Pontchartrain is dominated by cohesive sediments except for a small number of areas near the river mouths and tidal passes (Flocks et al. 2009). In response to wind and tidal induced flow shown in Fig. 15, sediment may transport/resuspend near shoreline and tidal passes, and deposit in the lake center.

During the BCS opening for flood release in 1997, large amount of sediment discharged from the Mississippi River into Lake Pontchartrain. Sediment transport in Lake Pontchartrain is dominated by the suspended sediment from the Mississippi River, and the effect of sediment cohesion can be ignored. In this period, the flow fields were majorly determined by the flow discharge at the BCS. They were also affected by the wind and tide induced flows. The simulated results and satellite imageries indicated that the suspended sediment moved eastward along the south shore first, gradually expanded northward, and was distributed throughout almost the entire lake about three weeks after the BCS opening. It was reported due to the flood release, a significant amount of nutrients were discharged into

Lake Pontchartrain, which may result in a massive algal bloom in the lake. However, the sediment and nutrients were simultaneously discharged into the lake, and the algae growth rate was restricted as a result of extremely high suspended sediment concentration in the lake (Chao et al 2007). So there was no algal bloom observed in the lake during the BCS opening. After the BCS was closed, the sediment derived from the Mississippi River gradually dispersed in the water column and deposited to the lake bed. In response to the low SS concentration, high nutrients, temperature and light intension, the algal bloom occurred in a large area of the lake, and the peak of the algal bloom observed in mid-June, about two months after the spillway closure (Dortch et al., 1998). In general, after the BCS was closed, it took about two to three months for the SS concentration in the lake recovered to the seasonal average level.

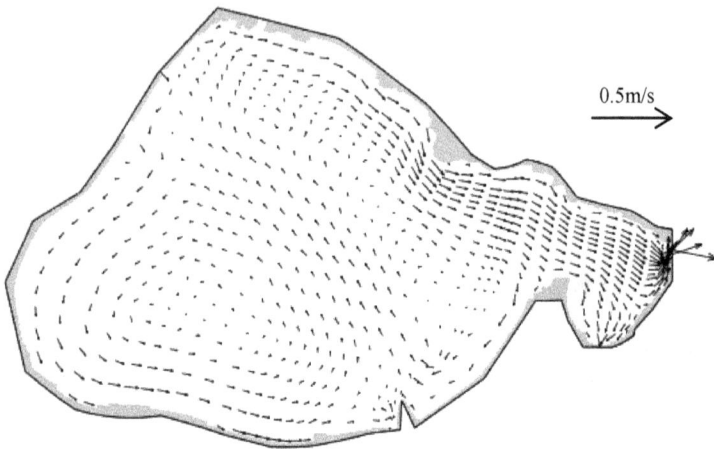

Figure 15. General flow circulations in Lake Pontchartrain due to tide and wind

7. Conclusions

A numerical model was applied to simulate the flow circulations and suspended sediment transport in Lake Pontchartrain in Louisiana, under tide, wind and flood release. It is one of the most significant real life problems we can find with reasonable field measurements obtained from USGS and USACE. Additional satellite imageries were obtained from NOAA for model validation. The results of these comparisons are in good agreement within the accuracy limitations of both the approximate numerical model solutions and the field measurements under the difficult conditions.

In the BCS flood release event, a vast amount of fresh water, sediment and nutrients were discharged into Lake Pontchartrain. The dispersion and transport processes of the suspended sediment in the lake were simulated successfully using the numerical model. The simulated SS concentrations are generally in good agreement with satellite imageries provided by NOAA. The differences of flow circulation and sediment transport in the lake under normal condition and BCS opening event were discussed. The simulated results and satellite imageries show that after the BCS opening, a large amount of sediment discharged into the lake, moved eastward and gradually expanded northward, eventually affecting the entire lake. After the BCS closure, the sediment derived from the Mississippi River gradually deposited to the lake bed and it took two to three months for the SS concentration in the lake recovered to the seasonal average level.

This research effort has positively demonstrated that the numerical model is capable of predicting free surface flow and sediment transport in Lake Pontchartrain under extreme natural conditions. It is a useful tool for providing information on hydrodynamics and sediment transport in such a big lake where the field measurements may not be sufficient. All the information obtained from the numerical model is very important for lake restoration and water quality management.

Acknowledgements

This research was funded by the US Department of Homeland Security and was sponsored by the Southeast Region Research Initiative (SERRI) at the Department of Energy's Oak Ridge National Laboratory. The authors would like to thank Rich Signell and David Walters of the USGS, and George Brown of the USACE for providing field measured data in Lake Pontchartrain. The technical assistance from Yaoxin Zhang, Weiming Wu and suggestions and comments provided by Yan Ding, Tingting Zhu and Kathy McCombs of the University of Mississippi are highly appreciated.

Author details

Xiaobo Chao, Yafei Jia and A. K. M. Azad Hossain

The University of Mississippi, USA

References

[1] Chao, X, Jia, Y, & Shields, D. (2004). Three dimensional numerical simulation of flow and mass transport in a shallow oxbow lake. Proc., World Water & Environmental Resources Congress 2004, ASCE, Resyon, Va. (CD-Rom).

[2] Chao, X, Jia, Y, & Zhu, T. (2006). CCHE_WQ Water Quality Module. Technical Report : NCCHE-TR-The University of Mississippi., 2006-01.

[3] Chao, X, Jia, Y, Shields, D, & Wang, S. S. Y. (2007). Numerical Modeling of Water
     Quality and Sediment Related Processes. *Ecological Modelling*, 201, 385-397.

[4] Department of Natural Resources(1997). Bonnet Carre Spillway Opening. Hydro-
     graphic Data Report, Final Report.

[5] Dortch, M. S, Zakikhani, M, Kim, S. C, & Steevens, J. A. (2008). Modeling water and
     sediment contamination of Lake Pontchartrain following pump-out of Hurricane Ka-
     trina floodwater. Journal of Environmental ManagementS: , 429-442.

[6] Dortch, Q, Peterson, J, & Turner, R. (1998). Algal bloom resulting from the opening of
     the Bonnet Carre spillway in 1997. Fourth Bi-annual Basics of the Basin Symposium.
     United States Geological Survey, , 28-29.

[7] Flocks, J, Kindinger, J, Marot, M, & Holmes, C. (2009). Sediment characterization and
     dynamics in Lake Pontchartrain, Louisiana. Journal of Coastal ResearchSpecial ,
     54(54), 113-126.

[8] Gulf Engineers & Consultants (GEC) ((1998). Biological and Recreational Monitoring
     of the Impacts of 1997 Opening of the Bonnet Carré Spillway Southeastern Louisiana.
     Final Report to U.S Army Corps of Engineers, New Orleans, Louisiana.

[9] Hamilton, G. D, Soileau, C. W, & Stroud, A. D. (1982). Numerical modeling study of
     Lake Pontchartrain. Journal of the Waterways, Port, Coastal and Ocean Division,
     ASCE. , 108, 49-64.

[10] Hossain, A, Jia, Y, Ying, X, Zhang, Y, & Zhu, T. T. (2011). Visualization of Urban Area
     Flood Simulation in Realistic 3D Environment, Proceedings of 2011 World Environ-
     mental & Water Resources Congress, May Palm Springs, California., 22-26.

[11] Huang, W, & Spaulding, M. (1995). 3D model of estuarine circulation and water
     quality induced by surface discharges. *Journal of Hydraulic Engineering*, 121(4),
     300-311.

[12] Ippen, A. T. (1966). Estuary and Coastline Hydrodynamics. McGraw Hill, New York.

[13] Kocyigit, M. B, & Kocyigit, O. (2004). Numerical study of wind-induced currents in
     enclosed homogeneous water bodies. Turkish J. Engineering & Environmental Sci-
     ence, , 28, 207-221.

[14] Koutitas, C, & Connor, O. B., ((1980). Modeling three-dimensional wind-induced
     flows. ASCE, Journal of Hydraulic Division, 106 (11), 1843-1865.

[15] Jia, Y, & Wang, S. S. Y. (1999). Numerical model for channel flow and morphological
     change studies. *Journal of Hydraulic Engineering*, 125(9), 924-933.

[16] Jia, Y, Wang, S. Y. Y, & Xu, Y. (2002). Validation and application of a 2D model to
     channels with complex geometry. *International Journal of Computational Engineering
     Science*, 3(1), 57-71.

[17] Jin, K. R, Hamrick, J. H, & Tisdale, T. (2000). Application of three dimensional hydro-
      dynamic model for Lake Okeechobee. J. Hydraulic Engineering, 126(10), , 758
      EOF-772 EOF.

[18] Manheim, F. T, & Hayes, L. (2002). Sediment database and geochemical assessment
      of Lake Pontchartrain Basin, in Manheim FT, and Hayes L (eds.), Lake Pontchartrain
      Basin: Bottom Sediments and Related Environmental Resources: U.S. Geological Sur-
      vey Professional Paper 1634.

[19] Mccorquodale, J. A, Georgiou, I, Chilmakuri, C, Martinez, M, & Englande, A. J.
      (2005). Lake Hydrodynamics and Recreational Activities in the South Shore of Lake
      Pontchartrain, Louisiana. Technical Report for NOAA, University of New Orleans.

[20] Mccorquodale, J. A, Georgiou, I, Retana, A. G, Barbe, D, & Guillot, M. J. (2007). Hy-
      drodynamic modeling of the tidal prism in the Pontchartrain Basin, Technical report,
      The University of New Orleans.

[21] Mccorquodale, J. A, Roblin, R. J, Georgiou, I, & Haralampides, K. A. (2009). Salinity,
      nutrient, and sediment dynamics in the Pontchartrain Estuary. Journal of Coastal Re-
      searchSpecial , 54(54), 71-87.

[22] Penland, S., Beall, A. and Kindinger, J. (2002). Environmental Atlas of the Lake
      Pontchartrain Basin. USGS Open File Report 02-206.

[23] Rueda, F.J. and Schladow, S.G. (2003). Dynamics of large polymictic lake. II: Numeri-
      cal Simulations. Journal of Hydraulic Engineering. 129(2): 92-101.

[24] Signell, R.P. and List, J.H. (1997). Modeling Waves and Circulation in Lake Pontchar-
      train. Gulf Coast Association of Geological Societies Transactions. 47: 529-532.

[25] Smagorinsky, J. (1993). Large eddy simulation of complex engineering and geophysi-
      cal flows, in Evolution of Physical Oceanography, edited by B. Galperin, and S. A.
      Orszag, pp. 3-36. Cambridge University Press.

[26] Stone, H.L. (1968). Iterative solution of implicit approximation of multidimensional
      partial differential equations. SIAM (Society for Industrial and Applied Mathematics)
      Journal on Numerical Analysis. 5: 530–558.

[27] Thomann, R.V. and Mueller, J.A. (1988). Principles of Surface Water Quality Model-
      ing and Control. Harper&Row Publication, New York.

[28] Thorne, C., Harmar, O., Watson, C., Clifford, N., Biedenham, D. and Measures, R.
      (2008). Current and Historical Sediment Loads in the Lower Mississippi River. Final
      Report, Department of Geology, Nottingham University.

[29] Turner, R.F., Dortch, Q. and Rabalais, N.N. (1999). Effects of the 1997 Bonnet Carré
      Opening on Nutrients and Phytoplankton in Lake Pontchartrain. Final report sub-
      mitted to the Lake Pontchartrain Basin Foundation, 117 p.

[30] United States Army Corps of Engineers (USACE) (2011). Bonnet Carre Spillway overview, Spillway pace (http://www.mvn.usace.army.mil/bcarre/2011opera-tion.asp).

[31] Wu, W. (2008). Computational River Dynamics, Taylor & Francis Group, London, UK.

[32] Wu, W., Wang, S.S.Y. and Jia, Y. (2000). Non-uniform sediment transport in alluvial river, Journal of Hydraulic Research, IAHR 38 (6) 427-434.

[33] Zhang, Q.Y. and Gin, K.Y.H. (2000). Three-dimensional numerical simulation for ti-dal motion in Singapore's coastal waters. Coastal Engineering. 39: 71–92.

[34] Zhang, R.J. and Xie, J.H. (1993). Sedimentation Research in China, Systematic Selec-tions. Water and Power Press. Beijing, China.

[35] Zhang, Y, & Jia, Y. (2009). CCHE Mesh Generator and User's Manual, Technical Re-port No. NCCHE-TR-University of Mississippi., 2009-1.

[36] Zhu, T, Jia, Y, & Wang, S. S. Y. (2008). CCHE2D water quality and chemical model capabilities. World Environmental and Water Resources Congress. Hawaii (CD-ROM).

Chapter 14

Numerical Modeling Tidal Circulation and Morphodynamics in a Dumbbell-Shaped Coastal Embayment

Yu-Hai Wang

Additional information is available at the end of the chapter

1. Introduction

Channel-shoal (ridge) system is a common morphological feature in wide, shallow coastal bays and estuarine mouth where tidal flow is relatively stronger. Sediment transport and morphological evolution is complex within such a system as constrained by interacting tidal force, river current, sediment source and characteristics, shoreline configuration, etc.

As the deeper tidal channels are usually utilized as navigation courses or water supply source for coal & nuclear power plants or other engineering purposes, it is vitally important to maintain the stability of these tidal channels, that is, they should not be allowed to migrate, merge or perish by siltation.

This chapter chooses the dumbbell-shaped Qinzhou Bay as the study site to investigate the sediment transport process and resultant morphological evolution of the channel-shoal system within the Bay using numerical simulations under the status quo situation. This is beneficial to the planned large-scale coastal engineering projects that might exert a profound long-term influence upon the stability of the channel-shoal system.

1.1. Research question

Qinzhou Harbor is one of the most important sea harbors connecting southwestern China's inland and the southeast Asian countries, it delivered up to 47.162 million ton cargos in 2011. The harbor has been using waterways as its navigational channel. Meanwhile, coal & nuclear power plants, industrial development zones, recreational parks, land reclamations and many other coastal projects have been or planned to be constructed around the Qinzhou Bay coast, all of which compete for the limited shoreline and water area resources.

For the optimum planning of the coastal engineering projects regarding size, location and the sustainable regional economic, social and environmental development, it is critically urgent to know how the stability of the channel-shoal system under the present coastal configuration and bathymetry. This might be answered by investigating the sediment transport processes and morphodynamics of the channel-shoal system.

1.2. Site description

Qinzhou Bay is located on Guangxi Province's coast facing South China Sea (Figure 1).

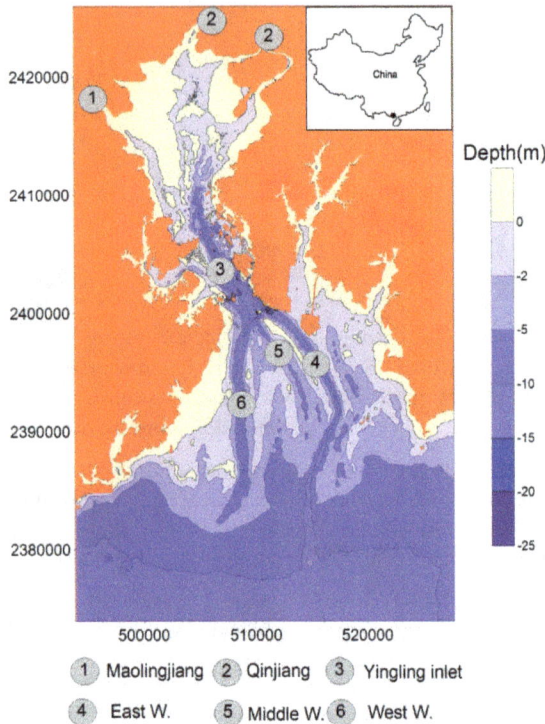

Figure 1. Sketch map of Qinzhou Bay coastal configuration (the axis is local coordinate, unit: m).

Appearing as a dumbbell shape it consists of three parts: the inner enclosed bay, also known as Maoweihai, the outer bay and the Yingling tidal inlet connecting two bays. Two rivers, i.e. Maolingjiang and Qinjiang flow into the inner bay, delivering annually 27.73×10^9 m^3 water and 86.4×10^3 t sediment; the outer bay is a show, trumpet-shaped bay, its area is nearly 2.55×10^8 m^2 with a mean depth of 4.67m (calculated by mean sea level). A complex channel-shoal system is present within the outer bay, consisting of three dominant tidal waterways, i.e. the east waterway, middle waterway and west waterway and sand ridges/shoals between them (Figure 1). The Yingling tidal inlet is 10.1 km long and 1.1~3.5 km wide with a water depth of 5~20 m; it is a rocky inlet with a total of 71 various-sized islands and 72 narrow, small waterways.

2. Dynamics

2.1. Tidal regime

The spring tide and middle tide in Qinzhou Bay are diurnal throughout most of the year but become irregularly diurnal in March and September each year, while the neap tide is usually semi-diurnal throughout the whole year. The mean tidal range is 2.51m, and the maximum tidal range is 5.27m. The flood tides last longer than the ebb tides in spring, middle and neap tides. It is 13 hours plus 14 minutes, 11 hours plus 18 minutes for spring tide; 14 hours plus 36 minutes and 10 hours plus 7 minutes for middle tide, and 6 hours plus 33 minutes and 5 hours plus 40 minutes for neap tide, respectively [1].

Tidal flow in the outer bay demonstrates as reciprocating flow in parallel to the major waterways; the mean flood velocity, ebb velocity of spring tide is 0.37m/s and 0.51 m/s, respectively, the mean flood velocity, ebb velocity of middle tide is 0.33m/s and 0.38 m/s, respectively, the mean flood velocity, ebb velocity of neap tide is 0.22m/s and 0.18 m/s; while flow velocity in the Yingling inlet becomes significantly larger, the mean flood velocity, ebb velocity of spring tide is 0.67m/s and 0.90 m/s, the mean flood velocity, ebb velocity of middle tide is 0.57m/s and 0.68 m/s, the mean flood velocity, ebb velocity of neap tide is 0.42m/s and 0.33m/s, respectively, and the maximum flood flow velocity reaches up to 1.40 m/s and the maximum ebb flow velocity is up to 1.32 m/s [1].

2.2. Wave climate

The Qinzhou Bay is influenced by subtropical monsoon and the waves within the Bay are mainly wind-driven with some surge waves traveled from the open sea.

The waves in winter season (October~Apirl) prevail in N-NE direction while they prevail in S-SW direction in summer season (May-September) and the stronger waves propagate in SSW, SSE directions; the mean wave height is 0.52m and mean wave period is 3.1s [2].

3. Sediment characteristics

3.1. Suspended sediment

Suspended sediment concentration within Qinzhou Bay water is generally low. In summer 2009, the mean full tidal concentration is 0.022 kg/m³, among which, it is 0.035 kg/m³ in spring tide, 0.020 kg/m³ in middle tide and 0.013 kg/m³ in neap tide; the maximum concentration is 0.081 kg/m³, occurred in spring ebb tide, the maximum middle tidal concentration is 0.034 kg/m³, occurred also in ebb tidal period, the maximum neap tidal concentration is 0.025 kg/m³, occurred also in ebb tidal period [1].

The medium diameters of suspended sediment vary within 0.0067~0.0152mm with a mean value of 0.0101mm. The suspended sediment is mainly clayey silt with 30.8% clay particles, 53.4% silt particles and 15.8% sand particles (Table 1). The sorting index is 1.90 [1].

Grading	sand	silt	clay	Sorting index
percentage%	15.8	53.4	30.8	1.90

Table 1. Grading of suspended sediment.

3.2. Bottom sediment

The bottom sediments in Qinzhou Bay mainly consist of gravel, coarse sand, medium sand, fine sand, silty clay and clayey silt, etc.

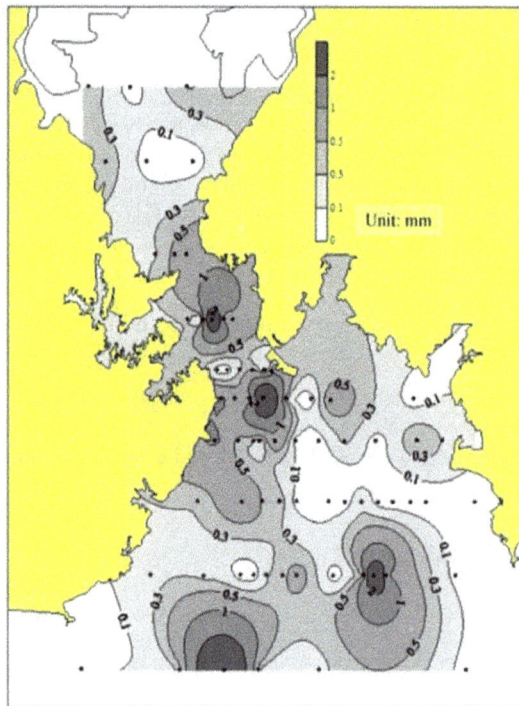

Figure 2. Bottom sediment grain size distribution in Qinzhou Bay.

The gain sizes vary remarkably in 0.0027~1.099mm (Figure 2). The spatial mean median diameter (D_{50}) in the inner bay is 0.334mm; it is 0.356mm in the deep channel but becomes 0.0041mm in the shallow parts in the Yingling inlet; in the outer bay, the spatial mean D_{50} is about 0.298mm with an overall deposition pattern: coarser in waterways but finer in shoals, and coarser in the western part than in the eastern part of the bay [1].

3.3. Sediment source

The Qinzhou Bay has been a drowned rocky valley by the last sea level transgression since 7,000-8,000 year before present [3]. Therefore, the huge amount of sand deposits in the outer bay has come from the deposits by paleo-Maolingjiang river and paleo-Qinjiang river, they have been reformed into the contemporary channel-shoal geomorphology by tidal dynamics. At the present day, sediments delivered by these two rivers are deposited within the inner bay with limited amount of fine particles transported into the outer bay and open sea; meanwhile, limited amount of sediment eroded from the adjacent slopes by storm rains also enter the outer bay. Generally speaking, sediment from the open sea into the Qinzhou Bay is very limited.

4. Numerical model

A 3D unstructured grid, finite-volume coastal ocean model (called FVCOM) has been developed in the Marine Ecosystem Dynamics Modeling Laboratory led by Dr. C. Chen at the University of Massachusetts–Dartmouth (UMASS-D) in collaboration with Dr. R. Beardsley at the Woods Hole Oceanographic Institute. FVCOM is a three-dimensional (3D) primitive equation ocean model, consisting of momentum, continuity, sediment, temperature, salinity, and density equations and is closed physically and mathematically using the Mellor and Yamada level-2.5 turbulent closure submodel; the irregular bottom slope is represented using a σ-coordinate transformation, and the horizontal grids comprise unstructured triangular cells; the finite-volume method used in the model combines the advantages of a finite-element method for geometric flexibility and a finite-difference method for simple discrete computation; current, sediment, temperature, and salinity in the model are computed in the integral form of the equations, which provides a better representation of the conservative laws for mass, momentum, and heat in the coastal region with complex geometry [4].

4.1. The primitive equations

The governing equations consist of the following momentum, continuity, temperature, salinity, and density equations:

(1) continuity equation

$$\frac{\partial u}{\partial x} + \frac{\partial v}{\partial y} + \frac{\partial w}{\partial z} = 0 \tag{1}$$

(2) x-direction momentum equation

$$\frac{\partial u}{\partial t} + u\frac{\partial u}{\partial x} + v\frac{\partial u}{\partial y} + w\frac{\partial u}{\partial z} - fv = -\frac{1}{\rho_0}\frac{\partial P}{\partial x} + \frac{\partial}{\partial z}(K_m\frac{\partial u}{\partial z}) + F_u \tag{2}$$

(3) y-direction momentum equation

$$\frac{\partial v}{\partial t} + u\frac{\partial v}{\partial x} + v\frac{\partial v}{\partial y} + w\frac{\partial v}{\partial z} + fu = -\frac{1}{\rho_0}\frac{\partial P}{\partial y} + \frac{\partial}{\partial z}(K_m\frac{\partial v}{\partial z}) + F_v \tag{3}$$

(4) temperature equation

$$\frac{\partial T}{\partial t} + u\frac{\partial T}{\partial x} + v\frac{\partial T}{\partial y} + w\frac{\partial T}{\partial z} = \frac{\partial}{\partial z}(K_h\frac{\partial T}{\partial z}) + F_T \tag{4}$$

(5) salinity equation

$$\frac{\partial s}{\partial t} + u\frac{\partial s}{\partial x} + v\frac{\partial s}{\partial y} + w\frac{\partial s}{\partial z} = \frac{\partial}{\partial z}(K_h\frac{\partial s}{\partial z}) + F_s \tag{5}$$

(6) pressure equation

$$\frac{\partial P}{\partial t} = -\rho g \tag{6}$$

$$\rho = \rho(T,s) \tag{7}$$

where x, y, and z are the east, north, and vertical axis of the Cartesian coordinate; u, v, and w are the x, y, z velocity components; T is the potential temperature; s is the salinity; P is the pressure; f is the Coriolis parameter; g is the gravitational acceleration; K_m is the vertical eddy viscosity coefficient; and K_h is the thermal vertical eddy diffusion coefficient. Here F_u, F_y, F_T, and F_s represent the horizontal momentum, thermal, and salt diffusion terms.

4.2. Numerical solutions

The momentum and continuity equations are solved using a 'model splitting' method [4], that is, the current is divided into external and internal modes that can be computed using two distinct time steps. The external mode is used to solve the 2D vertically integrated momentum and continuity equations while the internal mode is computed for the 3D equations, the latter is solved numerically using a simple combined explicit and implicit scheme, in which the local change of the current is integrated using the first-order accuracy upwind scheme; the advection terms are computed explicitly by a second-order accuracy Runge–Kutta time-stepping scheme [4].

4.3. Sediment computation

FVCOM adopts the Community Numerical Modeling System to simulate erosion, transport, deposition and the fate of sediments in the coastal ocean developed by experts from USGS [5]. The sediment-transport algorithms are implemented for an unlimited number of user-defined noncohesive/cohesive sediment classes. Each class has attributes of grain diameter, density, settling velocity, critical stress threshold for erosion, and erodibility constant. These properties are used to determine bulk properties of each bed layer. Suspended-sediment transport in the water column is computed with the same advection-diffusion algorithm used for all passive tracers and an additional algorithm for vertical settling that is not limited by the CFL criterion [5].

4.3.1. Suspended sediment

Suspended sediment transport equation is:

$$\frac{\partial C_i}{\partial t} + \frac{\partial u C_i}{\partial x} + \frac{\partial v C_i}{\partial y} + \frac{\partial (\omega - \omega_i) C_i}{\partial z} = \frac{\partial}{\partial x}(A_{s,h}\frac{\partial C_i}{\partial x}) + \frac{\partial}{\partial y}(A_{s,h}\frac{\partial C_i}{\partial y}) + \frac{\partial}{\partial z}(K_{s,h}\frac{\partial C_i}{\partial z}) \tag{8}$$

here C_i is the concentration of the ith sediment class, $A_{s,h}$ is the horizontal sediment diffusivity, $K_{s,h}$ is the vertical sediment diffusivity, ω_i is the settling velocity of the ith sediment class given by the user.

At the top boundary, the vertical diffusive flux is set to be zero:

$$K_{s,h}\frac{\partial C_i}{\partial t} = 0, z = \zeta \tag{9}$$

here ζ is surface elevation. At the bottom boundary, the vertical sediment flux is specified by:

$$K_{s,h}\frac{\partial C_i}{\partial t} = E_i - D_i, z = -D \tag{10}$$

here D is water depth. While the erosion flux of the ith sediment class is computed as:

$$E_i = \Delta t E_{0i}(1 - P_b)F_{bi}(\frac{\tau_b}{\tau_{ci}} - 1) \tag{11}$$

where E_{0i} is an empirical constant set by the user as the erosion rate of the ith sediment class, P_b is the porosity of the bed, F_{bi} is the fraction of the ith sediment class, τ_b is the bottom flow shear stress, τ_{ci} is the critical stress for the incipient motion of the ith sediment class.

When the bottom shear stress passes the critical stress erosion occurs on the bed. The sediment concentration profile in the water body is determined by horizontal convection, diffusion, vertical diffusion, settling and bottom erosion flux [5].

4.3.2. Bedload transport computation

Bedload transport rate is computed using established empirical formula, i.e. the Meyer-Peter and Muller formula or using formulas that the modeler considers appropriate, for example, the theoretical-based formula [6].

4.3.3. Sediment bed

The sediment bed consists of a constant number layers, and each layer is initialized with a thickness, sediment-class distribution, porosity, and age, the mass of each sediment class can be determined from these values and the grain density; the bed evolving properties include bulk properties of the surface layer (active-layer thickness, mean grain diameter, mean density, mean settling velocity, mean critical stress for erosion) [5].

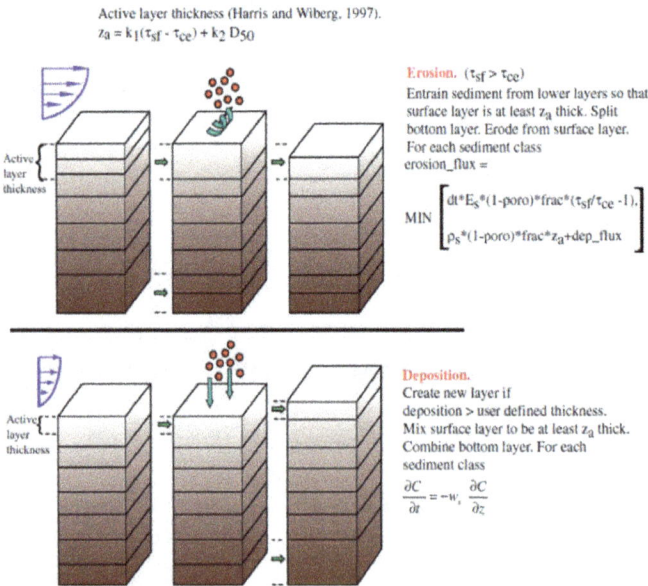

Active layer thickness (Harris and Wiberg, 1997).
$$z_a = k_1(\tau_{sf} - \tau_{ce}) + k_2 D_{50}$$

Erosion. $(\tau_{sf} > \tau_{ce})$
Entrain sediment from lower layers so that surface layer is at least z_a thick. Split bottom layer. Erode from surface layer. For each sediment class
erosion_flux =

$$\mathrm{MIN} \begin{bmatrix} dt*E_s*(1-poro)*frac*(\tau_{sf}/\tau_{ce} - 1), \\ \rho_s*(1-poro)*frac*z_a + dep\_flux \end{bmatrix}$$

Deposition.
Create new layer if deposition > user defined thickness. Mix surface layer to be at least z_a thick. Combine bottom layer. For each sediment class

$$\frac{\partial C}{\partial t} = -w_s \frac{\partial C}{\partial z}$$

Figure 3. Distribution of vertical layers in bed model (from [5]).

The bed layers are modified at each time step to account for erosion and deposition (Figure 3) and track stratigraphy; at the beginning of each time step, an active-layer thickness z_a is calculated based on the relation of Harris and Wiberg [7]:

$$z_a = \max\left[k_1(\tau_{sf} - \overline{\tau_{ce}})\rho_0, 0\right] + k_2 D_{50} \tag{12}$$

where τ_{sf} is bottom skin-friction stress; $\overline{\tau_{ce}}$ is the critical stress for erosion; and the overbar indicates this is averaged over all sediment classes; D_{50} is the median grain diameter of surface sediment; and k_1 and k_2 are empirical constants, 0.007 and 6.0, respectively.

Each sediment class can be transported by suspended-load and/or bedload; suspended-load mass is exchanged vertically between the water column and the top bed layer; mass of each sediment class available for transport is limited to the mass available in the active layer; bedload mass is exchanged horizontally between the top layers of the bed; mass of each sediment class available for transport is limited to the mass available in the top layer [5].

If continuous deposition results in a top layer thicker than a user-defined threshold, a new layer is provided to begin accumulation of depositing mass; the bottom two layers are then combined to conserve the number of layers; after erosion and deposition have been calculated, the active-layer thickness is recalculated and bed layers are readjusted to accommodate it [5].

5. Model setup

5.1. Model domain

Figure 4 shows the computation domain consisting of unstructured triangular grids.

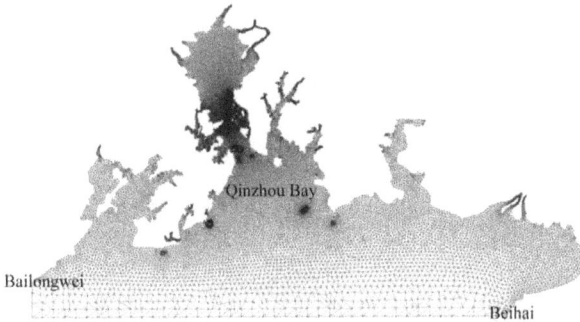

Figure 4. Unstructured grids of Qinzhou Bay.

It has 16466 nodes and 29722 elements in each horizontal layer and 7 sigma-levels in the vertical. The horizontal grid resolution varies from 2,000 m at the open boundary to 350m in the channel-shoal region to 150 m in the Yingling inlet, especially, down-to-30-m elements are interpolated around islands and in the estuarine channels.

5.2. Boundary conditions

The open boundary uses observed water level as input condition, for this purpose half-month water levels at each open boundary grid are interpolated from two tidal station, i.e. the Beihai Station and the Bailongwei Station (Figure 4). River boundaries use annually-mean discharges as input conditions.

Suspended-sediment concentrations at the open boundary grids are interpolated from adjacent observation sites, while those at river input grids are annually-mean suspended-sediment discharges.

The bathymetry in the computation domain consists of a local bathymetric survey in the Qinzhou Bay in 2008 and sea maps surveys in 2004 and 1997 to supplement other parts.

The time step for both of external mode and internal mode is 1 s.

6. Calibrations

TCZC [1] measured half month water level at three temporary tidal gauges, namely, Guozishan, Shabadun and Wulei, also measured 26-hour spring, middle and neap tidal flow velocity & direction and suspended-sediment concentrations at 10 sites (Figure 5).

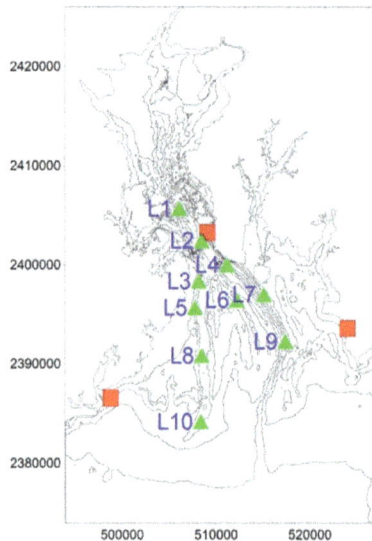

Figure 5. Tidal gauges (red square) and flow observations sites (green triangular) in Qinzhou Bay.

The present study firstly performs the calibrations of water level, flow velocity& direction and suspended-sediment concentrations.

6.1. Tidal water level

The Calibrations for water levels at three tidal gauges are shown in Figure 6.

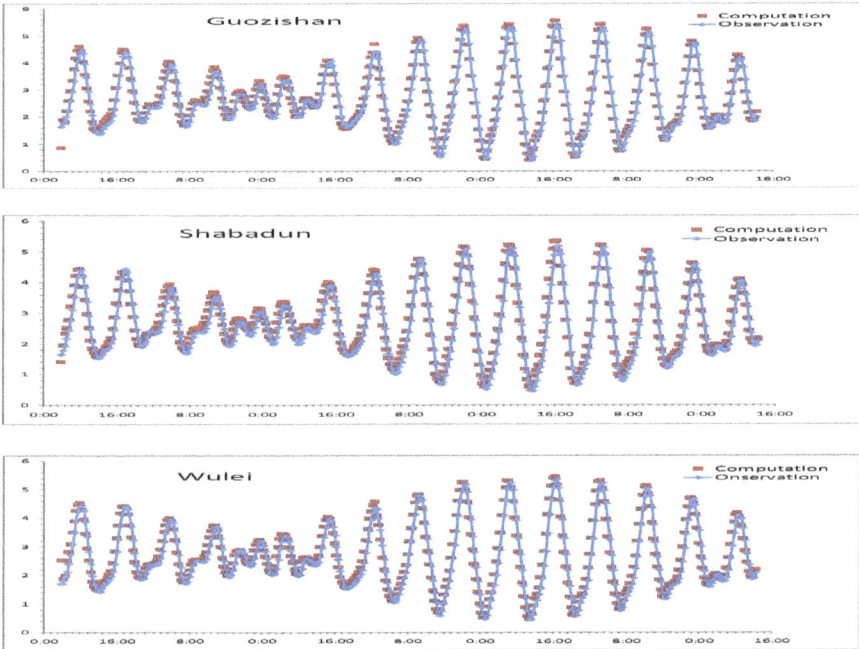

Figure 6. Tidal water level calibrations (the ordinate unit of is meter).

6.2. Flow calibrations

The flow calibrations are shown in Figure 7. For limited space only calibrations for spring tide are show here.

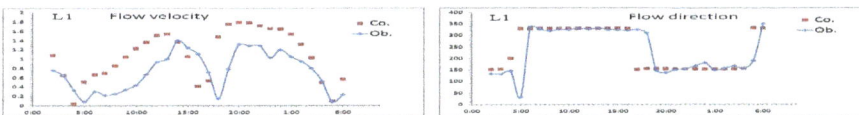

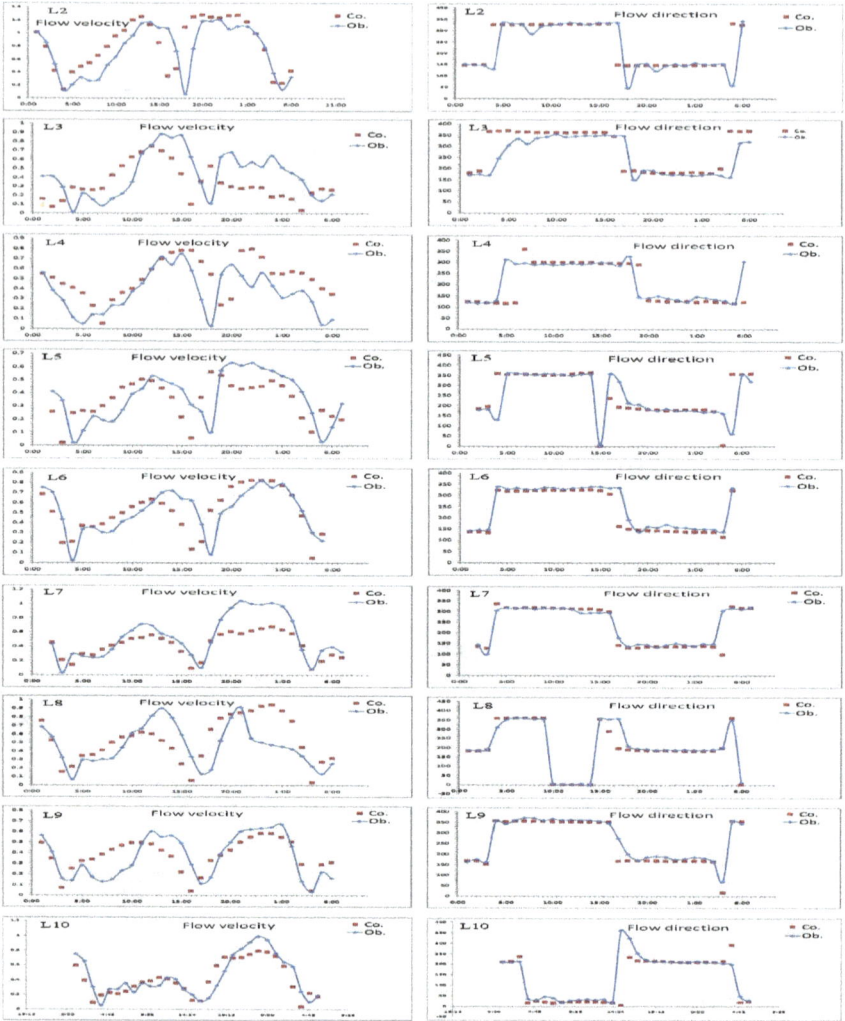

Figure 7. Flow calibrations.

6.3. Sediment calibration

Sediment calibrations for selected sites are shown on Figure 8. It needs to explain that the overall sediment calibrations are satisfactory, but results for some sites are not satisfactory enough due to observation and computation errors, ship activities and dredging during the observation period, etc.

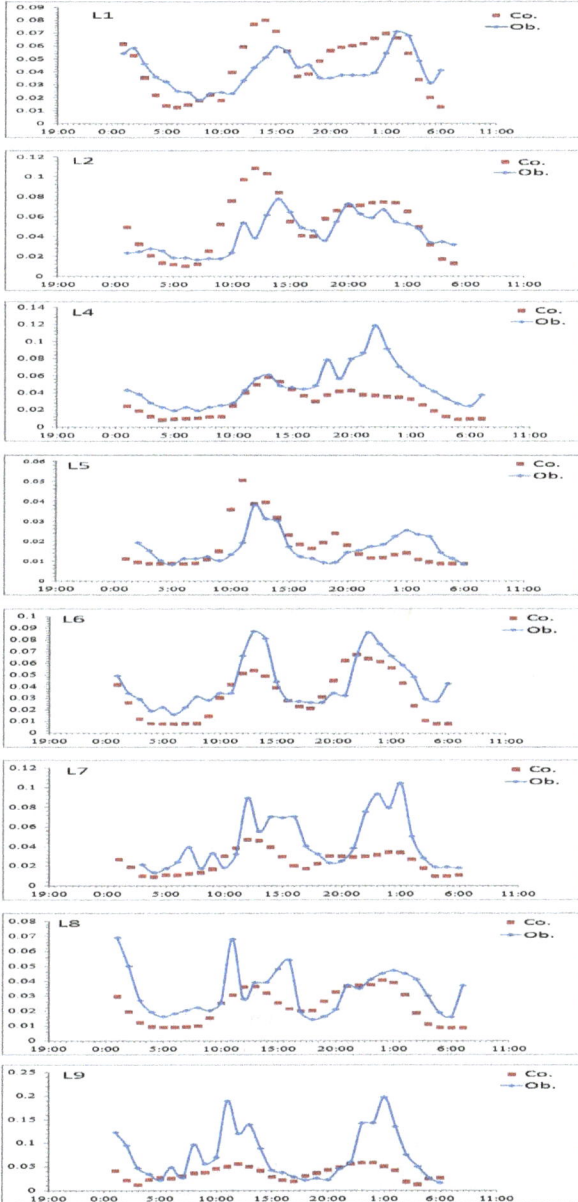

Figure 8. Suspended-sediment calibrations at selected sites for spring tide (the ordinate unit is kg/m³).

6.4. Morphological calibration

Based upon a local bathymetric survey conducted in 2008 summer and a sea map surveyed in 2004, statistics shows that the total siltation amount at this area is about 1.52 million m³with a spatial average value of 0.592m (the deposition volume is divided by the deposition area, the same hereinafter), the total erosion amount is nearly 3.35 million m³ with a spatial average value of 1.104m (the erosion volume is divided by the erosion area, the same hereinafter), the net eroded sediment amounts to 1.83 million m³ (Figure 9).

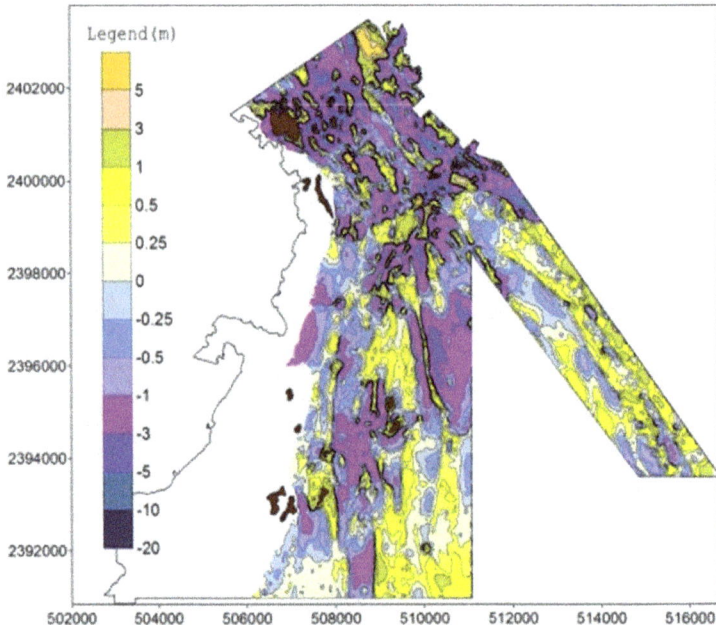

Figure 9. Erosion & deposition distribution map in a part of Qinzhou Bay.

The present study computes the morphological evolution using the 2004 bathymetry as the initial bathymetry, the computed four-year accumulative erosion & siltation distribution is shown in Figure 10. The computed total siltation amount is 1.244 million m³ with a spatial average value of 0.461m, amounting 81.89% and 77.87% of surveyed quantities, respectively; the computed total erosion amount is 2.97 million m³ with a spatial average value of 1.073m, amounting 88.53% and 97.19% of surveyed quantities, respectively; the computed net eroded sediment amounts to 1.72 million m³, amounting 94.03% of surveyed quantities.

In view of the discrepancies of computed results vs. surveyed quantities the present morphological calibration is quite satisfactory. This lays down very good basis for further morphodynamic study.

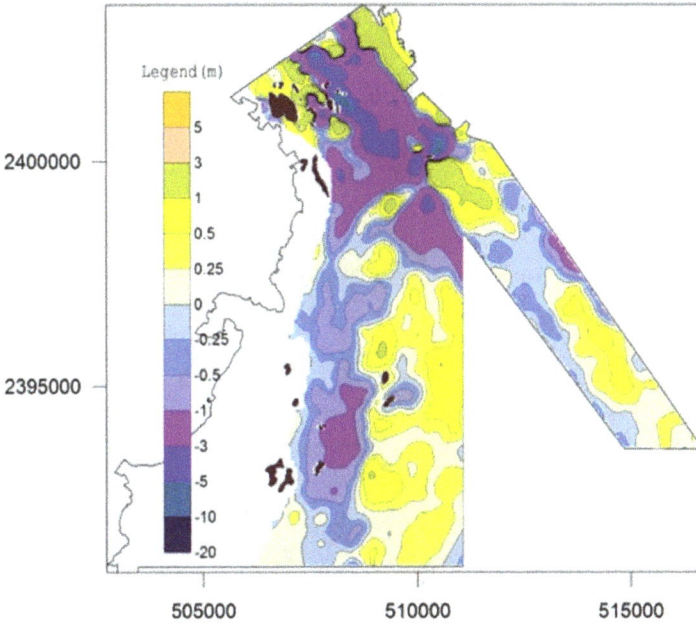

Figure 10. The computed 2004~2008 erosion & deposition distribution map.

7. Computation results

7.1. Tidal flow field

As for neap tide and middle tide, tidal water floods into the Qinzhou Bay in the northeastern direction while it floods into the Bay nearly in the northern direction during spring tide. As constrained by the shoreline and channel-shoal geomorphology tidal water propagates in the northwestern direction into the Yingling inlet and further into the inner bay, where it flows anticlockwise till stack water; then tidal water rushes into Yingling inlet and diverges among east, middle and west waterways; finally it leaves the Qinzhou Bay in the southwestern direction to the South China Sea (Figure 11).

Generally speaking, the tidal flow field in the Qinzhou Bay is characteristics of reciprocating flow in parallel to the major waterways, large scale eddies occur during flow reversal periods.

The computation results show that the ebb-mean velocities of spring tide and middle tide are all larger than the flood-mean velocities; while flood-mean velocity becomes larger than ebb-mean velocity during neap tide in the overall flow field of Qinzhou Bay.

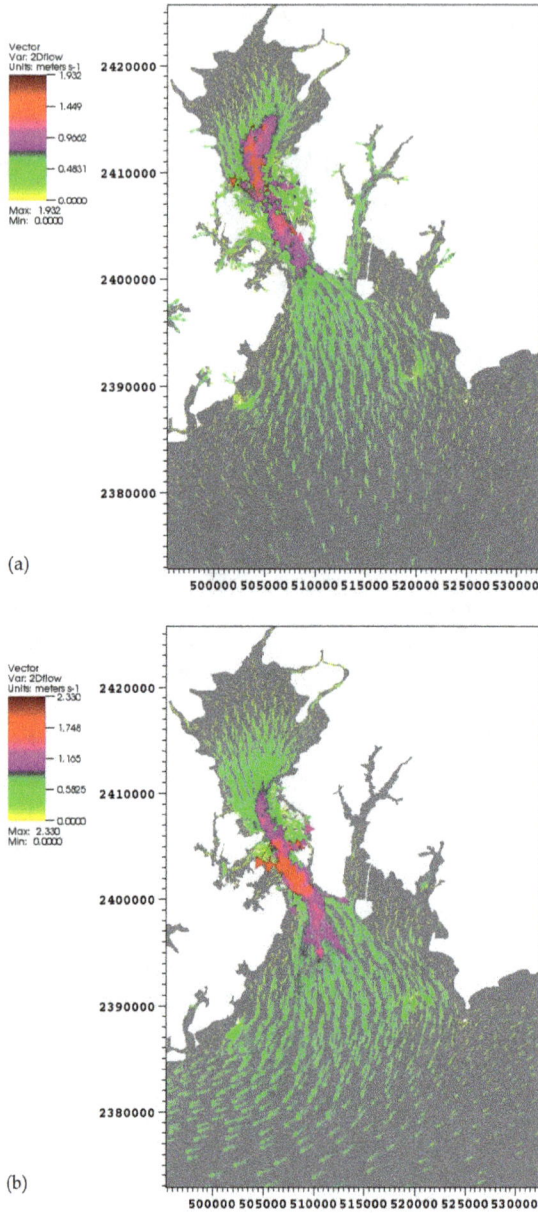

Figure 11. a) Flood peak flow of spring tide in Qinzhou Bay, b) Ebb peak flow of spring tide in Qinzhou Bay.

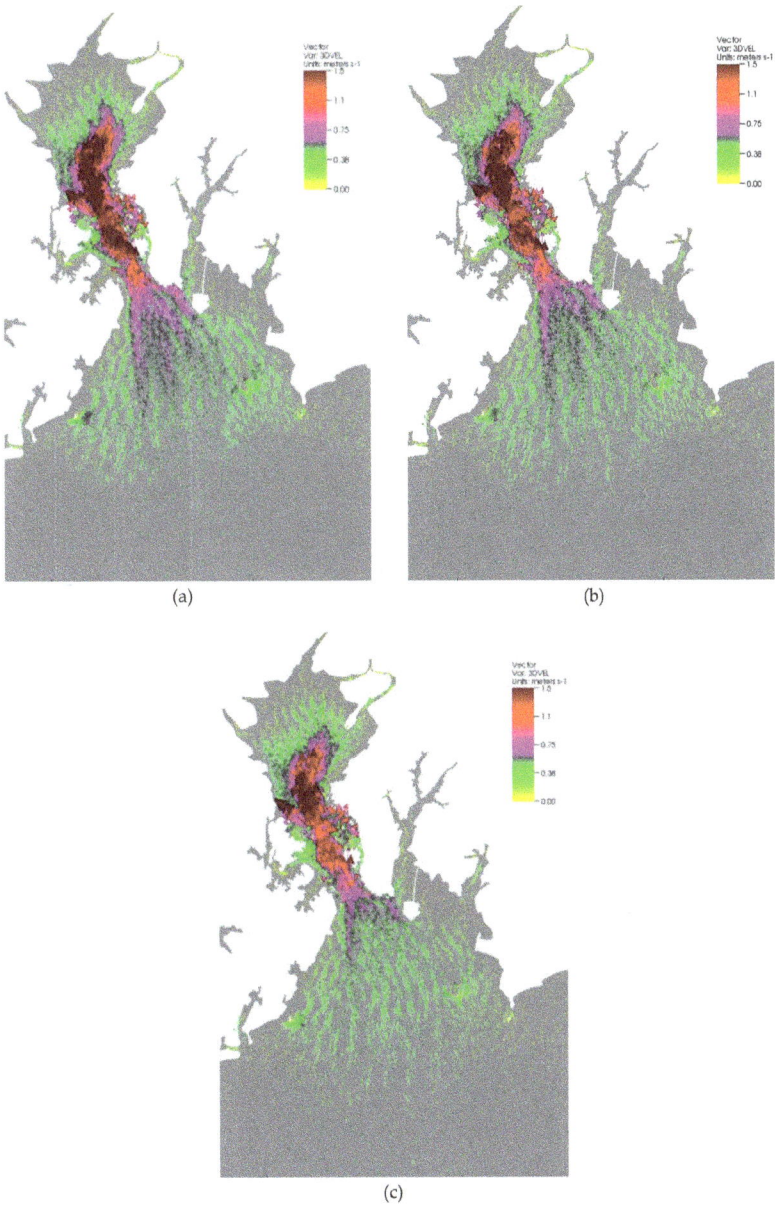

(a)

(b)

(c)

Figure 12. a) Upper-layer flood peak flow of spring tide in Qinzhou Bay, b) Middle-layer flood peak flow of spring tide in Qinzhou Bay, c) Lower-layer flood peak flow of spring tide in Qinzhou Bay.

Among three major tidal channels in the outer bay, the flood-mean and ebb-mean velocities of spring, middle, neap tides in the middle channel are all larger than those in the east and west channels; though flood-mean velocity in spring tide in the west channel is somewhat smaller than that in the east channel the flood-mean velocities in middle & neap tides in the west channel are all larger than those in the east channel; as for ebb-mean velocity, they are all larger in the west channel than those in the east channel. These data demonstrates that the west channel is the dominant channel for tidal water flowing into and out the Qinzhou Bay, the middle channel comes second and the east channel is third; ebb tide dominates in the west channel and middle channel but flood tide dominates in the east channel.

Generally speaking, flow velocity at the upper water layer is the largest and decreases from top to bottom (Figure 12). The depth-averaged residual flow is shown in Figure 13. Various-sized residual eddies occur in the Qinzhou Bay. Mean residual flow velocity in the Yingling inlet is around 0.15m/s with largest velocity of 0.489m/s, it is generally below 0.05m/s in other parts.

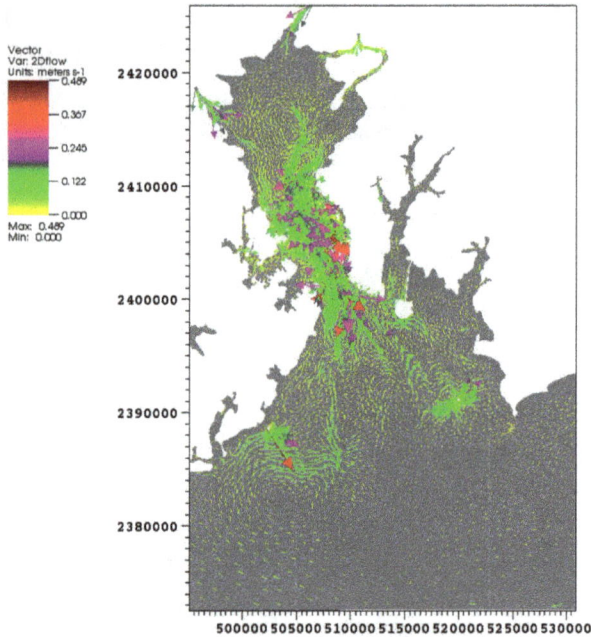

Figure 13. Depth-averaged residual flow in Qinzhou Bay.

7.2. Sediment transport

The suspended-sediment sources include those delivered by Maolingjiang river and Qinjing river and limited amount transported from the open sea, but the majority is eroded and resuspended in situ in the Bay. As a result, the spatial & temporal variations of suspended-sediment concentrations are in accordance with the processes of tidal flows.

Generally speaking, suspended-sediment concentrations are larger in major channels than those on shoals and intertidal zones (Figure 14), decrease from bottom to top. The majority of sediment delivered by Maolingjiang and Qinjiang rivers is deposited within the inner bay with limited amount of finer particles transported into the outer bay and deep water.

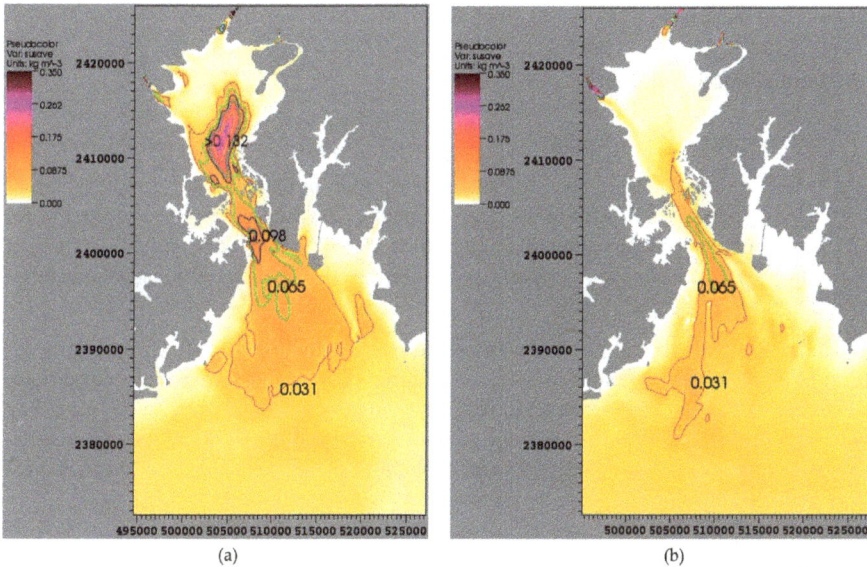

Figure 14. a) Suspended-sediment concentration field at flood peak of spring tide, b) Suspended-sediment concentration field at ebb peak of middle tide.

The computation results show that suspended-sediment concentration at spring tidal flood peak is 0.037 kg/m³, 0.021 kg/m³ at spring tidal flood stack, 0.034 kg/m³ at spring tidal ebb peak and 0.023 kg/m³ at spring tidal ebb stack, respectively, the mean spring-tidal concentration is 0.029 kg/m³; sediment concentration is 0.031 kg/m³ at middle tidal flood peak, 0.023 kg/m³ at middle tidal flood stack, 0.031 kg/m³ at middle tidal ebb peak and 0.018 kg/m³ at middle tidal ebb stack, respectively, the mean middle-tidal concentration is 0.026 kg/m³; the sediment concentration is 0.013 kg/m³ at neap tidal flood peak, 0.013 kg/m³ at neap tidal flood stack, 0.013kg/m³ at neap tidal ebb peak and 0.012 kg/m³ at neap tidal ebb stack, respectively, The mean neap-tidal concentration is 0.013 kg/m³. Generally speaking, suspended-sediment concentration is relatively low in the Qinzhou Bay.

The west channel has been the major channel to transport sediment from outer bay thought Yingling inlet to inner bay and vice versa; the middle channel comes second and the east channel contributes the least. The sediment discharges at ebb tide are all larger than those at flood tide in the three channels, demonstrating net sediment transport into the open deep water. The sediment discharges at spring tide in three channels are all larger than those at middle & neap tide, and the dividing ratio of west channel, middle channel and east channel is nearly 5:2:1 for spring flood tide and 7:5:1 for spring ebb tide, 4:2:1 for middle flood tide and 8:4:1 for middle ebb tide, respectively.

The sediment transport pattern is normally accordance with tidal flow asymmetry in three channels, that is, ebb flow strength and discharge are superior to flood flow strength and discharge.

7.3. Morphological evolution

7.3.1. 2009-year erosion and deposition

Due to lack of data on the deposit thickness distribution in Qinzhou Bay and considering that rock is exposed locally within the deep channel in Yingling inlet by strong tidal flows [8], the present study assumes the initial deposit thickness is 0.3m within the deep channel in Yingling inlet and 20m in other parts of the Qinzhou Bay. The morphological computation starts from year 2008.

The computed 2009 annual erosion & deposition distribution map is shown in Figure 15. It can be observed that erosions mainly occur within channels including three major channels in the outer bay, deep-water channel in the Yingling inlet and those in the inner bay while depositions occur at shoal & ridge area and at the end of channels. This asserts that tide flow is really the dominant force for maintaining and reforming the channel-shoal morphology in the Qinzhou Bay.

Generally speaking, eroded sediments exceed deposited sediment for the whole Qinzhou Bay with net erosion nearly up to 10.288 million m^3. Except for the inner bay where net deposition occurs with a quantity of 3.190 million m^3 net erosions all occur in the Yingling inlet and the outer bay, they are 2.999 million m^3 and 3.503 million m^3, respectively. Due to lack sediment supply, the offshore slope outside the Qinzhou Bay is also subjected to net erosion of 6.976 million m^3, where erosion mainly occurs at the middle and southeastern part while deposition occurs at the southwestern part.

The total deposition in the west channel (bounded by -5m bathymetric contour, the same for middle channel and east channel) in the outer bay is roughly 490,887.486 m^3, the total erosion is roughly 2,257,125.612 m^3, and the net erosion is about 1,766,238.1 m^3; the spatial mean deposition is 0.097m, the spatial mean erosion is -0.232m, the maximum deposition is 0.355m and the maximum erosion is -1.244m.

The total deposition in the middle channel in the outer bay is roughly 9,830.569 m^3, the total erosion is roughly 551,595.451 m^3, and the net erosion is about 453,285.88 m^3; the spatial mean deposition is 0.073m, the spatial mean erosion is -0.394m, the maximum deposition is 0.157m and the maximum erosion is -1.785m. The erosion mainly occurs at the channel mouth connecting with the Yingling inlet.

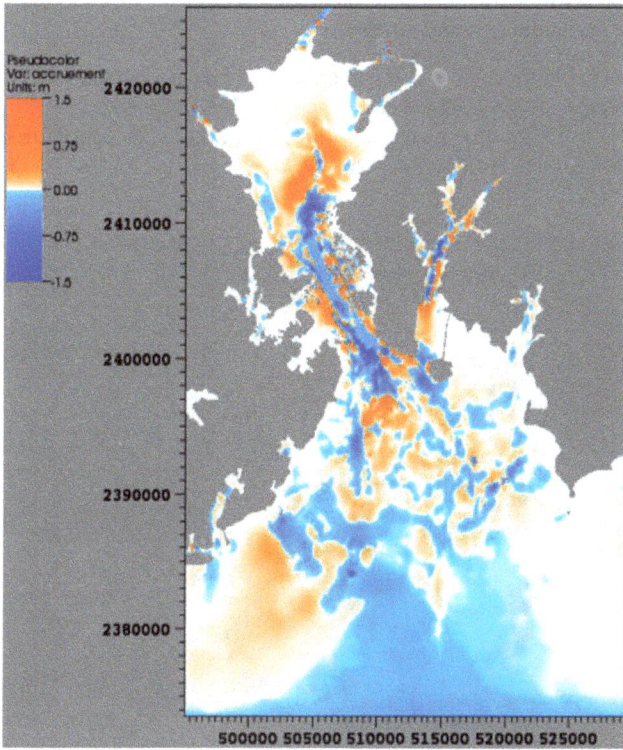

Figure 15. 2009-year annual erosion & deposition distribution map.

The total deposition in the east channel in the outer bay is roughly 499,259.028 m³, the total erosion is roughly 1,488,779.676 m³, and the net erosion is about 989,520.6 m³; the spatial mean deposition is 0.110m, the spatial mean erosion is -0.213m, the maximum deposition is 0.623m and the maximum erosion is -1.285m.

7.3.2. 2012-year erosion and deposition

The computed 2012-year annual erosion & deposition distribution map is shown in Figure 16. Eroded sediments still exceed deposited sediment for the whole Qinzhou Bay with net erosion nearly up to 10.469 million m³. The inner bay continues to accommodate net deposition of 2.832 million m³, net erosions still occur in the Yingling inlet and the outer bay, they are 0.809 million m³ and 4.161 million m³, respectively; the offshore slope outside the Qinzhou Bay is still subjected to net erosion of 8.331 million m³.

The total deposition in the west channel in the outer bay is roughly 240,510.501 m³, the total erosion is roughly 2,031,819.599 m³, and the net erosion is about 1,791,309.0 m³; the spatial mean deposition is 0.063m, the spatial mean erosion is -0.185m, the maximum deposition is 0.232m and the maximum erosion is -0.889m.

The total deposition in the middle channel in the outer bay is roughly 32,702.536 m^3, the total erosion is roughly 435,482.067 m^3, and the net erosion is about 402,779.53 m^3; the spatial mean deposition is 0.044m, the spatial mean erosion is -0.217m, the maximum deposition is 0.095m and the maximum erosion is -0.772m.

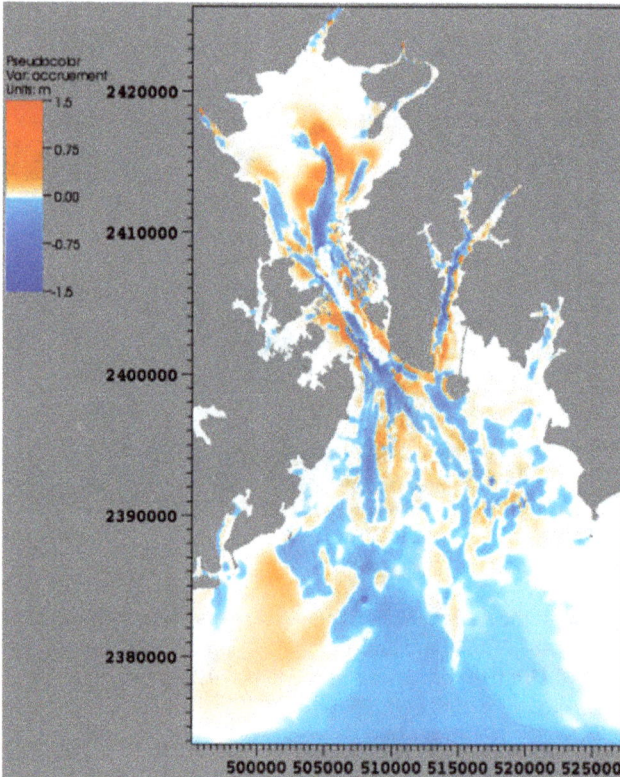

Figure 16. 2012-year annual erosion & deposition distribution map.

The total deposition in the east channel in the outer bay is roughly 356,241.632 m^3, the total erosion is roughly 1,160,837.006 m^3, and the net erosion is about 804,595.37 m^3; the spatial mean deposition is 0.084m, the spatial mean erosion is -0.159m, the maximum deposition is 0.527m and the maximum erosion is -0.705m.

These data in the three channels reflect that the erosion and deposition in the three major channels in the outer bay has steadily decreased. In particular, the erosion length in the west channel has increased substantially, leading almost to whole-channel erosion, and tidal channels have further developed in the inner bay.

7.3.3. 2020-year erosion and deposition

The computed 2020-year annual erosion & deposition distribution map is shown in Figure 17. Eroded sediments still exceed deposited sediment for the whole Qinzhou Bay with net erosion nearly up to 11.136 million m^3. The inner bay continues to accommodate net deposition of 2.601 million m^3, net erosions still occur in the Yingling inlet and the outer bay, they are 0.677 million m^3 and 3.095 million m^3, respectively; the offshore slope outside the Qinzhou Bay is still subjected to net erosion of 9.965 million m^3.

The overall morphological evolution trend is that total erosion & deposition amount have dropped steadily in all parts of Qinzhou Bay though net deposition might moderately increase or decrease in different parts of the Bay.

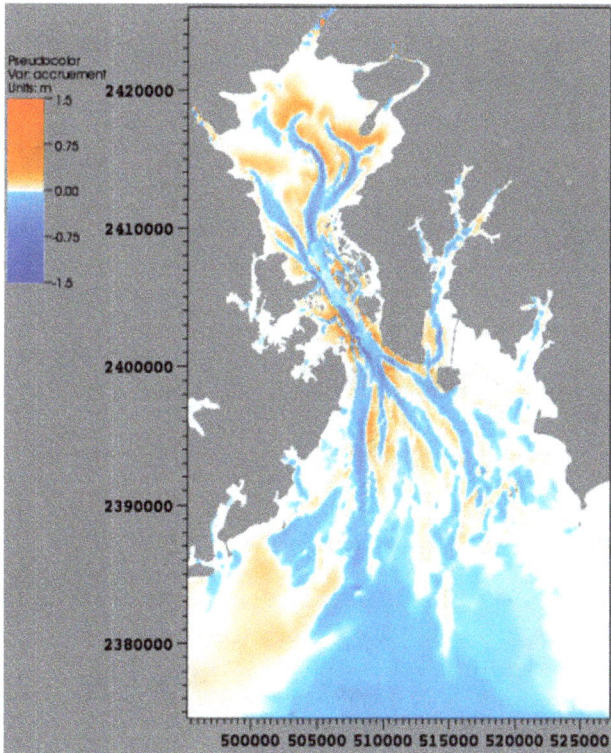

Figure 17. 2020-year annual erosion & deposition distribution map.

The total deposition in the west channel in the outer bay is roughly 207,835.846 m³, the total erosion is roughly 1,685,565.419 m³, and the net erosion is about 1,477,729.5 m³; the spatial mean deposition is 0.074m, the spatial mean erosion is -0.141m, the maximum deposition is 0.311m and the maximum erosion is -0.475m.

The total deposition in the middle channel in the outer bay is roughly 17,775.557 m³, the total erosion is roughly 296,870.467 m³, and the net erosion is about 279,094.91 m³; the spatial mean deposition is 0.025m, the spatial mean erosion is -0.145m, the maximum deposition is 0.049m and the maximum erosion is -0.405m.

The total deposition in the east channel in the outer bay is roughly 251,859.055 m³, the total erosion is roughly 929,903.211 m³, and the net erosion is about 678,044.15 m³; the spatial mean deposition is 0.069m, the spatial mean erosion is -0.118m, the maximum deposition is 0.431m and the maximum erosion is -0.412m.

7.3.4. 2040-year erosion and deposition

The computed 2040-year annual erosion & deposition distribution map is shown in Figure 18.

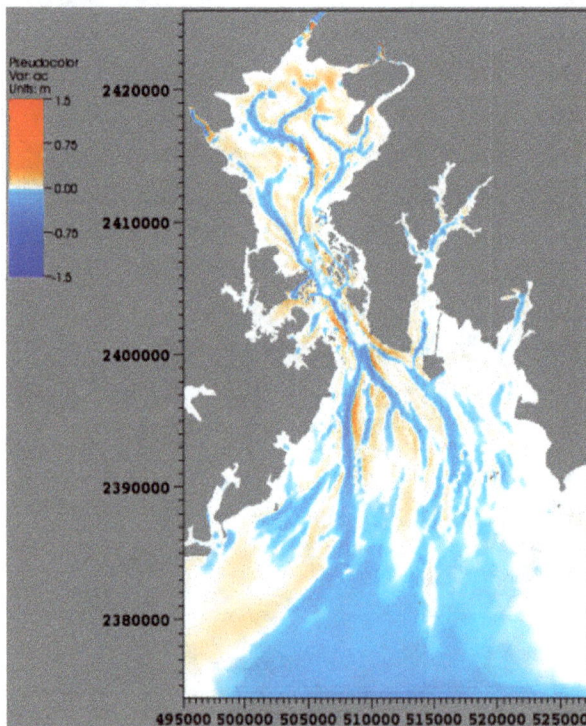

Figure 18. 2040-year annual erosion & deposition distribution map.

Eroded sediments still exceed deposited sediment for the whole Qinzhou Bay with net erosion nearly up to 13.915 million m^3. The inner bay continues to accommodate net deposition of 1.063 million m^3, but net deposition has occurred in the Yingling inlet with a value of 0.125 million m^3, net erosion in the outer bay is 1.592 million m^3, the offshore slope outside the Qinzhou Bay has been subjected to increased net erosion of 13.512 million m^3.

The total deposition in the west channel in the outer bay is roughly 387,349.439 m^3, the total erosion is roughly 1,114,549.259 m^3, and the net erosion is about 727,199.81 m^3; the spatial mean deposition is 0.098m, the spatial mean erosion is -0.103m, the maximum deposition is 0.335m and the maximum erosion is -0.429m.

The total deposition in the middle channel in the outer bay is roughly 37,682.474 m^3, the total erosion is roughly 202,893.855m^3, and the net erosion is about 165,211.38 m^3; the spatial mean deposition is 0.054m, the spatial mean erosion is -0.099m, the maximum deposition is 0.310m and the maximum erosion is -0.379m.

The total deposition in the east channel in the outer bay is roughly 170,181.197 m^3, the total erosion is roughly 420,256.804 m^3, and the net erosion is about 250,075.6 m^3; the spatial mean deposition is 0.051m, the spatial mean erosion is -0.051m, the maximum deposition is 0.271m and the maximum erosion is -0.259m.

The overall morphological evolution trend is that total erosion & deposition amount have continuously decreased, sub-channels haves occurred at the shoal between west channel and middle channel and new channel-shoal morphology has been developed within the inner bay (Figure 19).

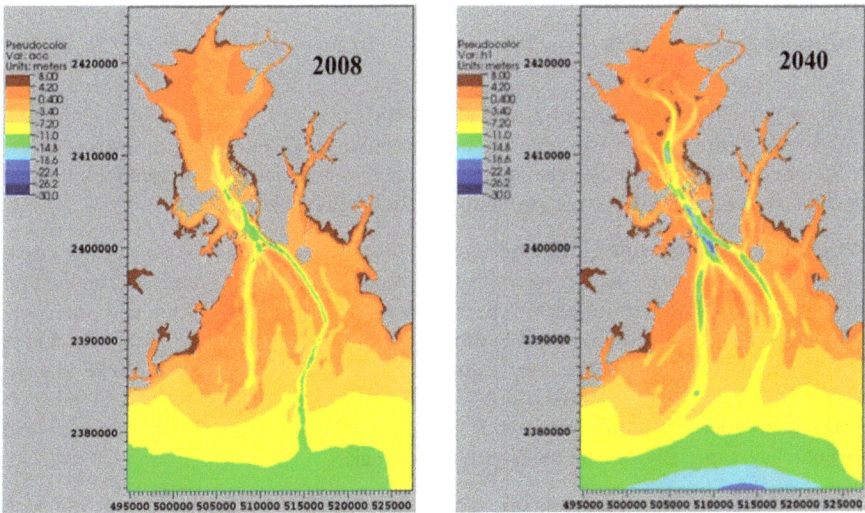

Figure 19. Bathymetry maps of 2008 vs. 2040.

8. Long-term morphological evolution

The Qinzhou Harbor has used the east channel as its major navigation channel to deliver cargos. Two large-scale dredgings were performed within this channel, i.e. 2009/9-2002/12 with a total dredged sediment of 8.234 million m³, 2004/2-2008/12 with a total dredged sediment of 45.307 million m³ [9]. These two dredging activities have exerted profound influences upon the morphological evolution of the channel-shoal system in the Qinzhou bay. The present computation has just reflected this evolution process and trend. During the adjusting process, the deeper channels have experienced further erosion while the shallower shoals (ridges) have accreted further higher, and the overall stability of the channel-shoal system has been maintained without horizontal migration or sign of merging or perishing; the inner bay has not only accepted sediments delivered by Maolingjiang river and Qinjiang river, but also sediments transported by flood tidal flows from the outer bay; the remaining part of the net eroded sediments from the outer bay has been transported into the offshore slope and deeper water by ebb tidal flows.

The magnitude of morphological adjustment by the above-mentioned channel dredging has been initially large but decreasing steadily with time. It could be estimated the morphological adjustment process at the outer bay would finished one hundred years later while other parts of the Qinzhou Bay might experience even longer adjustments. It should be clarified that such an estimation has assumed that no new engineering projects to be constructed and the computation conditions such as spatial bed thickness, horizontal & vertical sediment grading, shoreline configuration, tidal force and river discharges are unchanged.

9. Concluding remarks

Qinzhou Bay is characteristics of a unique dumbbell in shape, consisting of an inner enclosed bay, a trumpet-shaped outer bay and irregular rocky tidal inlet connecting them. Within the outer bay a complex channel-shoal (ridge) system has been present with the major channels serving as the navigation course for the Qinzhou Harbor.

FVCOM is a 3D unstructured grid, finite-volume coastal ocean model for the study of coastal oceanic and estuarine circulation, sediment transport and morphodynamics.. Having performed good calibrations of observed tidal water level, flow velocity and direction, suspended-sediment concentration at hydrographic sites [1] and morphological variation in the period of 2004 through 2008, the present study further investigates the diurnal tidal circulation including tidal asymmetry, residual eddy, and accompanying sediment transport processes in order to ascertain the water & sediment exchanges between the inner basin and the outer bay, especially, among the branching channels.

It is found that the inner basin has been acting as a sediment storage basin to accept sediments delivered by river flows and those by asymmetric tidal flows from the outer bay; Coriolis force together with rocks at the inlet mouth has controlled the dividing ratios of water and sediment among the branching channels, the west channel has been the dominant course for tidal flow and sediment to pass the outer bay, the middle channel comes secondly important and the east channel contribute the least.

Two large-scale dredging activities conducted in 2000/9-2002/12 and 2004/2-2008/12 in the east channel for deep navigation course development have exerted profound influence upon the morphodynamic evolution of the channel-shoal system. The erosion & deposition pattern, i.e. erosion in channels and deposition in shoals (ridges) has clearly demonstrated that tidal flow is the predominant force for maintaining and reforming the channel-shoal morphology; the dredging in the east channel has caused lasting erosions in the major channels in the outer bay, Yingling inlet and the inner bay as well as the offshore slope, meanwhile, depositions accumulate on shoals (ridges) and at the end parts of the channels.

Generally speaking, the overall channel-shoal system has been stable with channels becoming deeper and shoals becoming higher, and such a morphological adjustment process will probably finished over one hundred years later, if no new coastal engineering activity intervenes.

Acknowledgements

This study is supported by a grant from National Natural Science Foundation (No. 51179211) and a Young Researcher Fund of IWHR (No. NJ1009).

Author details

Yu-Hai Wang*

Address all correspondence to: wangyuhai-2166@126.com

Department of Sediment Research, China Institute of Water Resources and Hydropower Research, Beijing, China

References

[1] Tianjing Coastal Zone Engineering Company (TCZC). (2009). Fengchenggang Nuclear Power Plant Summer Full Tide Hydrological Survey and Analysis. *Technical Report*, in Chinese.

[2] Yan, X.-X., & Liu, G.-T. (2006). Study on deposition characteristics and channel siltation in offshore zone of Qinzhou Bay. *Journal of Waterway and Harbor*, 27(2), 79-83, in Chinese.

[3] Li, G.-Z., Liang, W., & Liu, J.-H. (2002). Discussion on the source and transport tendency of silt in the Qinzhou Bay in terms of the dynamic partition zones of heavy minerals in the sediments. *Marine Science Bulletin*, 21(5), 61-68, in Chinese.

[4] Chen, C., Liu, H., & Beardsley, R. (2003). An unstructured grid, finite-volume, three-dimensional, primitive equations ocean model: Application to coastal ocean and estuaries. *Journal of Atmospheric and Oceanic Technology*, 20(1), 159-186.

[5] Warner, J. C., Sherwood, C. R., Signell, R. P., Harris, C., & Arango, H. G. (2008). Development of a three-dimensional, regional, coupled wave, current, and sediment-transport model. *Computers and Geosciences*, 34, 1284-1306.

[6] Wang, Y. H. (2007). Formula for predicting bedload transport rate in oscillatory sheet flows. *Coastal Engineering*, 54(8), 594-601.

[7] Harris, C. K., & Wiberg, P. L. (1997). Approaches to quantifying long-term continental shelf sediment transport with an example from the northern California STRESS mid-shelf site. *Continental Shelf Research*, 17, 1389-1418.

[8] Li, G.-Z., Liang, W., & Liu, J.-H. (2001). Features of underway dynamic geomorphology of the Qinzhou Bay. *Geography and Territorial Research*, 17(4), 70-75, in Chinese.

[9] Wang, Y. H., Wang, C. H., Liu, D. B., & Lin, Y.-X. (2010). Preliminary study on channel stability in Qinzhou Bay. *Port & Waterway Engineering*, 8, 76-80, in Chinese.

Permissions

The contributors of this book come from diverse backgrounds, making this book a truly international effort. This book will bring forth new frontiers with its revolutionizing research information and detailed analysis of the nascent developments around the world.

We would like to thank Dr. Andrew J. Manning, for lending his expertise to make the book truly unique. He has played a crucial role in the development of this book. Without his invaluable contribution this book wouldn't have been possible. He has made vital efforts to compile up to date information on the varied aspects of this subject to make this book a valuable addition to the collection of many professionals and students.

This book was conceptualized with the vision of imparting up-to-date information and advanced data in this field. To ensure the same, a matchless editorial board was set up. Every individual on the board went through rigorous rounds of assessment to prove their worth. After which they invested a large part of their time researching and compiling the most relevant data for our readers. Conferences and sessions were held from time to time between the editorial board and the contributing authors to present the data in the most comprehensible form. The editorial team has worked tirelessly to provide valuable and valid information to help people across the globe.

Every chapter published in this book has been scrutinized by our experts. Their significance has been extensively debated. The topics covered herein carry significant findings which will fuel the growth of the discipline. They may even be implemented as practical applications or may be referred to as a beginning point for another development. Chapters in this book were first published by InTech; hereby published with permission under the Creative Commons Attribution License or equivalent.

The editorial board has been involved in producing this book since its inception. They have spent rigorous hours researching and exploring the diverse topics which have resulted in the successful publishing of this book. They have passed on their knowledge of decades through this book. To expedite this challenging task, the publisher supported the team at every step. A small team of assistant editors was also appointed to further simplify the editing procedure and attain best results for the readers.

Our editorial team has been hand-picked from every corner of the world. Their multi-ethnicity adds dynamic inputs to the discussions which result in innovative outcomes. These outcomes are then further discussed with the researchers and contributors who give their valuable feedback and opinion regarding the same. The feedback is then collaborated with the researches and they are edited in a comprehensive manner to aid the understanding of the subject.

Apart from the editorial board, the designing team has also invested a significant amount of their time in understanding the subject and creating the most relevant covers. They scrutinized every image to scout for the most suitable representation of the subject and create an appropriate cover for the book.

The publishing team has been involved in this book since its early stages. They were actively engaged in every process, be it collecting the data, connecting with the contributors or procuring relevant information. The team has been an ardent support to the editorial, designing and production team. Their endless efforts to recruit the best for this project, has resulted in the accomplishment of this book. They are a veteran in the field of academics and their pool of knowledge is as vast as their experience in printing. Their expertise and guidance has proved useful at every step. Their uncompromising quality standards have made this book an exceptional effort. Their encouragement from time to time has been an inspiration for everyone.

The publisher and the editorial board hope that this book will prove to be a valuable piece of knowledge for researchers, students, practitioners and scholars across the globe.

List of Contributors

Arnaud Héquette and Adrien Cartier
LOG - UMR CNRS 8187, University of Littoral Côte d'Opale, France

Philippe Larroudé
LEGI-UMR 5519 UJF, University of Grenoble, France

Alberto Sanchez
Centro Interdisciplinario de Ciencias Marinas, Instituto Politécnico Nacional, La Paz, Baja California Sur, México

Concepción Ortiz Hernández
El Colegio de la Frontera Sur, Unidad Chetumal, Chetumal, Quintana Roo, México

Rabin Bhattarai
Department of Agricultural and Biological Engineering, University of Illinois at Urbana-Champaign, Urbana, IL 61801, USA

X. H. Wang and F. P. Andutta
School of Physical, Environmental and Mathematical Sciences, University of New South Wales at Australian Defence Force Academy UNSW-ADFA, Australia

Jeremy R. Spearman, Richard J.S. Whitehouse and John V. Baugh
HR Wallingford, Howbery Park, Wallingford, Oxfordshire, UK

Emma L. Pidduck
School of Marine Science & Engineering, University of Plymouth, Plymouth, Devon, UK

Kate L. Spencer
Department of Geography, Queen Mary – University of London, Mile End Road, London, UK

Andrew J. Manning
HR Wallingford, Howbery Park, Wallingford, Oxfordshire, UK
School of Marine Science & Engineering, University of Plymouth, Plymouth, Devon, UK

Mohammed Achab
Department of Earth Sciences, Scientific Institute, University Mohammed V-Agdal, Rabat, Morocco

Ram Balachandar and H. Prashanth Reddy
Department of Civil and Environmental Engineering, University of Windsor, Canada

Levent Yilmaz
Hydraulic Division, Civil Engineering Department, Technical University of Istanbul, Maslak, Istanbul, Turkey

Yun-Chih Chiang
Tzu Chi University, Taiwan

Sung-Shan Hsiao
National Taiwan Ocean University, Taiwan

Vasileios Kitsikoudis and Vlassios Hrissanthou
Department of Civil Engineering, Democritus University of Thrace, Xanthi, Greece

Epaminondas Sidiropoulos
Department of Rural and Surveying Engineering, Aristotle University of Thessaloniki, Thessaloniki, Greece

Katerina Kombiadou and Yannis N. Krestenitis
Laboratory of Maritime Engineering and Maritime Works, Civil Engineering Department, Aristotle University of Thessaloniki, 54124, Thessaloniki

Charles Wilson Mahera and Anton Mtega Narsis
University of Dar-es-salaam, College of Natural and Applied Sciences, Department of Mathematics, Dar-es-salaam, Tanzania
Department of General Studies, Dar es Salaam Institute of Technology, Dar es Salaam, Tanzania

Xiaobo Chao, Yafei Jia and A. K. M. Azad Hossain
The University of Mississippi, USA

Yu-Hai Wang
Department of Sediment Research, China Institute of Water Resources and Hydropower Research, Beijing, China